THE WILEY GUIDE TO PROJECT, PROGRAM & PORTFOLIO MANAGEMENT

THE WILEY GUIDES TO THE MANAGEMENT OF PROJECTS

Edited by

Peter W. G. Morris and Jeffrey K. Pinto

The Wiley Guide to Project, Program & Portfolio Management
978-0-470-22685-8

The Wiley Guide to Project Control
978-0-470-22684-1

The Wiley Guide to Project Organization & Project
Management Competencies
978-0-470-22683-4

The Wiley Guide to Project Technology, Supply Chain &
Procurement Management
978-0-470-22682-7

THE WILEY GUIDE TO PROJECT, PROGRAM & PORTFOLIO MANAGEMENT

Edited by

Peter W. G. Morris and Jeffrey K. Pinto

JOHN WILEY & SONS, INC.

Library of Congress Cataloging-in-Publication Data:

ISBN: 978-0-470-22685-8

Printed in the United States of America

10 9 8 7 6 5 4 3 2 1

CONTENTS

THE WILEY GUIDE TO PROJECT, PROGRAM & PORTFOLIO MANAGEMENT: PREFACE AND INTRODUCTION

Peter W. G. Morris and Jeffrey Pinto

In 1983, Dave Cleland and William King produced for Van Nostrand Reinhold (now John Wiley & Sons) the *Project Management Handbook*, a book that rapidly became a classic. Now over twenty years later, Wiley is bringing this landmark publication up to date with a new series *The Wiley Guides to the Management of Projects*, comprising four separate, but linked, books.

Why the new title—indeed, why the need to update the original work?

That is a big question, one that goes to the heart of much of the debate in project management today and which is central to the architecture and content of these books. First, why "the management of projects" instead of "project management"?

Project management has moved a long way since 1983. If we mark the founding of project management to be somewhere between about 1955 (when the first uses of modern project management terms and techniques began being applied in the management of the U.S. missile programs) and 1969 / 1970 (when project management professional associations were established in the United States and Europe) (Morris, 1997), then Cleland and King's book reflected the thinking that had been developed in the field for about the first twenty years of this young discipline's life. Well, more than another twenty years have since elapsed. During this time there has been an explosive growth in project management. The professional project management associations around the world now have thousands of members—the Project Management Institute (PMI) itself having well over 200,000—and membership continues to grow! Every year there are dozens of conferences; books, journals, and electronic publications abound; companies continue to recognize project management as a core business discipline and work to improve company performance through it; and, increasingly, there is more formal educational work carried out in university teaching and research programs, both at the undergraduate and, particularly, graduate levels.

Yet, in many ways, all this activity has led to some confusion over concepts and applications. For example, the basic American, European, and Japanese professional models of project management are different. The most influential, PMI, not least due to its size, is the most limiting, reflecting an essentially execution, or delivery, orientation, evident both in its *Guide to the Project Management Body of Knowledge, PMBOK Guide, 3ʳᵈ Edition* (PMI, 2004) and its *Organizational Project Management Maturity Model, OPM3* (PMI, 2003). This approach tends to under-emphasize the front-end, definitional stages of the project, the stages that are so crucial to successful accomplishment (the European and Japanese models, as we shall see, give much greater prominence to these stages). An execution emphasis is obviously essential, but managing the definition of the project, in a way that best fits with the business, technical, and other organizational needs of the sponsors, is critical in determining how well the project will deliver business benefits and in establishing the overall strategy for the project.

It was this insight, developed through research conducted independently by the current authors shortly after the publication of the Cleland and King *Handbook* (Morris and Hough, 1987; Pinto and Slevin, 1988), that led to Morris coining the term "the management of projects" in 1994 to reflect the need to focus on managing the definition and delivery of *the project itself* to deliver a successful outcome.

These, at any rate, are the themes that we shall be exploring in this book (and to which we shall revert in a moment). Our aim, frankly, is to better center the discipline by defining more clearly what is involved in managing projects successfully and, in doing so, to expand the discipline's focus.

So, why is this endeavor so big that it takes four books? Well, first, it was both the publisher's desire and our own to produce something substantial—something that could be used by both practitioners and scholars, hopefully for the next 10 to 20 years, like the Cleland and King book—as a reference for the best-thinking in the discipline. But why are there so many chapters that it needs four books? Quite simply, the size reflects the growth of knowledge within the field. The "management of projects" philosophy forces us (i.e., members of the discipline) to expand our frame of reference regarding what projects truly *are* beyond of the traditional *PMBOK/OPM3*model.

These, then, are not a set of short "how to" management books, but very intentionally, resource books. We see our readership not as casual business readers, but as people who are genuinely interested in the discipline, and who seek further insight and information— the thinking managers of projects. Specifically, the books are intended for both the general practitioner and the student (typically working at the graduate level). For both, we seek to show where and how practice and innovative thinking is shaping the discipline. We are deliberately pushing the envelope, giving practical examples, and providing references to others' work. The books should, in short, be a real resource, allowing the reader to understand how the key "management of projects" practices are being applied in different contexts and pointing to where further information can be obtained.

To achieve this aim, we have assembled and worked, at times intensively, with a group of authors who collectively provide truly outstanding experience and insight. Some are, by any standard, among the leading researchers, writers, and speakers in the field, whether as academics or consultants. Others write directly from senior positions in industry, offering their practical experience. In every case, each has worked hard with us to furnish the relevance, the references, and the examples that the books, as a whole, aim to provide.

What one undoubtedly gets as a result is a range that is far greater than any individual alone can bring (one simply cannot be working in all these different areas so deeply as all these authors, combined, are). What one does not always get, though, are all the angles that any one mind might think is important. This is inevitable, if a little regrettable. But to a larger extent, we feel, it is beneficial for two reasons. One, this is not a discipline that is now done and finished—far from it. There are many examples where there is need and opportunity for further research and for alternative ways of looking at things. Rodney Turner and Anne Keegan, for example, in their chapter on managing innovation (*The Wiley Guide to Project Technology, Supply Chain & Procurement Management,* Chapter 8) ended up positioning the discussion very much in terms of learning and maturity. If we had gone to Harvard, to Wheelwright and Clark (1992) or Christensen (1999) for example, we would almost certainly have received something that focused more on the structural processes linking technology, innovation, and strategy. This divergence is healthy for the discipline, and is, in fact, inevitable in a subject that is so context-dependent as management. Second, it is also beneficial, because seeing a topic from a different viewpoint can be stimulating and lead the reader to fresh insights. Hence we have Steve Simister giving an outstandingly lucid and comprehensive treatment in *The Wiley Guide to Project Control,* Chapter 5 on risk management; but later we have Stephen Ward and Chris Chapman coming at the same subject (*The Wiley Guide to Project Control,* Chapter 6) from a different perspective and offering a penetrating treatment of it. There are many similar instances, particularly where the topic is complicated, or may vary in application, as in strategy, program management, finance, procurement, knowledge management, performance management, scheduling, competence, quality, and maturity.

In short, the breadth and diversity of this collection of work (and authors) is, we believe, one of the books' most fertile qualities. Together, they represent a set of approximately sixty authors from different discipline perspectives (e.g., construction, new product development, information technology, defense / aerospace) whose common bond is their commitment to improving the management of projects, and who provide a range of insights from around the globe. Thus, the North American reader can gain insight into processes that, while common in Europe, have yet to make significant inroads in other locations, and vice versa. IT project managers can likewise gather information from the wealth of knowledge built up through decades of practice in the construction industry, and vice versa. The settings may change; the key principals are remarkably resilient.

But these are big topics, and it is perhaps time to return to the question of what we mean by project management and the management of projects, and to the structure of the book.

Project Management

There are several levels at which the subject of project management can be approached. We have already indicated one of them in reference to the PMI model. As we and several other of the *Guides*' authors indicate later, this is a wholly valid, but essentially delivery, or execution-oriented perspective of the discipline: what the project manager needs to do in order to deliver the project "on time, in budget, to scope." If project management professionals cannot do this effectively, they are failing at the first fence. Mastering these skills is the *sine qua non*—the 'without which nothing'—of the discipline. Volume 1 addresses this basic view of the discipline—though by no means exhaustively (there are dozens of other books on the market that do this excellently—including some outstanding textbooks: Meredith and Mantel, 2003; Gray and Larson, 2003; Pinto, 2004).

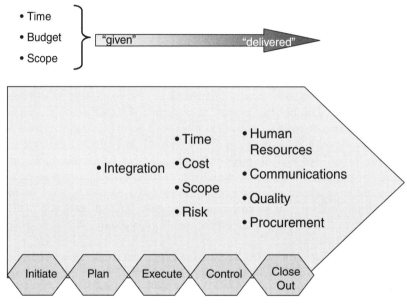

PROJECT MANAGEMENT:
"On time, in budget, to scope" execution/delivery

The overriding paradigm of project management at this level is a control one (in the cybernetic sense of control involving planning, measuring, comparing, and then adjusting performance to meet planned objectives, or adjusting the plans). Interestingly, even this model—for us, the foundation stone of the discipline—is often more than many in other disciplines think of as project management: many, for example, see it as predominantly oriented around scheduling (or even as a subset, in some management textbooks, of operations management). In fact, even in some sectors of industry, this has only recently begun to change, as can be seen towards the end of the book in the chapter on project management in the pharmaceutical industry. It is more than just scheduling of course: there is a whole range of cost, scope, quality and other control activities. But there are other important topics too.

Managing project risks, for example, is an absolutely fundamental skill even at this basic level of project management. Projects, by definition, are unique: doing the work necessary to initiate, plan, execute, control, and close-out the project will inevitably entail risks. These need to be managed.

Both these areas are mainstream and generally pretty well understood within the traditional project management community (as represented by the PMI *PMBOK*® '*Guide*' (PMI, 2004) for example). What is less well covered, perhaps, is the people-side of managing projects. Clearly people are absolutely central to effective project management; without people projects simply could not be managed. There is a huge amount of work that has been done on how organizations and people behave and perform, and much that has been written on this within a project management context (that so little of this finds its way into *PMBOK* is almost certainly due to its concentration on material that is said in *PMBOK* to be "unique" to project management). A lot of this information we have positioned in Volume 3, which deals more with the area of competencies, but some we have kept in the other volumes, deliberately to make the point that people issues are essential in project delivery.

It is thus important to provide the necessary balance to our building blocks of the discipline. For example, among the key contextual elements that set the stage for future activity is the organization's structure—so pivotal in influencing how effectively projects may be run. But organizational structure has to fit within the larger social context of the organization—its culture, values, and operating philosophy; stakeholder expectations, socioeconomic, and business context; behavioural norms, power, and informal influence processes, and so on. This takes us to our larger theme: looking at the project in its environment and managing its definition and delivery for stakeholder success: "the management of projects."

The Management of Projects

The thrust of the books is, as we have said, to expand the field of project management. This is quite deliberate. For as Morris and Hough showed in *The Anatomy of Major Projects* (1987), in a survey of the then-existing data on project overruns (drawing on over 3,600 projects as well as eight specially prepared case studies), neither poor scheduling nor even lack of teamwork figured crucially among the factors leading to the large number of unsuccessful projects in this data set. What instead were typically important were items such as client changes, poor technology management, and poor change control; changing social, economic, and environmental factors; labor issues, poor contract management, etc. Basically, the message was that while traditional project management skills are important, they are often not *sufficient* to ensure project success: what is needed is to broaden the focus to cover the management of external and front-end issues, not least technology. Similarly, at about the same time, and subsequently, Pinto and his coauthors, in their studies on project success (Pinto and Slevin, 1988; Kharbanda and Pinto, 1997), showed the importance of client issues and technology, as well as the more traditional areas of project control and people.

The result of both works has been to change the way we look at the discipline. No longer is the focus so much just on the processes and practices needed to deliver projects "to scope, in budget, on schedule," but rather on how we set up and define the project to deliver stakeholder success—on how to manage projects. In one sense, this almost makes the subject impossibly large, for now the only thing differentiating this form of management from other sorts is "the project." We need, therefore, to understand the characteristics of the project development life cycle, but also the nature of projects in organizations. This becomes the kernel of the new discipline, and there is much in this book on this.

Morris articulated this idea in *The Management of Projects* (1994, 97), and it significantly influenced the development of the Association for Project Management's Body of Knowledge as well as the International Project Management Association's Competence Baseline (Morris, 2001; Morris, Jamieson, and Shepherd, 2006; Morris, Crawford, Hodgson, Shepherd, and Thomas, 2006). As a generic term, we feel "the management of projects" still works, but it is interesting to note how the rising interest in program management and portfolio management fits comfortably into this schema. Program management is now strongly seen as the management of multiple projects connected to a shared business objective—see, for example, the chapter by Michel Thiry (*The Wiley Guide to Project, Program & Portfolio Management*, Chapter 6.) The emphasis on managing for business benefit, and on managing projects, is exactly the same as in "the management of projects." Similarly, the recently launched *Japanese Body of Knowledge, P2M* (*Program and Project Management*), discussed *inter alia* in Lynn Crawford's chapter on project management standards (*The Wiley Guide to Project Organization & Project Management Competencies*, Chapter 10), is explicitly oriented around managing programs and projects to create, and optimize, business value. Systems manage-

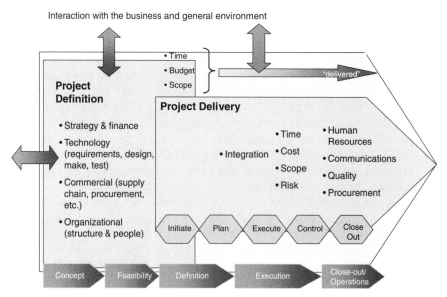

THE MANAGEMENT OF PROJECTS involves managing the definition and delivery of the project for stakeholder success. The focus is on the project in its context. Project and program management – and portfolio management, though this is less managerial – sit within this framework.

ment, strategy, value management, finance, and relations management for example are all major elements in *P2M:* few, if any, appear in *PMBOK*.

("The management of projects" model is also more relevant to the single project situation than *PMBOK* incidentally, not just because of the emphasis on value, but via the inclusion of design, technology, and definition. There are many single project management situations, such as Design & Build contracts for example, where the project management team has responsibility for elements of the project design and definition.)

Structure of *The Wiley Guide to Project, Program & Portfolio Management*

The Wiley Guides to the Management of Projects series consists of four distinct, but interrelated, books:

* *The Wiley Guide to Project, Program & Portfolio Management*
* *The Wiley Guide to Project Control*
* *The Wiley Guide to Project Organization & Project Management Competencies*
* *The Wiley Guide to Project Technology, Supply Chain & Procurement Management*

This book, *The Wiley Guide to Project, Program & Portfolio Management,* is based on the "meta" level of management, understanding and exploiting the strategic management of projects, portfolios, and program management and the linkage with context and strategy.

Strategy represents the fundamental goals and objectives that drive the organization and which, if well understood and delineated, should affect the manner in which projects are selected, shaped, and executed. The organization's strategy encompasses the way in which it makes sense of its external environment, identifies opportunities, and evaluates its performance. In this manner, projects become, in a term David Cleland coined, "the building blocks of strategy," allowing the organization to operationalize its goals in meaningful, measurable ways (Cleland, 1990). The organization's use of its strategic portfolio of projects and the manner in which it shapes and maintains its programs, reflects its commitment to a proactive, rather than reactive, means of achieving its goals.

Acknowledging the links between strategy at the corporate, portfolio, program, and project levels allow organizations to focus on improving their portfolio and program management. These are themes explored in several different chapters in this volume; the result is that our basic "management of projects" model can now be expanded to reflect our increased knowledge of program management and its concerns—managing for business benefit, managing products (brands, technology), resource allocation, etc.—along with portfolio management and its special challenges.

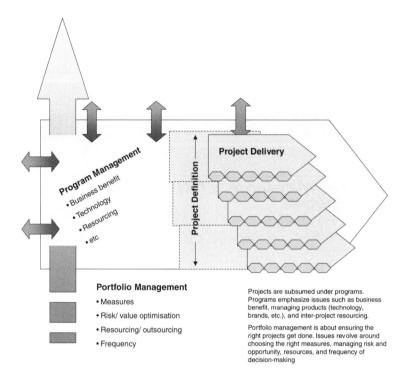

Projects are subsumed under programs. Programs emphasize issues such as business benefit, managing products (technology, brands, etc.), and inter-project resourcing.

Portfolio management is about ensuring the right projects get done. Issues revolve around choosing the right measures, managing risk and opportunity, resources, and frequency of decision-making

1. In Chapter 1, Karlos Artto and Perttu Dietrich offer a comprehensive and very thorough overview on "Strategic Business Management through Multiple Projects." Covering a wealth of academic work in the area, Karlos and Perttu examine the way that companies manage the relationships between portfolios, programs, and projects in different situations. They then generalize this into an overall theoretical model which they illustrate from a series of research projects they have undertaken with industry.

2. Ashley Jamieson and Peter Morris, in Chapter 2, similarly survey the literature on moving from corporate strategy to project strategy, again emphasizing the sequence of moving, via portfolios and programs, into projects (and subprojects/tasks). Like Artto and Dietrich, their chapter introduces fresh research evidence to substantiate their findings, this time from a project funded largely by PMI calling on case study data as well as evidence from a questionnaire survey of PMI members showing that most of those who replied routinely work in programs and projects, and use techniques to value optimize the strategy (as will be later discussed in Chapter 9).

3. David Cleland builds on this argument in Chapter 3, drawing on a wealth of experience to extend the arguments regarding corporate planning and programs with more detail at the project planning end, for example explaining how project strategic planning feeds into work packages via the work breakdown structure.

4. Joe Lampel and Pushkar Jha, in Chapter 4, in their chapter on project orientation, contend that many projects fail as a result of poor strategic orientation—of not achieving a proper fit between the enterprise and the project. They hypothesize three types of

project/enterprise orientation—project based, project led, and core operations led—and propose that project scoping, programming, and autonomy shape the interaction between the project and its corporate parent. They conclude by presenting research findings that explore these interactions.

Having reviewed aspects of the linkages between strategy at the corporate, portfolio, program and project levels, the next two chapters focus more purposefully on, firstly, portfolio management, and secondly, program management.

5. Norm Archer and Fereidoun Ghasemzadeh are two of the world's leading authorities on portfolio management (in the project, as opposed to the financial, sense). Their chapter demonstrates the importance particularly of managing risk and outsourcing options at the portfolio level, and the need for a framework for classifying project type. They then proceed to look at the different characteristics that affect portfolio choice and develop a generic process model for portfolio selection.

6. Michel Thiry, in Chapter 6, develops a number of interesting perspectives to better understand the characteristics of program management. Building on the ideas already presented in the previous chapters on strategy and portfolio management, Michel emphasizes the importance of learning in developing a strategic response to evolving conditions—in his process model of program management, learning needs understanding as clearly as performance delivery (this leads him to define project management in the more specific sense of being primarily about uncertainty reduction). He then elaborates this into a two-dimensional phase model linking strategy, programs, projects, and operations cyclically around the activities of formulation, organization, deployment, appraisal, and dissolution.

7. Ali Jaafari, in Chapter 7, focuses on the characteristics of large (engineering) projects, emphasizing in particular how they are subject to environmental uncertainty and may need much more front-end, strategic management than the smaller project. Ali then walks us through a high-level process model of the major things that need doing in the management of large projects.

8. An issue one comes up against in looking at the discipline of managing projects across a broad range of contexts is how to categorize the application area. Aaron Shenhar and Dov Dvir have done as much work as anyone in this area and they provide a stimulating discussion in Chapter 8 showing that there are several different categorizations which are valid and which work well under different circumstances. Based on their research, Aaron and Dov believe that in order to select the appropriate management style, managers should first assess the environment, the product, and the task; second, classify the project by the levels of uncertainty, complexity, and pace; and third, select the right style to fit the specific project type.

9. Value Management (VM), the process of formally optimising the overall approach to the project (including whether or not it should be done) is addressed in Chapter 9 by Michel Thiry. He begins by discussing what is meant by value and by defining the various terms used in VM (Value Engineering, Value Analysis, etc.). VM is positioned as a strategic process comprising sense-making, ideation, elaboration, choice, and mas-

tery. Techniques within each of these, such as Function(al) Analysis, are described. Overall Michel takes an ambitious view of VM, positioning it as "the method of choice to deal with the ambiguity of stakeholders' needs and expectations and the complexity of changing business environment at program level and project initiation."

10. Project success is often quoted as the measure on which the project should be assessed. The trouble is this is a portmanteau term covering many different issues. Terry Cooke-Davies in Chapter 10 reviews the studies that have been carried out in this area since the landmark work by Baker, Murphy, and Fisher in 1974. He concludes that there are essentially three levels of success measure: was the project done right (what he calls "project management success"), was the right project done ("project success"), and were the right projects done right time after time ("consistent project success"). While warning that there are no silver bullets, Terry nevertheless identifies the half dozen or so key factors that he believes, from his research and that of his colleagues, are critical at each level.

11. Roland Gareis, in Chapter 11, discusses the characteristics of organizations that are project (and program) oriented. Again building on years of original research as well as practical consulting, Roland encapsulates most of the ideas this resource book addresses, although using his own distinctive "management by projects" framework (which is just slightly different, as the wording would suggest, from our "management of projects").

12. Graham Winch, in Chapter 12, provides an enormously useful practical account of the importance of stakeholder management in achieving successful project outcomes. Taking a systems development project as an example (the computerization of share dealing on the London Stock Exchange) Graham shows how the failure to identify and manage different parties' expectations can not only lead to "academic" discussions of whether and for whom the project was or was not a success, but can in a very real sense lead to loss of control and ultimately project disaster. Graham concludes by drawing out nine key lessons for managing stakeholders effectively.

Finance is an important dimension to the strategic development of projects. The availability of money will affect what can be done, and when. In the public area particularly, changes in the way projects are funded have had a huge affect on the whole way projects are set up and carried out. The next two chapters, by Rodney Turner and Graham Ive, illustrate this.

13. In Chapter 13 Rodney Turner gives a masterful overview of the characteristics and means of financing projects as well as the process of financial management on projects.

14. One of the newer forms of project finance to have developed over the last 20 or so years (coming out of the oil sector in the 1970s) has been that of basing the project's funding solely on the revenues generated specifically by the project itself, with no other security from other parties. This form, strictly termed "project financing," has become very important in many parts of the world in bringing ways of using private funds to finance public infrastructure projects. It is no exaggeration to say that this has had a revolutionary impact on the management of public sector projects where it has been applied. Graham Ive, in Chapter 14, discusses this form of project financing in detail,

with particular reference to its application in the British public sector, which is widely regarded as leading the field in this area. He outlines the origin of this form of financing, known in the UK as PFI (for Private Finance Initiative), and shows how it impacts on the management of projects, for example by requiring increased clarity on project objectives, risk management, value management, securing stakeholder consent, capturing users' requirements, and on procurement and bidding practices. (All issues we have either looked at already or will be doing later in the book.) Interestingly, with regard to the procurement challenges, Graham uses an economic tool (agency theory) to analyze the problems of devising the reward structure, selection of resources, and moral hazard. Throughout, he illustrates his points with reference to a real PFI project—a new hospital.

About the Authors

Karlos A. Artto

Dr. Karlos Artto is Professor at the Department of Industrial Engineering and Management at the Helsinki University of Technology (HUT), Finland, where he is responsible for developing research and education in the field of project-oriented activities in business. His current research interests cover the management of project-based organizations (including the area of project portfolio management and strategic management of multiple projects in organizations); the management of innovation, technology, R&D, new product development and operational development projects in different organizational contexts; project networks and project delivery chains; and risk management (with the emphasis on management of business opportunities and considerations of project success and related criteria).

Perttu H. Deitrich

Perttu Dietrich works as a project manager and researcher at the Helsinki University of Technology (HUT), Finland. He has pursued research and development in several companies in the area of project portfolio management. His research interests include strategic management, multi-project management, and organizational development. His doctoral research will explore these areas by investigating the linkage between projects and strategy.

Ashley Jamieson

Ashley Jamieson worked for many years as a business manager, senior program manager, and project manager with global aerospace and defense companies on British, European, North American, Middle East, South East Asia, and Australasian aircraft programs and projects. He then took up a career in research and academia. He has worked at the Centre for Research in the Management of Projects (CRMP) at UMIST where he carried out research into design management in major construction projects, and more recently at University College London, where through a PMI funded research project, he investigated how strategy is moved from corporate to projects. He holds an M.Sc. in Engineering Business Management and is currently studying for a Ph.D. in project strategy at The Bartlett School of Construction and Project Management, University College London. He is a visiting lecturer in project management at the University of Manchester and UMIST and is the co-author (with Peter Morris) of *Moving Strategy from Corporate to Projects* (2004) published by PMI.

Peter W. G. Morris

Peter Morris is Professor of Construction and Project Management at University College London, Visiting Professor of Engineering Project Management at UMIST, and Director of the UCL/UMIST based Centre for Research in the Management of Projects. He is also Executive Director of INDECO Ltd, an international projects oriented management consultancy. He is a past Chairman and Vice President of the UK Association for Project Management and Deputy Chairman of the International Project Management Association. His research has focussed significantly around knowledge management and organizational learning in projects, and in design management. Dr. Morris consults with many major companies on developing enterprise-wide project management competency. Prior to joining INDECO, he was a Main Board Director of Bovis Limited, the holding company of the Bovis Construction Group. Between 1984 and 1989 he was a Research Fellow at the University of Oxford and Executive Director of the Major Projects Association. Prior to his work at Oxford, he was with Arthur D Little in Cambridge, Massachusetts and previously with Booz, Allen & Hamilton in New York, and Sir Robert McAlpine in London. He has written approximately one hundred papers on project management as well as the books: *The Anatomy of Major Projects* (Wiley, 1988) and *The Management of Projects* (Thomas Telford, 1997). He is a Fellow of the Association for Project Management, Institution of Civil Engineers, and Chartered Institute of Building and has a Ph.D., M.Sc. and B.Sc., all from UMIST.

David Cleland

David I. Cleland is currently Professor Emeritus in the School of Engineering at the University of Pittsburgh. He is the author/editor of thirty-five books in the fields of project management, engineering management, and manufacturing management. He has served as a consultant for both national and foreign companies, and has been honored for his original and continuing contributions to his disciplines. Dr. Cleland is a Fellow of the Project Management Institute (PMI), and has received the Distinguished Contribution to Project Management Award from PMI in 1983, 1993, and again in 2001.

Joseph Lampel

Joseph Lampel is a Professor of Strategy at Cass Business School, City University, London. He obtained his undergraduate degree in Physics from McGill University, Canada, and later pursued his M.Sc. in Technology Policy at the Institut d'Histoire et Sociopolitique des Sciences at Université de Montréal, Canada. After working for the Science Council of Canada and the Ontario government he returned to McGill University to pursue doctoral studies in Strategic Management. His dissertation "Strategy in Thin Industries" won the Best Dissertation Award from the Administrative Science Association of Canada in 1992. He was Assistant Professor at the Stern School of Business, New York University from 1989–1996. Subsequently, he was Reader at the University of St. Andrews from 1996 to 1999, and Professor of Strategic Management at University of Nottingham Business School from 1999 to 2001. He has also taught at McGill University, Concordia University, Montreal, and the University of Illinois at Urbana-Champaign. Joseph Lampel is the author with Henry Mintzberg and Bruch Ahlstrand of the *Strategy Safari*, Free Press & Prentice-Hall, 1998. He is also the editor with Henry Mintzberg, James Brian Quinn, and Sumantra Ghoshal of the fourth edition of *The Strategy Process*, Prentice-Hall (2003). He edited a Special Issue of *International Journal of Project Management*, Special Issue on Strategic Project Management (2001, vol. 19, No. 8). He has published in *Strategic Management Journal, Sloan Management Review, Organization Science, Journal of Management, Journal of Management Studies, R&D Management, International Journal of Technology Management,* and *Fortune Magazine*.

Pushkar P. Jha

Pushkar Jha holds an M.Phil. in process engineering from the University of Newcastle-upon-Tyne and degrees in management and commerce. He is currently a research fellow at Cass Business School, London, working in the area project-based organizational learning. He was a member of the Advanced Process Control Group at the University of Newcastle upon Tyne, and prior to that he worked on development projects in India.

Norman P. Archer

Dr. Norm Archer is Professor Emeritus in the Management Science and Information Systems Area of the Michael G. DeGroote School of Business at McMaster University, and is Special Advisor to the McMaster eBusiness Research Centre (MeRC). Dr. Archer consults, teaches, and supervises graduate student research projects. He is active in the study of organizational problems relating to the implementation of eBusiness approaches in existing organizations, and the resulting impacts on processes, employees, customers, and suppliers. Current research projects include the study of knowledge transfer in network organizations, supply chain management issues, change management in organizations, and management of eBusiness projects. Together with his students, he has published more than seventy papers in refereed journals and conference proceedings, primarily on project management, business-to-business eCommerce, intelligent agents, and the human-computer interface, in *Decision Support Systems, Internet Research, International Journal of Management Theory and Practice, IEEE Transactions on Systems, Man, and Cybernetics, International Journal of Human-Computer Studies, Journal of the Operational Research Society, Communications of the ACM,* and many others.

Fereidoun Ghasemzadeh

Dr. Fereidoun Ghasemzadeh is an assistant Professor in the Management and Economics School at Sharif University of Technology. Dr. Ghasemzadeh has been involved in teaching MIS, DSS, and E-Commerce courses to MBA students, and supervises graduate students in their research projects. He is the co-founder and currently the CEO of Afranet, a leading Internet and E-Commerce company in Iran, and is heavily involved in advanced E-Commerce and E-Government research and applied projects. Articles by, or interviews with, Dr. Ghasemzadeh have appeared more than fifty times in the media, and he has organized many conferences, symposia, and workshops on these topics. With his colleagues and students Dr. Ghasemzadeh has published ten academic papers in refereed journals and conference proceedings, primarily on project management, electronic commerce, and electronic government.

Michel Thiry

Michel Thiry is a Brussels-based organizational consultant and facilitator with thirty years experience in project management in North America and Europe. He has worked in Canada, the United States, Australia, the UK and continental Europe. He holds an M.Sc. in Organizational Behaviour from the School of Management and Organizational Behaviour, University of London (UK) and regularly speaks and publishes at the international level. Currently he is a seminar leader for PMI SeminarsWorld and associate professor for the ISGI (France)/UTS (Australia) joint *Program and Project Management* MSc; he also lectures at Reading University (UK). He has also provided value, project or program management expertise to major organizations, in various fields, including construction, pharmaceutical, IT and IS, telecom, water treatment, transportation (air and rail), and others. He has authored the book *Value Management Practice* and co-authored the "Managing Programmes of Projects" chapter in the *Gower Handbook of Project Management,* 3rd Ed. He also authored the Program Management and Value Management Chapters in *Project Management Pathways,* published by the Association for Project Management (UK) in early 2003. In addition, he regularly writes and reviews for the *International Journal of Project Management.* Mr. Thiry is also past Director of the Project Management Institute's Montreal and UK Chapters and past President of the European Governing Board for Value Management Certification and Training, based in Paris.

Ali Jaafari

Professor Jaafari received his Ph.D. in Business Economics (Quantitative Modeling & Forecasting) from Surrey University in the UK (Joint SSRC-SRC scholarship holder and Swan Award) in 1977, and his Master of Science (Distinction) in Highway Engineering and Transportation Management in 1974 from the same university. He has a Master of Engineering (Distinction) from Tehran University awarded in 1968. He has acted as an expert consultant to industry and governments worldwide for more than fifteen years. In 1994, he acted as a special consultant to the European Community on the management of the Productivity Initiative Programme as part of TACIS. He has, to date, authored over one hundred and thirty publications in project and program management, including strategy-based project management, whole-of-life framework & philosophy, concurrency, management of technology and innovations, information management systems, TQM, risk, opportunity and uncertainty analysis & management, and education of professionals. Since 1982, he has conducted courses and seminars for over 3,000 executives, managers and professionals in

Australia, Asia, and Europe. He specializes in graduate education and professional development, and has developed innovative on-line graduate programs that have won three Excellence Awards. He has been a member of the International Project Management Association since 1984, and a regular contributor to the World Congresses on Project Management. Professor Jaafari has chaired many functions, both nationally and internationally, and is the winner of many prizes and awards. Professor Jaafari has held visiting professorial appointments at a number of universities, in the United States, the UK, Europe, and Asia.

Aaron J. Shenhar

Dr. Aaron J. Shenhar is the Institute Professor of Management and the founder of the Project Management Program at Stevens Institute of Technology. He is also a visiting professor at Tel-Aviv University and the Technion in Israel. He was named, "Engineering Manager of the Year," by the Engineering Management Society of IEEE in 2000. Prior to his academic career, he has been involved in managing projects, innovation, R&D, and high technology businesses for almost twenty years. Working for the Israel defense industry, he participated in all phases of engineering and management—from project manager up through the highest executive posts. As executive at Rafael, the Armament Development Authority of Israel, he was appointed Corporate Vice President, Human Resources, and later, President of the Electronic Systems Division. In his second career in academia, Dr. Shenhar's work focuses on research, teaching, and consulting in project management; strategic project leadership, technology, and innovation management; product development and the leadership of professionals in technology-based organizations. He is serving as a consultant to several major corporations. With more than 150 publications to his credit, his writings have influenced project management research and education throughout the world. Dr. Shenhar holds five academic degrees in engineering, statistics, and management, including a Ph.D. in Electrical Engineering from Stanford University. In 2003, he became the first recipient of the PMI Research Achievement Award.

Dov Dvir

Dr. Dov Dvir is Senior Lecturer at the School of Management, Ben Gurion University, Israel. Formerly, he was the Head of the Management of Technology (MOT) department at the Holon Center for Technological Education. He holds a B.Sc. in electrical engineering from the Technion—Israel Institute of Technology, M.Sc. in operations research and an MBA from Tel Aviv University; and Ph.D. in management (specialization in MOT) from Tel Aviv University. His research interests include project management, technology transfer, technological entrepreneurship and the management of technological organizations. Dr. Dvir has accumulated over 25 years of technical, management and consulting experience in government and private organizations.

Terry Cooke-Davies

Terry has been a practitioner of both general and project management continuously since the end of the 1960s. He is the Managing Director of Human Systems Limited, which he founded in 1985 to provide services to organizations in support of their innovation projects and ventures. Through the family of project management knowledge networks created and supported by Human Systems, he is in close touch with the best project management practices of more than seventy leading organizations globally. The methods developed in support of the networks are soundly based in theory, as well as having practical application to members, and this was recognized by the award of a PhD to Terry by Leeds Metropolitan University in 2000 for a thesis entitled, "Towards Improved Project Management Practice: Uncovering the evidence for effective practices through empirical research." He is now an Adjunct Professor of Project Management at the University of Technology, Sydney. Terry is a regular speaker at international project management conferences in Europe, North America, Australia and Asia and has published more than thirty book chapters, journal and magazine articles, and research papers. He has a bachelor's degree in Theology, and qualifications in electrical engineering, management accounting and counselling in addition to his doctor's degree in Project Management.

Roland Gareis

Dr. Roland Gareis graduated from the University of Economics and Business Administration, Vienna; had habilitation at the University of Technology, Vienna, Department of Construction Industry; was Professor at the Georgia Institute of Technology in Atlanta; and was Visiting professor at the ETH, Zurich, at the Georgia State University, Atlanta and at the University of Quebec, Montreal. He is currently Professor of Project Management at the University of Economics and Business Administration, Vienna and Director of the postgraduate program "International Project Management." He owns the firm Roland Gareis Consulting.

Graham Winch

Graham Winch is Professor of Construction Project Management at the Manchester Centre for Civil and Construction Engineering, UMIST, where he is head of the Project Management Division. He taught for ten years at the Bartlett School, University College London after a career in management research in business schools, and managing construction projects. He is author of *Managing Construction Projects, an Information Processing Approach* (Blackwell, 2002). He is the author of three other books and over thirty refereed journal articles, complemented by numerous book chapters, conference papers, and research reports. His research currently focuses on the strategic management of projects, and on innovation in the construction industry. In addition to the work on stakeholder management, he works also investigating the processes of risk identification using cognitive mapping.

Rodney Turner

Rodney Turner is Professor of Project Management at Erasmus University Rotterdam, in the Faculty of Economics. He is also an Adjunct Professor at the University of Technology Sydney, and Visiting Professor at Henley Management College, where he was previously

Professor of Project Management, and Director of the Masters program in Project Management. He studied engineering at Auckland University and did his doctorate at Oxford University, where he was also for two years a post-doctoral research fellow. He worked for six years for ICI as a mechanical engineer and project manager, on the design, construction and maintenance of heavy process plant, and for three years with Coopers and Lybrand as a management consultant. He joined Henley in 1989 and Erasmus in 1997. Rodney Turner is the author or editor of seven books, including *The Handbook of Project-based Management,* the best selling book published by McGraw-Hill, and the *Gower Handbook of Project Management.* He is editor of *The International Journal of Project Management,* and has written articles for journals, conferences and magazines. He lectures on and teaches project management world wide. From 1999 and 2000, he was President of the International Project Management Association, and Chairman for 2001–2002. He has also helped to establish the Benelux Region of the European Construction Institute as foundation Operations Director. He is also a Fellow of the Institution of Mechanical Engineers and the Association for Project Management.

Graham Ive

At the Bartlett since 1977, Graham Ive is responsible for the overall direction of the M.Sc. Construction Economics and Management (CEM) course, and for economics teaching on the masters program. His research has focused on aspects of the industrial economics of the construction sector, embracing the structure of the construction industries, the strategies, behavior and performance of construction firms; other research includes the complex and specific economic institutions of the construction process, specifically contracting and procurement systems. Current research within this theme focuses on the Private Finance Initiative (Build-Own-Operate-Transfer projects). Much of his research has been undertaken with partners from the UK construction sector, coordinated through the UK's Construction Industry Council, to whom Graham is an economic advisor. He is author / co-author of two studies of PFI for the Construction Industry Council (*The Constructors' Key Guide to PFI;* and *The Role of Cost-Saving and Innovation in PFI Projects'*), published by Thomas Telford; and two books, *The Economics of the Modern Construction Sector* and *The Economics of the Modern Construction Firm,* published by Palgrave Macmilllan.

The Wiley Guides to the Management of Projects series offers an opportunity to take a step back and evaluate the status of the field, particularly in terms of scholarship and intellectual contributions, some twenty-four years after Cleland and King's seminal *Handbook.* Much has changed in the interim. The discipline has broadened considerably—where once projects were the primary focus of a few industries, today they are literally the dominant way of organizing business in sectors as diverse as insurance and manufacturing, software engineering and utilities. But as projects have been recognized as primary, critical organizational forms, so has recognition that the range of practices, processes, and issues needed to manage them is substantially broader than was typically seen nearly a quarter of a century ago. The old project management "initiate, plan, execute, control, and close" model once considered the basis for the discipline is now increasingly recognized as insufficient and inadequate, as the many chapters of this book surely demonstrate.

The shift from "project management" to "the management of projects" is no mere linguistic sleight-of-hand: it represents a profound change in the manner in which we approach projects, organize, perform, and evaluate them.

On a personal note, we, the editors, have been both gratified and humbled by the willingness of the authors (very busy people all) to commit their time and labor to this project (and our thanks too to Gill Hypher for all her administrative assistance). Asking an internationally recognized set of experts to provide leading edge work in their respective fields, while ensuring that it is equally useful for scholars and practitioners alike, is a formidable challenge. The contributors rose to meet this challenge wonderfully, as we are sure you, our readers, will agree. In many ways, the *Wiley Guides* represent not only the current state of the art in the discipline; it also showcases the talents and insights of the field's top scholars, thinkers, practitioners, and consultants.

Cleland and King's original *Project Management Handbook* spawned many imitators; we hope with this book that it has acquired a worthy successor.

References

Christensen, C. M. 1999. *Innovation and the General Manager.* Boston: Irwin McGraw-Hill.

Cleland, D. I. and King, W. R. 1983. *Project Management Handbook.* New York: Van Nostrand Reinhold.

Cleland, D. I. 1990. *Project Management: Strategic Design and Implementation.* Blue Ridge Summit, PA: TAB Books.

Gray, C. F., and E. W. Larson. 2003. *Project Management.* Burr Ridge, IL: McGraw-Hill.

Griseri, P. 2002. *Management Knowledge: a critical view.* London: Palgrave.

Kharbanda, O. P., and J. K. Pinto. 1997. *What Made Gertie Gallop?* New York: Van Nostrand Reinhold.

Meredith, J. R., and S. J. Mantel. 2003. *Project Management: A Managerial Approach,* 5th Edition. New York: Wiley.

Morris, P. W. G. and G. H. Hough. 1987. *The Anatomy of Major Projects.* Chichester: John Wiley & Sons Ltd.

Morris, P. W. G. 1994. *The Management of Projects.* London: Thomas Telford; distributed in the USA by The American Society of Civil Engineers; paperback edition 1997.

Morris, P. W. G. 2001. "Updating The Project Management Bodies Of Knowledge," *Project Management Journal* 32(3):21–30.

Morris, P. W. G., H. A. J. Jamieson, and M. M. Shepherd. 2006. "Research updating the APM Body of Knowledge 4th edition," *International Journal of Project Management* (24):461–473.

Morris, P. W. G., L. Crawford, D. Hodgson, M. M. Shepherd, and J. Thomas. 2006. "Exploring the Role of Formal Bodies of Knowledge in Defining a Profession—the case of Project Management" *International Journal of Project Management* (24):710–721.

Pinto, J. K. and D. P. Slevin. 1988. "Project success: definitions and measurement techniques," *Project Management Journal* 19(1):67–72.

Pinto, J. K. 2004. *Project Management.* Upper Saddle River, NJ: Prentice-Hall.

Project Management Institute. 2004. *Guide to the Project Management Body of Knowledge.* Newtown Square, PA: PMI.

Project Management Institute. 2003. *Organizational Project Management Maturity Model.* Newtown Square, PA: PMI.

Wheelwright, S. C. & Clark, K. B. 1992. *Revolutionizing New Product Development.* New York: The Free Press.

CHAPTER ONE

STRATEGIC BUSINESS MANAGEMENT THROUGH MULTIPLE PROJECTS

Karlos A. Artto, Perttu H. Dietrich

Effective management of single projects does not suffice in today's organizations. Instead, the managerial focus in firms has shifted toward simultaneous management of whole collections of projects as one large entity, and toward effective linking of this set of projects to the ultimate business purpose. This approach is contained in concepts of project-based management, programs, and portfolios. Portfolios of different project types are typically positioned under the governance of organizational units or responsibility areas (see Figure 1.1). Management processes above projects must link projects to business goals and assist in reaching or exceeding the expectations set by company strategy.

One major starting point for the development of business-oriented management of projects in a company context was introduced in the end of 1980s in an expert seminar in Vienna, where the contribution of project management to the general world of management was discussed as contained in the concept of "management by projects" (see the chapter by Gareis). Since that time there have been an increasing number of studies on the broader role available for project-based management, project-based organizations, and project business. Recent examples of such studies include Turner (1999), Turner and Keegan (1999, 2000), Turner et al. (2000), Gareis (2000a, 2000b), Artto (2001), Artto et al. (2002), and Elonen and Artto (2003).

Early theories of organizational strategy saw "strategy as an action of intentionally and rationally combining selected courses of action with the allocation of resources in order to carry out organizational goals and objectives in order to achieve strategic fit and thereby obtain competitive advantage" (Hatch, 1997). This is based on the idea that strategy involves creating a match between organization and environment (Ansoff, 1965). Galbraith (1995) proposed that strategy establishes the criteria for choosing among alternative organizational forms. Each organizational form enables some activities to be performed well while hinder-

FIGURE 1.1. TWO COMPANIES (OR TWO BUSINESS UNITS WITHIN ONE COMPANY) WITH NETWORKED PROJECTS AND PORTFOLIOS. THERE ARE CROSS-ORGANIZATIONAL PROCESSES IN THE SHARED NETWORK AT STRATEGIC, PORTFOLIO, AND PROJECT LEVELS.

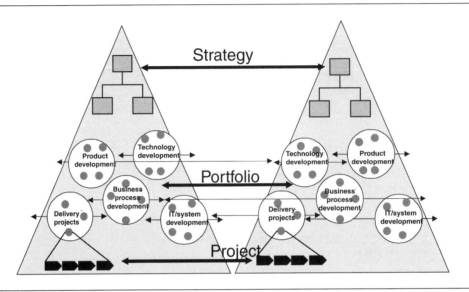

Source: Artto et al. (2002).

ing others. Choosing between organizational alternatives involves trade-offs. Strategy can help with this by pointing to those activities that are most necessary, thereby providing a basis for making the best trade-offs.

The purpose of this chapter is to introduce managerial practices relevant to strategic business management in multiple-project environments. Multiproject environments are introduced in terms of different project types, programs, and portfolios and their management. Based on the knowledge from this, we introduce issues that serve as guidelines to the theme of strategic business management of multiple projects. We conduct an analysis of content and process of strategy and how these relate to the setting of goals and objectives, and to effective decision making with multiple projects. Based on this, the chapter identifies effective managerial practices for the strategic business management in multiple-project environments. We also combine strategy with management from an applications viewpoint by looking at four case organizations.

Different Project Types and Their Different Strategic Importance

Different project types have different strategic importance; each type typically requires different management approaches. Crawford et al. (2002), Shenhar et al. (2002), and Youker

(1999) are studies of project classification that attempt to address this issue. (See Shenhar and Dvir's chapter in this book.) These are valuable in understanding not only different project types and their characteristics but also the different success criteria and respective strategic importance, and accordingly, different successful managerial practices associated with each project type.

Shenhar et al. classify projects into external and internal types, where the position or closeness of the customer (external or internal) provides the basis for the classification. This classification also considers the ultimate customer in the external markets in relation to how direct or indirect the relationship of the ultimate customer is to the project deliverable. Their starting point is innovation management literature that makes a distinction between incremental and radical innovation. Thus, according to Shenhar et al., projects can be either strategic or operational in their nature, depending on the project type.

External projects typically relate to developing products for customers in the market. Shenhar et al. distinguish between derivative, platform, and breakthrough projects, all as external projects. Wheelwright and Clark (1992) call these three project types *commercial development projects*. Based on Shenhar et al.'s considerations, derivative projects relate to extending, improving, or upgrading existing products. They typically aim at short-term benefits, and they are thus more operational than strategic in their nature. Platform and breakthrough projects relate to new product development or production processes where there is a longer-term perspective, and, accordingly, a reaching for a more strategic nature. Another interpretation of an external project is that of a delivery project where the project is in a commercial setting, and where an organization is running projects for other organizations (Turner and Keegan, 1999). Such external delivery projects are often mere production or manufacturing devices that run more or less predetermined work for an organization according to a contract between the customer and project supplier (Artto, 2001). The similarity of project-based operations with both external and internal customers is demonstrated by Turner and Keegan (1999), who defined a project-based organization as a stand-alone entity that makes products for external customers, or a subsidiary of a business unit of a larger firm that makes products for internal or external customers.

Shenhar et al. (2002) divide internal projects into problem solving, utility, maintenance and research projects. Wheelwright and Clark (1992) distinguish between internal projects based on research and development, which are a precursor to commercial development, and alliances and partnerships, which can be commercial or basic research directed. Figure 1.2 describes Wheelwright and Clark's view on different types of development projects (the figure includes four types; the fifth type—alliances and partnerships—can include any of the other four types). Mikkelsen et al. (1991) define internal projects as organizational or operational development projects, such as systems planning and implementation, the introduction of new manufacturing technology, and organizational change. Shenhar et al.'s utility and research projects usually come from a long-term perspective and can be considered as strategic projects. Problem-solving and maintenance projects usually focus on the shorter term, typically aim at performance improvements, and can be seen as operational projects (Shenhar et al., 2002).

We appreciate the consideration of strategic importance now given to project types but consider that the "strategic versus tactical" importance given to these also depends on parameters other than project type as defined by existing project classification literature. Fur-

FIGURE 1.2. MAPPING THE TYPES OF DEVELOPMENT PROJECTS.

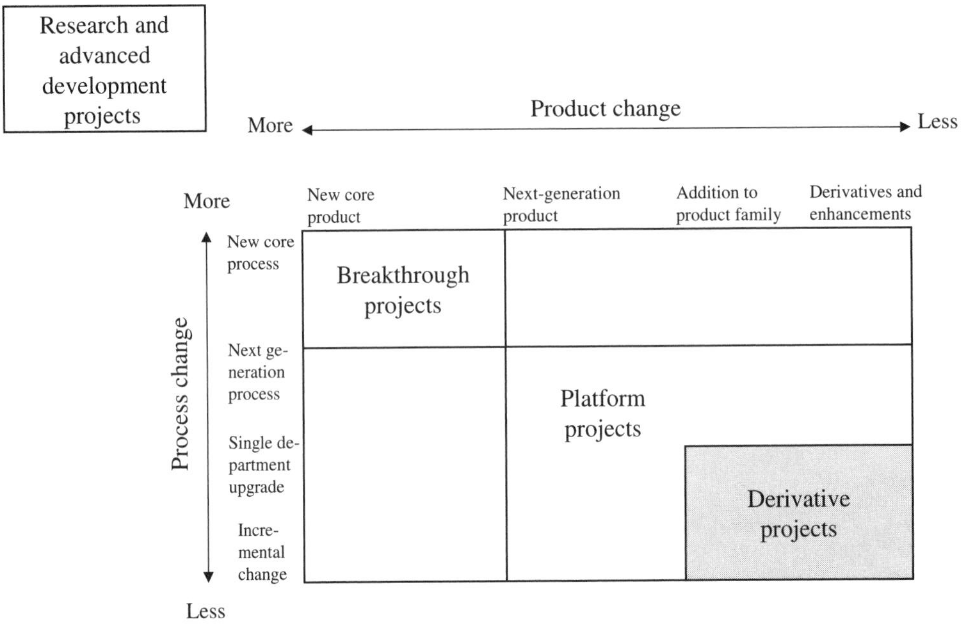

Source: Wheelwright and Clark (1992).

thermore, the strategic importance cannot be evaluated in a straightforward manner, such as presuming that long-term projects are always more strategic, as is widely argued in the literature.

Programs and Portfolios

Guidance for the management of multiple projects in organizations can be derived from several different theoretical and practically oriented discussion arenas. The program management and project portfolio management contents are outlined in the following section.

Archer and Ghasemzadeh (1999) define a project portfolio as a group of projects that are conducted under the sponsorship or management of a particular organization (see their chapter in this book). They point out that these projects compete for scarce resources. The three well-known objectives of portfolio management are as follows (Cooper et al., 1998):

- Maximizing the value of the portfolio
- Linking the portfolio to the strategy
- Balancing the portfolio

Dye and Pennypacker (1999) define project portfolio management as the art and science of applying a set of knowledge, skills, tools, and techniques to a collection of projects to meet or exceed the needs and expectations of an organization's investment strategy. In PMBOK (2000), project portfolio management refers to the selection and support of project investments or program investments that are guided by the organization's strategic plan and available resources. A strategic task of project portfolio management is to maintain corporate identity and ensure linkages between projects and constrain the impact of individually implemented projects with no links to other projects (Lundin and Stablein, 2000). According to Platje et al. (1994), a portfolio is a set of projects that are managed in a coordinated way to deliver benefits that would not be possible if the projects were managed independently. This definition is similar to many definitions introduced for a project program. For example, Turner (1999) and Poskela et al. (2001) emphasize that projects in a program are a coherent group that is managed in a coordinated way for added benefit. Murray-Webster and Thiry (2000) define a program as a collection of change actions (projects and operational activities) purposefully grouped together to realize strategic and/or tactical benefits. (See the chapter by Thiry on program management.)

From the strategic management point of view, the main driver for the management of multiple projects in different forms—for instance, programs—is the change in the business environment of an organization (OGC, 2002). Changes in the environment imply system or organizational changes (Ackoff, 1999). In these changes, program management provides a framework for the management of complexity and risk with the general intent of implementing business strategies and initiatives, or large-scale change (OGC, 2002).

The management of risk and uncertainty can appear in different ways. For example, in the R&D area, the important task of a business manager may be to increase risk to balance the portfolio of projects for business benefit. We can see this from findings illustrating how radical projects with high risk have the highest business potential (Loch, 2000).

Programs usually represent entities that have a determined purpose, predefined expectations related to the benefits scheme, and an organization, or at least a plan for organizing the effort. A program is set up to produce a specific outcome that may be defined at a high abstraction level of a "vision." According to PMBOK (2000), a program consists of several associated projects that will contribute to the achievement of a strategic plan. Many programs also include elements of ongoing operations. Program management helps to organize, manage, accommodate, and control adaptation and changes such that the eventual outcome meets the objectives set by the business strategy (OGC, 2002). Program management includes the management of interfaces between projects, prioritization of resources, and a reduction in overall management effort (Turner, 1999). The objectives of projects under the same project program are interdependent (Platje et al., 1994). Turner (1999) emphasizes the importance of the overall strategic resource sharing scheme related to program management. Such strategic resource sharing is implemented through a well-organized balance of responsibility, where the program directors' responsibility is to link programs with corporate objectives, the overall corporate plan, and corporate resource plan. OGC (2002) defines program management as the coordinated management of a portfolio of projects that change organizations to achieve benefits that are of strategic importance.

Constructing a Theoretical Framework

The previous sections introduced aspects of existing knowledge on multiproject environments and attempted to show the need for new knowledge in the area of strategic business management of multiple-project environments. Based on this analysis, and the needs reflected by it, we can identify the following issues:

1. How can multiple projects be collectively aligned with business strategy in a manner that generates enhanced benefits for the whole business?
2. What is the role of specific projects in implementing, creating, and renewing business strategies?
3. How best can strategic business management be applied in organizations with multiple projects, and what are the relevant managerial practices for accomplishing this?

The preceding three questions are addressed in this remainder of this chapter via current strategy, business administration and project management literature, as well as findings from four case organizations.

Strategy and Strategic Management

In ancient military terminology in Athens, *Strategos* referred initially to attributes of the general commander in the army. The word strategy later was expanded to include the art of managerial skills for employing forces to overcome opposition (Mintzberg et al., 1995). Ancient military terminology and early strategic management literature emphasize the relative position of an organization to its external competitive environment, with emphasis on activities necessary to achieve a desired position (Chaffee, 1985). The concept of strategy has also used contributions from other disciplines, such as industrial organizational economics approach, resource-based approaches, ecologist-evolutionary approaches, and systems thinking approaches (Pavón, 2002). These emphasize, among others, the importance of rational decision making, and learning as an issue that shapes strategies.

The variety of attempts to express the specific nature of the strategy has led different authors to create different strategic schools (see, for example, Chaffee, 1985; Mintzberg et al., 1995). Mintzberg et al. (1995) introduced five Ps of strategy as a means to show the complex nature of strategy, where the Ps include definition of strategy as a Plan, Ploy, Pattern, Position, and Perspective. This last P—perspective—emphasizes sharing visions and mental images inside the organization to form a common understanding and culture, as strategies are abstractions in individuals' minds. This is consistent with Chaffee's (1985) *interpretative view*, which focuses on corporate culture and symbolic management as essential means to motivate participants and potential participants in ways that can favor an organization. This view makes a clear distinction with traditional strategic literature (see, for example, Chandler, 1962; Andrews, 1971; Hofer, 1975; Mintzberg, 1978) by suggesting that organizations' behavior is rather irrational in nature.

Early studies on strategic management focused on the content of strategy. Later literature distinguished between the content of the strategy and the process of strategy formulation and implementation (Chaffee, 1985). A distinction was made between an analytically objective strategy formulation process and a behavioral implementation process (see Andrews, 1971; Fredrickson, 1984; Pettigrew, 1992). Organization theorists tended to emphasize the meaning of *human processes (e.g., decision styles) in strategy making* (Burgeois, 1985), which started with rationality as a principal assumption of strategy process (see, for example, Andrews, 1971; Ansoff, 1965; Porter, 1980). From this, strategy management research introduced ideas of bounded rationality as a means to circumvent the reality of aspects of "organizational anarchy" within an organization (Simon, 1957; Cohen et al., 1972). This emerging recognition of an existing imperfect rationality in organizations has shifted toward emphasizing the extent and type of involvement of individuals in the organization or its environment (stakeholders) in strategy process (Hart, 1992). For example, Chaffee's (1985) emerging school of interpretative perspective on strategic management is an example of seeing the importance of individuals' involvement in strategy making. We can conclude from this that strategy is, and is accepted as, an important concern of the whole organization, not just its top management, and that motivation arises as a more crucial element of strategy realization.

Strategy Formulation and Implementation

Strategic processes comprise both strategy formulation and implementation. The strategic management literature mainly focuses on the strategy formulation aspect, with less attention given to strategy implementation (Aaltonen and Ikävalko, 2001). Andrews (1995) identifies organizational structures as requirements for the efficient implementation of intended actions. These structures include elements such as information systems and relationships enabling execution and management of subdivided activities. Moreover, Andrews (1995) states that one critical requirement for the successful implementation of strategy is to ensure that decisions made by managers and senior managers are consistent with the organization's goals and objectives. Leading the organization to intended goals and objectives requires measurement of the current state or performance of actions, analyzing the gap between the current and intended state, and making corrective actions. Diagnostic control systems are traditionally recognized as an important means of controlling the intended performance of the organization (see, for example, Simons, 1995). These systems are supported by defined performance characteristics called *critical performance variables* or *key success factors* that serve as indicators for achievement of organizational goals in the means of efficiency and effectiveness (Simons, 1995).

The balanced scorecard method introduced by Kaplan and Norton (1992) is a good example of a way to measure the performance of an organization in enhancing the achievement of organizational goals and strategy implementation. The scorecard can be used to derive objectives and measures related to company vision and strategy that can be derived to further project-specific objectives that are well aligned with business strategy. The strategic objectives to be measured fall into four perspectives:

- Customer
- Financial

- Internal business process
- Learning and growth

Employee capabilities, technology, and corporate climate contribute to the organization's capability for learning and growth (Kaplan and Norton 2001).

The success domains/dimensions in some project success studies are analogous to the four perspectives of balanced scorecard introduced by Kaplan and Norton (1992, 1996). For example, Shenhar et al. (1997) introduce the following four dimensions of project success:

- Project efficiency
- Impact on customer
- Business success
- Preparing for the future

In general, project success studies contribute to definition of requirements for decision-related information used, for instance, in project selection criteria or in performance measures.

Another contribution of project success studies is their indication of the most relevant managerial areas and even managerial practices that can serve as enablers for success (see the chapter by Cooke-Davies). Although many project success studies still limit their views to the success and successful management of one single project only, they can also introduce the important aspect of the overall context where a single project occurs. This extends the evaluation of success toward strategic issues that take a viewpoint of the whole business. According to Saravirta (2001) and Kotsalo-Mustonen (1996), the relevant success domains are related to the following:

- Strategy (e.g., new competitive advantage, reference value)
- Relationship (e.g., client satisfaction)
- Situation (e.g., learning by doing, unlearning)
- Product/service (e.g., commercial success, quality)
- Project implementation (e.g., cost, time, process quality)

Furthermore, evaluation of success depends on the stakeholder and its perspective on the project. From Morris and Hough (1987) and Rouhiainen (1997), we can derive the following synthesis of what the important success domains are:

1. Technical performance, project functionality, client satisfaction, and technical and financial performance of the deliverable for the sponsor/customer
2. Project management: on budget, on schedule, and to technical specification
3. Supplier's commercial performance: commercial benefit for the project service providers
4. The learning that project stakeholders acquire

Emergent Strategies

Mintzberg (1978) examined the relation between an organization's intended strategy and its realized strategy. Mintzberg showed that in addition to intentional strategies, strategies can also include unintentional, emergent components. Strategies emerge from different sources and from different levels of organization. Mintzberg proposes that the concept of realized strategy consists of intentions that lead to deliberate strategy, intentions that lead to unrealized strategy, and emergent strategies that develop in the organization without a priori intentions. Simons (1995) explains that an emergent strategy process consists of actions of individuals at all organizational levels to seize the opportunities and deal with the problems.

The emergent perspective of the strategy process seems to focus now on organizational learning (see the chapter by Lampel and Jha) and works to identify strategy as the cumulative impact of operative decisions taken by management (Christensen, 2000). Lindblom (1959) explained strategic management from the policy formation viewpoint, by seeing policies as consisting of small, politically acceptable, disjointed decisions. Moreover, Quinn's (1995) logical incrementalism proposed that strategies should rely on flexibility and experimental applications to move from broad concepts toward specific commitments, and strategic decisions should be made at the last possible moment in order to utilize the most topical and available information for minimizing risks. Quinn's argument is based on recognition of the biases that are found in reality among the formal "systems planning" and "power-behavioral" approaches of strategy formation in organizations. Good strategies are not formulated in a comprehensive master plan. According to Quinn, the formal systems planning approach relies on quantitative data ignoring vital qualitative, organizational, and power-behavioral factors, which often tend to represent the dynamic, time-related attributes of organizational success. Power-behavioral perspectives focus on psychological issues, trying to understand the influence of human dynamics, power relationships, and organizational processes in strategy formation. However, power-behavioral approaches can introduce drawbacks associated with ignoring the normative component of rationality in strategic decision making. Quinn thus emphasizes the importance of "process limits" in strategic decision making and management.

Process limits deal with issues such as timing and sequencing, building comfort levels, developing consensus, and selecting and training people. These imperatives can become the determinants of the system itself, and they finally determine the outcome of the decisions. This resembles Mintzberg and Waters' (1985) umbrella strategy perspective, where top-managers define boundaries and guidelines for the organization to operate, and where within these boundaries individuals in the organization can take initiatives. Mintzberg and Waters's study illustrates that even if the goals and objectives for the organization are predetermined at the top level of the organization, managers at the middle level can, by their actions and decisions, affect the formation of strategy. Burgelman (1983) supports this while proposing that in addition to induced strategic behavior, there exists also an autonomous strategic behavior within the organizations, and that behavior develops outside of the strategic umbrella defined by top management. This autonomous behavior appears when people at the operational level notify the resources provided by the organization as a means to utilize new opportunities (Floyd and Wooldridge, 2000). In his later study, Burgelman (1991) reported evidence from a longitudinal case study of Intel Corporation. The findings indicate that successful firms are characterized by both top-down strategic intent and bottom-up experimentation and selection process. Hart (1992) further developed the idea of organization-wide involvement in strategy formation and claimed that strategy making is an organizational capability that determines an organization's success or failure.

The preceding can be summarized as confirming that the role of individuals can be extremely important in viable strategy formulation and implementation. Projects and the individuals who work on them are particularly important. This is supported in the literature concerning product development and internal development projects, which emphasizes the project manager's role as a champion, gatekeeper, facilitator, or coach, and the top management representative's involvement and supporting role (Loch, 2000; Terwiesch et al., 1998, Brown and Eisenhardt, 1995; Eisenhardt and Tabrizi, 1995; and Mikkelsen et al. 1991). An important managerial problem is to encourage projects and individuals in their role in emerging strategies to create new ideas and renew existing strategies.

Thus, the challenge of successful strategic management may lie in managing the tension between creative innovation and predictable goal achievement. This tension occurs by

- reconciling unlimited opportunities with managers' limited attention;
- implementing top-down strategies while allowing bottom-up strategies to emerge;
- creating predictable environments while maintaining innovativeness; and
- controlling actions while simultaneously allowing the organization to learn new ones (Simons, 1995).

The ability to learn is raised as one major sources of sustainable competitive advantage in many companies. The study by De Geus (1988) provides a good example of the impact of learning to the success of companies. He examined the survival of Fortune 500 companies and found that one-third of the companies listed in 1973 had vanished by 1983. A key source of the success of the survivors was their ability to learn by continuously exploring opportunities for new business and organizational development. The emphasis should be placed on focusing that organizations are doing the right things, rather than doing things right. This capability of an organization to question its underlying policies and goals is called *double-loop learning*. Senge (1990) proposes creative tension in organizations as a principal building block of learning organizations. Creative tension is created by integrating pictures of desired future and current reality. However, this creative tension differs from solving existing problems in an undesired state of current reality. Rather, it comes from individuals' intrinsic motivation and generative learning with its emphasis on continuous experimentation and feedback. Brown and Eisenhardt (1997) argue that managers learn from possible futures. Small losses through experimental products that fail, or futurists' predictions that do not come true, are probably the most effective learning devices. A variety of probes creates hands-on experiences (experimental products and experimental strategic alliances) and indirect experiences (meetings). Eisenhardt's 1997 study suggested semi-structures that would ensure responsibilities, ownership, prioritization, and communication. Semi-structures relate to quasi-formal structures (committees, teams, task forces, information exchange relationships and arrangements) introduced by Schoonhoven and Jelinek (1996). Hence, in board meetings that represent gates or reviews, practical issues such as agendas, visual aids, and other decision support mechanisms, together with chairing, coaching, facilitating, and communication issues may play an important role as knowledge-sharing meetings and meetings for learning.

Organizational Design and Decision Making from a Strategy Perspective

As we have already indicated, any individual, and especially managers at the middle level (e.g., project managers), can, by their actions and decisions, affect the formation of strategy. The early strategic literature suggested that this approach of strategy formation by individuals at the lower organizational levels may not be effective. Instead, the early strategic literature suggested that strategic issues must be placed as part of a higher-level strategy process at the top level of the organization (e.g., Mintzberg, 1978; Ansoff, 1965). Shendel and Hofer (1979) extended this executive-focused view of strategic management to include other organizational levels. They specified three distinct organizational levels where strategic consideration should happen. First, at the corporate level, the main question is what business the organization should be in. Second, at business unit level, the focus is more on how to compete in that given business. Third, there is the integration of subfunctional activities and the integration of functional areas with the environment. The focus and perspective on strategy thus changes by levels.

Hart (1992) studied different models of strategy-making processes and classified five principal models of the strategy-making process according to the distribution of power in the organization: command, symbolic, rational, transactive, and generative modes. The command mode represents one extreme, where the role of top management is dominant and the participation of other members of the organization is limited to strategy implementation. At the other extreme, in a generative mode, the role of the top manager is to sponsor new ideas—for instance, project proposals emerging from the bottom of the organization—and guide those initiatives to a strategic direction. Moreover, Hart (1992) proposed that the three middle modes of strategy making (symbolic, rational, and transactive), characterized by better use of resources and organizational capabilities, led to higher levels of performance than the two extreme modes. He concluded that the strategy process should be considered as an issue that concerns the whole organization. Moreover, Hart (1992) proposed that strategy making is a capability of an organization that influences its overall performance, and organizational success requires multiple modes in strategy making.

Loch's (2000) study of a European technology manufacturer provides an excellent example of how the organizational setting is arranged in a multiproject environment in terms of distribution of power. It also emphasizes the importance of decision making as an important part of organizational design. Loch identified three different project clusters that defined how the manufacturer initiated and executed product development. An interesting finding was that there was no actual difference in success among the three clusters. Each of the three approaches had its strength. The first cluster, "formal process" projects, used the company's institutionalized product development process and relied on the Stage-Gate process recommended by Cooper (1994), and the formal process supports professional execution of the majority of all new product development projects (Cooper and Kleinschmidt 1987; Cooper, 1994); The second cluster, "under-the-table-projects," represented small teams or "skunk works" (Wolff, 1987) that supported organizational experimentation for new and unstructured ideas and flexibility (Quinn, 1985). The third cluster, comprising "pet projects," or "sacred cows," (projects determined by a powerful senior manager; see, for example, Meredith and Mantel, 1999), can be effective for difficult actions that need management support from a high level, and patience.

Two important weaknesses of undifferentiated process use were what Loch called "rigidity" and "lack of linkage." First, rigidity appears as the formal process where a company follows a relatively rigid Stage-Gate process and is perceived as inflexible in adjusting to specific project needs. Employees resorted to under-the-table projects because the formal process was too rigid and no alternative structure was available. Loch argued that the formal process may be too heavy-handed for incremental projects and too structured for radical projects. Second, lack of linkage occurs where there is a lack of structure for feeding unofficial under-the-table projects into the formal process. Loch argues that many companies suffer from the problem of new-product development not being integrated with strategy. He suggests that the company should develop a customized project portfolio with strategic positioning of projects, and a corresponding mixture of processes to meet its strategic innovation needs. Moreover, Loch considers that the lack of training of business unit managers in general strategy and technology management limit their ability to link strategic context and new-product development approach.

Our analysis of the role of managerial boards, and project and other teams pointed to an emphasis on meetings and reviews that relate to appropriate cross-organizational communication and decision-making processes. From an organizational design viewpoint, Ackoff (1978, 1981) introduces boards and board meetings as major organizational vehicles for participation and communication in what he calls a *circular organization.* McGrath (1996) provides an example of how cross-organizational cooperation is organized through teams, boards, or committees in a managerial model with practical orientation for product management and new product development. A product development project is conducted by a cross-functional core team. The core team is directly responsible for the success of the project, and the team is empowered with full authorization. The core team generally consists of five to eight individuals with different skills and a core team leader. The core team does not have the classical hierarchical approach to organization. Product development decisions are made by the product approval committee designated with the authority and responsibility to make them. The committee members are representatives of senior management representatives. Because the committee is a decision-making group, it should remain small. Four to five executives is an appropriate size. In some cases the committee is the company's executive committee. The decisions are made at phase reviews that are decision-making sessions that occur at specific milestones of the product development. Specifically, the product approval committee initiates new product development projects, cancels and reprioritizes projects, ensures that products being developed fit the company's strategy, and allocates development resources. While the core teams and the product approval committee are for short-term product development, the mid-term technology development is organized in a similar manner through technology development teams and a senior review committee. The senior review committee is a decision-making body of senior scientists and business managers that oversees technology development projects via technology phase reviews. Technology transfer teams with evolving team membership transfer the technology to product development projects (McGrath, 1996).

Important factors—or enablers—for project success often represent issues that are significant from the viewpoint of organizational design. For example, Mikkelsen et al. (1991) studied internal organizational and operational development projects and reported that the characteristics and roles of project managers and top managers were important drivers for project success. Furthermore, according to Brown and Eisenhardt (1995), important success factors of product development include cross-functional teams enabling cross-organizational integration, effective internal and external communication, powerful project leader, and senior management support. Brown and Eisenhardt also discuss the important role of team tenure that reflects the effectiveness of the pattern of working together, the important role of gatekeepers who are individuals that supply external information to the team, and the important role of a team group process that enables effective internal and external communication within the team and with customers, suppliers, and other individuals in the organization. Loch (2000) investigated a larger body of work on new product development and concluded that the following success drivers would represent good management practices: customer orientation and demand pull, cross-functional cooperation, top management support, existence of a champion, good planning and execution with a strong project manager, and the use of a well-defined process with formal measures. The success factors of new product development have slight differences according to the industry, though (e.g., Eisenhardt and Tabrizi, 1995; Terwiesch et al., 1998).

Goal Setting in Time and Aspects of Timing in Relation to Doing the Right Thing

Recent project management and business management literature has raised various aspects of managing time as one important issue in determining how overall efficiency can be achieved (Yeo and Ning, 2002; Steyn, 2002; DeMarco, 2001; Perlow, 1999; Goldratt, 1997). This literature, however, often argues that efficiency of timely performance would contribute to other indirect benefits in terms of efficiency and even effectiveness in overall performance. However, when discussing the management of time, the literature too often emphasizes the aspect of just doing the work efficiently instead of the more strategic dimension of doing the right things. This issue is introduced by Rämö (2002) by focusing on different notions of chronological and nonchronological time in organizational settings. He refers to Drucker's (1974) well-known discussion on efficiency and effectiveness, arguing that efficiency is concerned with doing things right. This is reflected in managerial approaches such as Taylor's scientific management or Deming's Total Quality Management, which both are concerned with doing things right and [just] in time (Rämö, 2002; Drucker, 1999). Such approaches emphasize exact clock time—*chronos*. They require efficiency and doing things right, which requires management and improvement of what is already known. Effectiveness, instead, is doing the right things (Drucker 1974). Rämö (2002) suggests that Drucker's discussion on the difference between efficiency and effectiveness also implies a dualism of time: clock time (*chronos*) emphasizes the chronological sequences of activities and, accordingly, rules efficiency, while the nonchronological aspect of time (*kairos*) relates to right timing and, accordingly, is essential for effectiveness. Seizing windows of opportunities requires a good sense of timing. *Chronos*, or clock time, does not govern such a sense of timing which, instead, it is based on a *kairic* feeling for the right moment (Rämö 2002).

Ackoff (1999) discussed the introduction of different types of systems, with particular attention to organizations as systems. A system may have a memory that can increase its efficiency over time in producing the outcome that is its goal. A purposeful system changes its goals under constant conditions; it selects ends and means and thus displays a will. An ideal-seeking system is a purposeful system that, on attainment of its goals or objectives, then seeks another goal and objective that more closely approximates its ideal. An ideal-seeking system is thus one that has a concept of "perfection" or the "ultimately desirable" and pursues it systematically—that is, in interrelated steps. The time that it would take to reach the ideal could be considered as infinite. Ackoff (1999) introduces the concept of "ends planning" that takes an approach to different perspectives in terms of three types of desired outcomes, each related to different timely perspective. First, "goals" represent ends that are expected within the period of a plan. The goals may be related to entities like, for instance, projects. Second, "objectives" represent ends that provide right directions but are not expected to be obtained until after the period planned for. Our interpretation is that such objectives may be achieved through a collection of projects, such as programs or portfolios, during a longer time span. Third, "ideals" are ends that are believed to be unattainable but toward which continuous progress is thought to be possible and is expected. Ideals provide strategic directions that enable good portfolio decisions for selecting the right projects—and indeed whole collections of well-balanced projects—with the right strategic intent. Goals are means with respect to objectives, and objectives are considered with respect to ideals. Ackoff's ends planning includes four steps: first, selecting a mission; second, specifying the desired properties of the system planned for; third, idealized redesign of that system; and fourth, selecting the gaps between this design and the reference scenario that planning will try to close. One additional related issue that Ackoff (1994) introduces is backward planning, and within backward planning, working backward from the present—that is, from where one wants to be right now to where one is right now.

Aalto et al. (2003) provide an example of what different timely perspectives mean in the R&D context for the management of projects and their portfolios, and the linkage between projects and portfolios. This is illustrated in Figure 1.3, as adapted from Aalto et al. (2003), with modifications to the figures presented by Groenveld (1997, 1998) and Kostoff and Schaller (2001). Aalto et al. use the term R&D to include research, technology development, and product development. Product development is the shortest-term activity. Technology development is more volatile by nature and the projects are typically focused on producing certain technologies or their combinations in the medium term. Such technologies are used in short-term product development. Research is the longest-term activity. It provides technology development with a potential for paradigm shifts and, thus, new points of departure. The interrelatedness of different projects with different time spans and purposes introduces challenges to successful R&D management in terms of how projects and project portfolios are managed.

Summary of the Theoretical Framework Construction

Table 1.1 summarizes the preceding theoretical analysis on strategic business management through multiple projects. The right column of the table presents existing artifacts that are

FIGURE 1.3. INTERRELATEDNESS OF DIFFERENT PROJECTS AND THEIR PORTFOLIOS WITH DIFFERENT TIMELY PERSPECTIVES IN THE R&D FIELD.

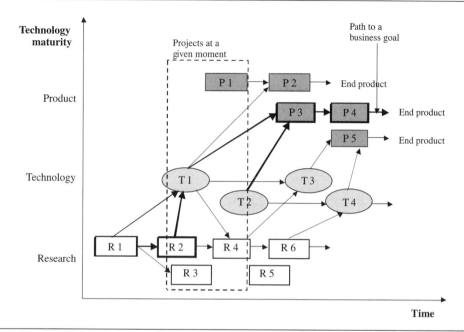

Source: Adapted from Aalto et al. (2003).

relevant to this topic. The left column of the table groups the artifacts by their content in the focal areas. These focal areas can be considered as important prerequisites for successful strategic business management in multiple projects environment. The focal areas are as follows:

1. Categorizing projects by their type
2. Supporting structured and flexible decision making
3. Ensuring effective communication and information transparency
4. Linking projects and strategy process
5. Establishing an organizational design to support strategic management in the multiple-project environment
6. Setting and measuring goals for different time spans in the future
7. Evaluating strategic contents, distinguishing between effectiveness and efficiency

Table 1.1 is self-explanatory. Concerning the table, only two additional explanations are raised here. First, concerning categorizing projects, we argue that the strategic content is partly specific to single projects. This occurs as the project itself is a fundamental managerial entity that interacts with its environment by producing, transferring, and receiving strategic

TABLE 1.1. EXISTING ARTIFACTS FROM THE THEORETICAL FRAMEWORK CONSTRUCTION GROUPED TO SEVEN FOCAL AREAS.

Focal Area	Existing Artifacts
1. Categorizing projects by their type	• Different project types are of different strategic importance. • Different project types require different managerial approaches.
2. Supporting structured and flexible decision making	• Structured decision-making practices (e.g., in board meetings with project-specific decisions) are important for adopting a view on whole portfolios and for linking strategy with projects. • Structured decision-making practices support the realization of strategic intentions of the organization. • Flexible decision-making practices enhance the emergence of innovative ideas and learning. Use of various types of processes in decision making (both formal and informal) increase an organization's ability to succeed. • Decision-making structures such as meetings are important for communication among top management, middle management, and project management. In addition to communicating the intended strategy component top-down, the meetings and their communication serve to foster grounds for the bottom-up emergent strategy component.
3. Ensuring effective communication and information transparency	• Information transparency, both vertical transparency across organizational levels and horizontal transparency across projects and organizational boundaries, and open communication help with building linkages. • Effective communication and information transparency enhances creativity and appearance of new strategic ideas. • Effective communication and information transparency enhances quality and optimality of decisions. • Communication enables learning. • Open sharing of information and information transparency results in better commitment and involvement among individuals and groups in organization.
4. Linking projects and strategy process	• Linking projects and the strategy process enables top management to acquire a holistic picture of ongoing project activities and new innovative ideas emerging from different organizational levels. This holistic picture of project activities and new ideas increase top management's ability to manage organization in a concrete manner toward desired direction. • Linking projects and the strategy process ensures that projects positioned at lower levels become aware of their status in the whole picture of implementing business strategy. This means that project managers are aware of why each of their projects exists and what should be accomplished in the end. A prerequisite for this is that the project manager understands clearly the intended strategy and the ways the project manager is capable of adjusting his or her project's direction. • Linking projects and the strategy process ensures that strategic initiatives are introduced both top-down and bottom-up. • Linking projects and the strategy process ensures that resources are allocated to "strategically right activities." • Linking projects and the strategy process ensures that those activities as a whole contribute in an optimal manner to the whole business.

TABLE 1.1. (*Continued*)

Focal Area	Existing Artifacts
5. Establishing an organizational design to support strategic management in the multiple-project environment	• Because organizational design and related structures partly determine the strategy and strategic capabilities (e.g., controllability and innovativeness) of the organization, it is essential to establish an organizational design that supports successful management schemes. • Hierarchy and boundaries of portfolios in the organization determine: which project activities must be viewed as a whole, how different kinds of portfolios contribute the strategic aims of the organization, and what is the relationship between different portfolios in the organization. • Power structures in the organization determine the organizational decision-making practices. • Management culture and project culture in the organization are important issues. The managerial practices must match the culture, and on the other hand, the culture can be changed by introducing new managerial practices. • An important enabler for the emergent strategic component is that there is a fluent interaction between different organizational levels, that projects are put into strategic perspective by being viewed as whole entities throughout the organization, and that there is communication about how these entities contribute to new strategic dimensions.
6. Setting and measuring goals for different time spans in the future	• Long-term strategic objectives of any organization differ from how short-term objectives are set. Further, different projects may be established simultaneously for different time spans. Accordingly, these objectives and their associated projects must be managed simultaneously by taking both the long term and the short term into account. • Especially in the long perspective, the future is uncertain, and the basic aim is to take advantage for different possible futures. This occurs through managing options toward the uncertain future. • In the case of different planning horizons, concepts such as risk, uncertainty, imperfect knowledge, and ambiguity become important parameters for how projects are managed successfully. • Organizational levels may relate to the length of projects, at least in that top management must have a long-term view of the future. Thus, at higher organizational levels, many projects may be established to pave the way for the long-term mission of the organization.
7. Evaluating strategic contents, distinguishing between effectiveness and efficiency	• For the successful management of multiple projects, it is important to distinguish whether the projects are established for effectiveness or for efficiency. Effectiveness refers to doing the right thing, and efficiency refers to doing the thing right. Effectiveness often means creating something new; efficiency means perfecting something that is already known.

information. Second, concerning support for structured and flexible decision making, flexible practices are needed to allow freedom to adjust the project management approach to fit the project type or its strategic importance. For example, creativity and the emergence of new strategic directions should be allowed in innovative project schemes. This could be achieved by avoiding too centralized and/or too formal management schemes in such projects.

Finally, the following sections of this chapter discuss empirical examples, and the conclusion section at the end of this chapter introduces a framework for strategic business management with suggestions for managerial practices as derived from theoretical and empirical reasoning.

Empirical Examples from Four Case-Study Organizations

The empirical examples discussed in the following pages are based on a study carried out with four case organizations—two private and two public. The investigated project environments included organizational and operational development projects and product development projects. Depending on the case study corporation, our empirical study focused either on organization-unit-/business-unit-specific project portfolios, or on cross-functional project portfolios of a certain project type (e.g., projects with strong IT orientation) across the whole corporation. We call the four organizations C-service, D-engineering, E-maintenance, and M-service. C-service is a large public organization delivering services locally for society and individuals. D-engineering is a large public organization delivering services for society. E-maintenance is a medium-sized private organization with engineering services, systems, and equipment deliveries for industrial customers. M-service is a large private organization with mainly service product deliveries for individual and industrial customers. The empirical examples discussed in the following are partly derived from the extensive empirical analysis and documentation of company-specific data, produced by our colleagues in our research team. Lindblom (2002), Elonen (2002), Hongisto (2002), and Nurminen (2003) are examples of such material. The majority of the documents are proprietary. The object of the research was the broad area of the management of project portfolios in the case study organizations. The methodology was modified from the developmental workshop scheme suggested by Järvinen et al. (2000). The empirical data gathering occurred in 2001 to 2004. We and our colleagues acted in the role of university researchers representing external change agents and facilitators for workshops, piloting efforts, and other developmental schemes. This way, our empirical study represented an organizational and operational change project from the company representatives' viewpoint.

Decision-Related Processes

Figures 1.4 and 1.5 and the following discussion illustrate a multiproject management process as generalized from case organization-specific processes. In the processes with all the case study organizations, the role of making decisions both at the single-project level and at the level of multiple projects was central. In the organization-specific processes, however,

FIGURE 1.4. DECISION-MAKING FROM THE PERSPECTIVE OF ONE SINGLE PROJECT.

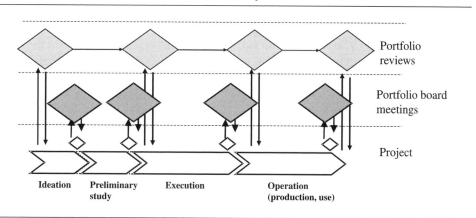

FIGURE 1.5. DECISION-MAKING WITH MULTIPLE PROJECTS.

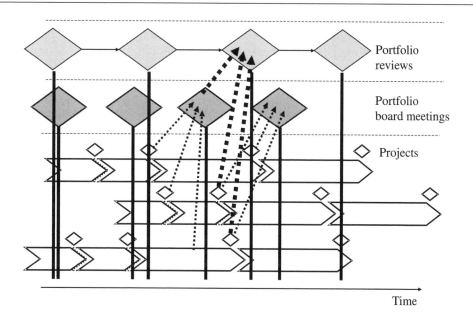

the extent and role of top-down and bottom-up components of the strategy varied depending on the organization. Our process model divides decision-making power at three different levels and emphasizes the communication and information sharing between those levels (see Figure 1.4).

The portfolio board consists of managers from different organizational units or responsibility areas. The cross-organizational composition of the board increases the variety of perspectives on organization within the board and enhances communication and discussion between different organizational areas. This develops a holistic view not only for selecting and prioritizing the right projects, but also for appropriate decisions on resource allocation and timing issues. Thus, the portfolio board is responsible for go/no-go decisions and decisions on major project-specific modifications. The strategic considerations at this decision level concentrate on ensuring that the set of projects under execution and new project ideas provide the best possible basis to achieve organizational objectives and goals. The portfolio board decision meetings strongly support the implementation of intended strategies through projects. The decision flow must be structured in a way that the meeting enhances the possibility of new ideas being introduced effectively, and in that sense also enforces the emergence of new strategies.

While the portfolio board meetings are strongly focused on making project-specific decisions project by project, whole project portfolios are considered in portfolio review meetings (see Figure 1.5). Portfolio reviews adopt a view on whole portfolios rather than single projects as strategic vehicles of action to link current reality and intended positions. Portfolio review meetings are held by boards/bodies consisting of managers representing a higher organizational level than the members of portfolio boards. The objective of the portfolio review meeting is to create a strategic snapshot of the portfolio of projects under execution and new project ideas, and to use this whole entity as a roadmap in planning guidelines for future objectives. The aim of the strategic decisions at this level is to create a feasible match between the organization's capabilities and the resources and opportunities and risks of today and the future.

The vertical arrows up from the level of single projects and down from upper-level decision-making points in Figure 1.4 indicate information and communication flows that are essential for the whole decision-oriented process for the strategic management of multiple projects. The arrows describe information and communication flows that are essential inputs and outputs for respective decision points. Figure 1.5 develops this scheme further by illustrating how decisions made at the level of single projects, and information derived from lower levels, serve as important triggers for decision making at upper levels and simultaneously for strategy implementation and emergence. Figure 1.5 illustrates also how portfolio board meetings and portfolio reviews occur at discrete points in time. Projects and project ideas enter into those upper-level meetings and reviews. The strategic management of multiple projects toward the achievement of business benefits/advantage requires the dynamic comparison of portfolios of project ideas, ongoing projects, and already completed projects. New project ideas, ongoing projects, and completed projects (e.g., internal IT systems, or existing offering of products) potentially affect decision making on any portfolio or any project. Thus, they should all be considered as belonging to the same pool of interlinked activities and opportunities.

Decision-Making Flows in Board Meetings

The following examples of board meetings in M-service and C-service describe the content of decision-making practices at the multiple-projects level. In M-service, the two management boards (i.e., portfolio board for gate decisions, and top management board for reviews) are preexisting boards, but the portfolio focus and related systematic managerial practices introduced new responsibilities and tasks for these managerial bodies. The portfolio review meetings take place two or three times a year and serve as a forum for top management members to discuss and determine the strategic guidelines and objectives for the portfolio. Our discussion here concerns the monthly portfolio board meetings that provide a vital basis for strategy implementation and the emergence of new ideas. The decision flow of the portfolio board meeting in M-service starts by reviewing the information of project reports (including project reports for new project ideas) provided by responsible project managers or owners of the projects/ideas. There must be adequate information that is considered as sufficiently valid and reliable to proceed to a scoring discussion aimed at achieving consensus against a variety of decision criteria. The meeting proceeds by comparing the project's score with the average score of all projects in the portfolio. This prepares the next step of accepting or rejecting projects, and for allocating priorities and resources among projects. The aim of the balancing is to compare ongoing projects to new project ideas by taking into account at least the most important parameters related to strategic importance, benefits, risk, and resources. The actual balancing considerations may use visual graphs as inputs for discussion (for an example of such graph, see Figure 1.6). Decisions on resources are the most important outputs of the discussion. The final step of the meeting includes deriving feedback to be delivered to projects. The feedback includes both written information and information to be explained orally to responsible project individuals. The information comprises the most relevant decision issues related to progress of the project and an explanation on reasons for the decision.

As the focus in C-service's application is on IT project portfolios, the portfolio board consists of managers responsible for IT projects from different functional areas. New project ideas and ongoing projects are evaluated and prioritized in the meeting, and decisions on resource allocations are made. The meeting agenda is not as structured/formal as in the case of M-service, but the discussions are stimulated by using visual Web-based IT tool as an catalyst to capture different views of the current situation of the portfolio of projects. The chair of the portfolio board facilitates the decision-making situation in the meeting. The Web-based application includes important parameters of projects recorded by the responsible project managers or project owners prior to the meetings. The IT tool provides semistructured project-related information as an input for the decision making. This information enforces discussion, and with the help of the facilitator, consensus is achieved and decisions made. An important advantage of structuring decision meetings around the Web-based IT tool is that it provides a shared communication channel to enhance two-way information sharing between projects and the management levels above them. This channel is used to integrate the organizational vision, strategies, and the actual project work.

Information Contents and Decision Criteria

The important information content for the information flows between the project and portfolio board levels was investigated separately for each decision point through the project

FIGURE 1.6. VISUAL GRAPH ON SCORE PROFILES OF MULTIPLE PROJECTS.

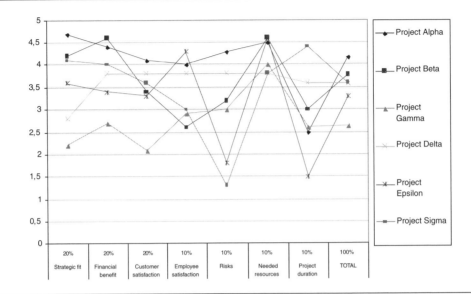

process. The strategic information contents in decision points change during the project process. In the early ideation phase, the information is strongly focused on result-related issues of the project from the business perspective. In the execution phase, the information relates more to monitoring pre-estimated strategic issues and updating estimates. Furthermore, in the post-project phase, the information relates more to the start of the application (going operational or production start) and to gathering feedback information from the operation phase for the purpose of learning for future projects.

The practical applications in organizations with explicit information displays are simplified schemes from contents of strategic information. For example, M-service applies a condensed set of simple criteria for project selection, covering the following categories of criteria: strategic importance, benefits, risk, and resources. As the major current challenge in M-service is to deal with the problems of organizing and resourcing its multiproject efforts vis-à-vis its extensive number of ongoing projects and scarce resources, the criteria set in M-service includes also many issues that relate to the project execution phase. This orientation is reflected by criteria with a major aim of monitoring issues that relate to execution and the successful implementation of the project work.

The portfolio board meetings in M-service are provided with additional structure by allowing the individuals to encode their opinions and beliefs with many criteria in terms of estimating quantitative scores for each criteria prior to the meetings. The scores and weights for criteria allow wide possibilities for preparatory calculations and visual illustrations that can be used as catalysts for communication and decision making that occurs in the actual portfolio board meeting.

Figure 1.6 is a visual look-alike sample graph from M-service's application that illustrates how weighted scores can be used for creating a simultaneous view of many projects at one point of time. This kind of figure can also serve as a tool for evaluating whether intended projects as a whole are effective for fulfilling the strategic objectives. Scoring, quantitative data, and visual illustrations can be used to structure the discussion among meeting attendants on how each project and the project portfolios as a whole contribute to the overall achievement of business benefits, and how well they fulfill strategic business objectives. Such aggregate visual information is helpful especially in portfolio review meetings among top management representatives, where the focus should be more on the information that relates to portfolios as an entity, rather than just to project-specific information. Illustrations with aspects of timing and timely perspectives, for example, roadmap-like presentations, are effective in many decision situations. However, in addition to recording scores for each criteria in M-service, the explicit information recorded for each criterion also includes qualitative and other quantitative information (e.g., monetary figures from economic/financial calculations, resource usage/needs, timely units indicating the schedule). The importance of qualitative information (i.e., documentation, explanations) is emphasized in the M-service's decision process. Finally, we believe that the true content for decision making and related important communication and learning occurs in board meeting events themselves rather than in explicit information contents in practical applications.

The explicit information contents of the project-specific parameters in C-service's application do not represent in a straightforward manner decision criteria as such, but rather represent the relevant information contents to be communicated for decision making and other management purposes. Furthermore, as compared to M-service, C-service emphasizes even more the importance of recording explicit qualitative information for each parameter. Such qualitative information is recorded in the Web-based IT tool. However, for enhanced communication and for increased clarity in comparing projects in a portfolio, many qualitatively expressed parameters (e.g., status, or priority, benefit, and risk) are categorized into classes. Such subdivision into classes, marked with integer figures in C-service, has some analogies with the application of scores in M-service. The qualitative and quantitative information contents in C-service's www tool include, among others, the following important themes: priority, benefit, risk, interconnectedness and linkages to other projects, contact data of responsible project manager and project owner, start date and complete date, cost, resource usage (person-hours), percent complete, and status. Much of the information in C-service's current application reflects the need to organize effectively ongoing projects and manage project progress, resources, cost, and time in a multiproject scheme, where the constraint of scarce resources and the interrelatedness of projects play an important role.

Conclusion: A Framework for Strategic Business Management

Figure 1.7 summarizes the preceding analysis by presenting a framework for strategic business management in multiple-projects environment. For successful multiproject management, it is essential that the managers and decision makers understand the sphere of ultimate

FIGURE 1.7. FRAMEWORK FOR STRATEGIC BUSINESS MANAGEMENT THROUGH PROJECTS.

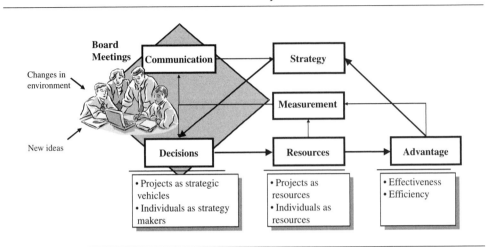

potential purposes of any project or idea. This occurs only if a mature understanding is in place that makes a clear distinction between effectiveness and efficiency.

Meetings, reviews, workshops, or other communication platforms where a group of individuals are collected together are central elements in strategy formation. Such occasions and situations are not only fostering grounds for communication and creative implicit or explicit decisions but also for new ideas. New ideas often arise simultaneously while a well-planned structure is applied for decision making in a group of individuals. However, such a structure should leave enough room for communication and/or expressing new and even radical ideas that fall outside the scope of the actual and concrete decision-making situation at hand.

Figure 1.7 illustrates the central role of a board meeting when managing strategies through multiple projects. When the issue concerning a set of multiple projects is brought to such a meeting through structured information and/or appropriate visual display of such information, and through a well-structured meeting agenda or well-facilitated meeting flow, both explicit and implicit decisions over multiple projects occur in an effective manner. Such information, visualization, agendas, and flows serve as structures that guarantee effective decisions and support for the realization of strategic intentions of the organization, produced in a creative manner, while simultaneously allowing the appearance of creative and innovative new ideas that reformulate strategies and strategic directions.

The corporate strategy (or business strategy) provides the individuals with guidelines, goals, and objectives for decision making. The dynamic nature of strategy implementation is supported by measuring both the achievement of advantages and the resources as the organization's internal capabilities, in relation to the requirements set by the project initia-

TABLE 1.2. MANAGERIAL PRACTICES FOR STRATEGIC BUSINESS MANAGEMENT THROUGH MULTIPLE PROJECTS BY FOCAL AREAS.

Focal Area	Managerial Practice
1. Categorizing projects by their type	• Form specific portfolios or buckets based on strategic guidelines. Consider strategic goals and responsibility areas while doing this. • Establish specific and tailored management models both at the project level and at levels above projects for each portfolio. Ensure that these models enable strategic management in an appropriate manner.
2. Supporting structured and flexible decision making	• Establish meetings, reviews, or workshops where a group of individuals are collected together, to serve as central elements for decision making. • Define specific levels where decision making is expected to occur. Distinguish the different roles of top management, middle management, and project management. • Differentiate operational single project decisions from the strategic ones. Authorize project managers or middle managers to make most operational project decisions. Furthermore, extend strategic considerations to include simultaneous consideration of multiple projects. • Distinguish between two types of portfolio-level decision making. Portfolio board meetings serve as a frequent forum for active monitoring and decision making to ensure that the structure of portfolio aligns with intended strategic guidelines. Portfolio reviews are typically organized few times in a year, and their focus is in strategic future-oriented planning and monitoring the overall situation of the portfolio. • Establish clear and limited roles and responsibilities for decision making. Assign a responsible individual (e.g., portfolio coordinator) for each portfolio, to take the overall responsibility of introducing the situation of the portfolio. • Avoid unnecessary rigidity while following the intended strategy. • Appropriate visual display of structured information, a well-structured meeting agenda, and well-facilitated meeting flows enhance not only effective explicit and implicit decisions but also the appearance of creative and innovative new ideas that reformulate strategies and strategic directions. • Establish criteria that enable comparison, selection, and prioritization of projects. Include strategic issues and project-success-related issues in those criteria. • Organize for measurement of projects, activities, and portfolios. Ensure that measurement is in line with established criteria and strategic guidelines. • Leave room for interactive discussion as principal element of decision making in meetings and boards.

TABLE 1.2. (*Continued*)

Focal Area	Managerial Practice
3. Ensuring effective communication and information transparency	• Support communication by establishing and using systematic structures such as: • Structured meeting agendas for board meetings • Systematic follow-up and measurement of portfolios of projects • Project type specific criteria and decision tools for project prioritization, and for stimulus for discussion • Top-down flow of feedback information down to projects • IT tools that enable the availability of project-related information vertically and horizontally in the organization. • Enhance learning, both directly by allowing experimental schemes or small probes that may fail and indirectly by exchanging experiences in meetings. • Use projects themselves as structured communication platforms, similar to meetings: Both meetings and projects as such bring individuals together to the same structured sphere of communication, decision making, and fostering of new ideas.
4. Linking projects and strategy process	• Make strategic portfolio review meetings timely (e.g., meetings related to the annual strategy process). Use effectively the advantage of knowing the situation of the current portfolio of projects when forming strategic guidelines for the organization. • Ensure a fluent interaction between different organizational levels, and make sure that projects are put into strategic perspective by looking at project entities as a whole (e.g., in portfolio review meetings) and by looking at how these entities contribute to new strategic dimensions. The interaction between organizational levels can be achieved through a cascade of meetings across lower and higher organizational levels. • Introduce top-management-originated intended strategies to lower-level boards and organizational bodies. Furthermore, emphasize the importance of feedback by providing the lower level bodies and boards with top management's decisions concerning projects or portfolios. • Use visualization methods in portfolio review meetings among top management representatives to reflect the project portfolios' role in the adaptation of new strategic directions through the set of projects and their strategic content as a whole. The visualization can, for example, be a time phased, roadmap-like view that paves the potential paths for the future.

TABLE 1.2. (*Continued*)

Focal Area	Managerial Practice
5. Establishing an organizational design to support strategic management in the multiple-project environment	• Define the role of different bodies and individuals at different organizational levels, especially the responsibilities and authorization for decision making. • Use (preferably) existing board structures that are assigned new tasks and responsibilities. • Pay attention to appropriate level of openness, trust, and encouragement of individuals. Top management support is an important factor. • Create clear ownership for each functional activity, cross-organizational process activity, and portfolio of projects. Recognize the overlapping areas of responsibility, and deal with such complexity by organizing for effective communication and information sharing. • Establish project-office-like organizational bodies or responsibilities for such supportive activities that ensure effective support for the overall complex setting of managing multiple projects. • Plan carefully how centralized or how decentralized different decision-making-related activities are. Match the level of centralization/decentralization to fit the organizational culture.
6. Setting and measuring goals for different time spans in the future	• Use roadmap-like presentations that put the projects and their mutual interrelations into timely perspective. • Use supportive illustrations that emphasize the life cycle perspectives (both product life cycle and project life cycle), in order to understand the relationship of new or existing projects to current products or systems and their life cycles. • Analyze stated assumptions carefully, and make different scenarios of the business environment in the uncertain future. This is especially important for understanding the potential outcomes in the long-term future. • Establish effective risk management or uncertainty management procedures for coping with the imperfect knowledge, ambiguity, and uncertainty. Manage options in an effective manner; sometimes one important strategy is to keep options open as long as possible.
7. Evaluating strategic contents, distinguishing between effectiveness and efficiency	• Make a clear distinction between those projects that are driven by improving effectiveness and those that are driven by improving efficiency. • Evaluate the strategic importance of each project for understanding the type of strategic impact produced by the project. Furthermore, estimate the managerial challenge and need for a specific management style that relates typically to newness and risk dimensions in the project.

tives and the external environment. In our framework, projects are used for strategy implementation and emergence. The framework also emphasizes the role of individuals as strategy makers. Individuals' commitment and motivation often guarantee that the intended strategies are in fact realized. Furthermore, the role of projects and their individuals are important strategy makers and remarkable introducers of new ideas. This occurs as projects serve as structured communication platforms with similar impacts to what was discussed previously regarding meetings and group sessions. Both meetings and projects as such bring individuals together to the same structured sphere of communication.

Although we emphasize the role of projects and individuals both at the project level and at levels above projects as *strategy makers*, our framework shows that projects and individuals also can be seen in another role: *resources*. It may be clear that projects in the execution phase can be interpreted as resources, but we emphasize here that even ideas at a very early pre-execution phase are important resources that carry important issues related to the future, often in terms of the potential or strategic business content embedded in the idea. When projects are thought of as resources, measuring resources can be seen as measuring current project-based activities and new projects ideas. Having a clear picture of the current situation of the organizational realities at the project level (capabilities), and comparing that with the desired state of the future, provides a frame for successful decision making in the multiple-projects environment.

Our framework emphasizes the role of face-to-face discussions and communications that takes place in certain specific contexts—for example, meetings or projects. An important managerial challenge is to avoid unnecessary rigidity and sometimes even too much discipline while following the intended strategy. Discussions characterized by various perspectives and stimulated by measures that support flexibility and creativity, are the necessary components for emerging of new strategies. Decisions—small and big—then determine future outcomes in terms of effectiveness and efficiency. The framework suggests that advantages and benefits result primarily from individuals' (company and business unit managers') decisions on projects and project ideas, and individuals' (project managers' and project team members') decisions made in single-project contexts. Finally, Table 1.2 concludes this chapter with what we perceive to be the most important managerial practices in the framework of strategic business management in a multiproject environment.

Acknowledgments

Professor David L. Hawk participated in meetings in the early stages when writing this chapter. Furthermore, he provided valuable comments to the early versions of the manuscript. Our research colleague M.Sc. (Eng.) Merja I. Nurminen helped the writing process with fruitful discussions, while simultaneously conducting her own research scheme on strategic management with project portfolios. In the editor's role, Professor Peter W. G. Morris put in extensive effort by providing us with helpful suggestions and constructive comments for how to make the chapter much better. Professor David L. Hawk, M.Sc. (Eng.) Merja I. Nurminen, and Professor Peter W. G. Morris all deserve our greatest thanks for their most valuable help.

References

Aalto, T., M. Martinsuo, K. A. Artto. 2003. Project portfolio management in telecommunications R&D: Aligning projects with business objectives. 99–147. In *Handbook of product and service development in communication and information technology*, ed. T. O. Korhonen and A. Ainamo. Boston: Kluwer Academic Publishers.

Aaltonen, P., and K. Ikävalko. 2001. Implementing strategies successfully. *XII World Productivity Congress. Track One: Cultivating Innovation*. Hong Kong and Beijing, November 5–10.

Ackoff, R. L. 1978. *The art of problem solving*. New York: Wiley.

———. 1981. *Creating the corporate future*. New York: Wiley.

———. 1994. *The democratic corporation: A radical prescription for recreating corporate America and rediscovering success*. New York: Oxford University Press.

———. 1999. *Ackoff's best: His classic writings on management*. New York: Wiley.

Andrews K. R. 1971. *The concept of corporate strategy*. Homewood, IL: Irwin.

———. 1995. The concept of corporate strategy. In *The strategy process*. H. Mintzberg, J. B. Quinn, and S. Ghoshal. Prentice Hall London Exerpted from K. R. Andrews. 1980. *The concept of corporate strategy*. Copyright Richard D. Irvin, Inc.

Ansoff, H. I. 1965. *Corporate strategy*. New York: McGraw-Hill.

Archer, N. P., and F. Ghasemzadeh. 1999. An integrated framework for project portfolio selection. *International Journal of Project Management* 17(4):207–216.

Artto, K. A. 2001. Management of project-oriented organisation: Conceptual analysis. In *Project portfolio management: Strategic management through projects*, ed. K. A. Artto, M. Martinsuo, and T. Aalto, 5–20. Helsinki, Finland: Project Management Association Finland.

Artto, K. A., P. H. Dietrich, and T. Ikonen. 2002. Industry models of project portfolio management and their development. In *Proceedings of the PMI Research Conference 2002*, ed. D. P. Slevin, J. K. Pinto, and D. I. Cleland. 3–13. Seattle, July 14 to 17. Newtown Square, PA: Project Management Institute.

Bourgeois, L. J. 1985. Strategic goals, perceived uncertainty, and economic performance in volatile environments. *Academy of Management Journal* 28(3):548–573.

Brown, S. L., and K. M. Eisenhardt. 1995. Product development: Past research, present findings, and future direction. *Academy of Management Review* 20(2):343–378.

———. 1997. The art of continuous change: Linking complexity theory and time-paced evolution in relentlessly shifting organizations. *Administrative Science Quarterly* 42(1):1–34.

Burgelman, R. A. 1983. A model of the interaction of strategic behavior, corporate context, and the concept of strategy. *Academy of Management Review* 8(1):61–70.

———. 1991. Intraorganizational ecology of strategy making and organizational adaptation: Theory and field research. *Organizational Science* 2(3):239–262.

Chaffee, E. E. 1985. Three models of strategy. *Academy of Management Review* 10(1):89–98.

Chandler, A. D. 1962. *Strategy and structure: Chapters in the history of the American industrial enterprise*. Cambridge, MA: MIT Press.

Cohen, M. D., J. G. March, and J. P. Olsen. 1972. A garbage can model of organizational choice. *Administrative Science Quarterly* 17:1–25.

Cooper, R. G. 1994. Debunking the myths of new product development. *Research Technology Management* (July–August): 40–50.

Cooper, R. G., and E. J. Kleinschmidt 1987. New products: What separates winners from losers. *Journal of Product Innovation Management* 4.

Cooper, R. G., S. J. Edgett, and E. J. Kleinschmidt. 1998. *Portfolio management for new products*. New York: Perseus Books.

Crawford, L., J. B. Hobbs, and J. R. Turner. 2002. Investigation of potential classification systems for projects. *PMI Research Conference 2002.* pp. 181–190. Seattle, July 14–17. Newtown Square, PA: Project Management Institute.

De Geus, A. P. 1988. Planning as learning. *Harvard Business Review* (March–April): 70–74.

DeMarco, T. 2001. *Slack.* New York: Random House.

Drucker, P. F. 1974. *Management. tasks, responsibilities, practices.* London: Heinemann.

———. 1999. *Management challenges for the 21st century.* New York: Harper Business.

Dye, L. D., and J. S. Pennypacker. 1999. An introduction to project portfolio management. In *Project portfolio management: Selecting and prioritizing projects for competitive advantage,* ed. L. D. Dye and J. S. Pennypacker xi–xvi. West Chester, PA: Center for Business Practices.

Eisenhardt, K. M., and S. L. Brown. 1998. Time pacing: Competing in markets that won't stand still. *Harvard Business Review* (March–April):59–69.

Eisenhardt, K. M., and B. N. Tabrizi. 1995. Accelerating adaptive processes: Product innovation in the global computer industry. *Administrative Science Quarterly* 40:84–110.

Elonen, S. 2002. Project portfolio management: Managerial problems and solutions for business development portfolios. Master's thesis, Helsinki University of Technology.

Elonen, S., K. A. Artto. 2003. Problems in managing internal development projects in multi-project environments. *International Journal of Project Management* 21(6):395–402.

Floyd, S. W., and B. Woodridge. 2000. *Building strategy from the middle: Reconceptualizing strategy process.* Thousands Oaks, CA: Sage Publications.

Fredrickson, J. 1984. The competitiveness of strategic decision processes: Extension, observations, future directions. *Academy of Management Journal* 27:445–466.

Galbraith, J., 1995. *Designing organizations: An executive briefing on strategy, structure and process.* San Francisco: Jossey-Bass.

Gareis, R. 2000a. Programme management and project portfolio management: New competences of project-oriented companies. *Fourth International Conference of the International Research Network on Organising by Projects IRNOP IV.* Sydney, January 10th–12th.

———. 2000b. Competences in the project-oriented organization. In *Project Management Research at the Turn of the Millennium: Proceedings of the PMI Research Conference, ed.* D. P. Slevin, D. L. Cleland, and J. K. Pinto. 17–22. Newtown Square, PA: Project Management Institute.

Goldratt, E. M. 1997. *Critical chain.* Great Barrington, MA: North River Press.

Groenveld, P. 1997. Roadmapping Integrates Business and Technology. *Research-Technology Management* (September–October):48–55.

———. 1998. The roadmapping creation process, Presentation at the Technology Roadmap Workshop, Washington, D.C., October 29.

Hart, S. L. 1992. An integrative framework for strategy making process, *Academy of Management Review* 17(2):327–351.

Hatch, M. 1997. *Organization theory: Modern, symbolic and postmodern perspectives,* New York: Oxford University Press.

Hofer, C. W. 1975. Toward a contingency theory of business strategy. *Academy of Management Journal* 18:784–810.

Hongisto, J. 2002. Project selection and decision making for successful project portfolio management. Master's thesis, Helsinki University of Technology.

Järvinen, P., K. A. Artto, and P. Aalto. 2000. Explorations on the integration of fractured process improvement: The 3A-workshop procedure. *Project Management* 6(1):77–83.

Kaplan, R. S., and D. P. Norton. 1992. The balanced scorecard: Measures that drive performance. *Harvard Business Review* (January–February).

———. 1996. *The balanced scorecard: Translating strategy into action.* Boston: Harvard Business School Press.

————. 2001. Transforming the balanced scorecard from performance measurement to strategic management: Part I. *Accounting Horizons* 15 (1, March): 87–104

Kostoff, R. N., and R. R. Schaller. 2001. Science and technology roadmaps. *IEEE Transactions on Engineering Management* 48(2):132–143.

Kotsalo-Mustonen, A. 1996. Diagnosis of business success: Perceptual assessment of success in industrial buyer-seller business relationship. PhD. diss., Helsinki School of Economics and Business Administration, Publications A-117, Helsinki, Finland.

Lindblom, C. E. 1959. The science of muddling through. *Public Administration Review*, pp. 79–88.

Lindblom, L., 2002. A decision support system for project prioritization and selection: A case study. Seminar study, unpublished company report, Helsinki University of Technology.

Loch, C. 2000. Tailoring product development to strategy: Case of a European technology manufacturer. *European Management Journal* 18 (3, June): 246–258

Lundin, R., and R. Stablein. 2000. Projectisation of global firms: Problems, expectations and meta-project management. Fourth International Conference of the International Research Network on Organising by Projects IRNOP IV, Sydney, Australia, January 10–2th.

McGrath, M. E., ed. 1996. *Setting the PACE in product development: A guide to Product and cycle-time excellence.* Boston: Butterworth-Heinemann.

Meredith, J. R., and S. J. Mantel, Jr. 1999. Project selection. In *Project portfolio management: Selecting and prioritizing projects for competitive advantage*, ed. L. D. Dye and J. S. Pennypacker. 135–167. West Chester, PA: Center for Business Practices.

Mikkelsen, H., W. Olsen, J. O. Riis. 1991. Management of internal projects. *International Journal of Project Management* 9(2):77–81.

Mintzberg, H. 1978. Patterns in strategy formation. *Management Science* 24:934–948.

Mintzberg, H. and Waters, J. A. 1985. Of strategies, deliberate and emergent. *Strategic Management Journal* 6:257–272.

Mintzberg, H., J. B. Quinn, and S. Ghoshal. 1995. *The strategy process.* Prentice Hall London.

Morris, P. W. G., and G. H. Hough. 1987. *The anatomy of major projects: A study of the reality of project management.* Chichester, UK: Wiley.

Murray-Webster, R., and M. Thiry M. 2000. Managing programmes of projects. In *Gower handbook of project management*, ed. J. R. Turner and S. J. Simister. 3rd ed. 47–63. Aldershot, UK: Gower.

Nurminen, M. 2003. Strategic management with project portfolios. Master's thesis, Helsinki University of Technology.

Office of Government Commerce (OGC). 2002. *Managing successful programmes.* 3rd ed. London: The Stationery Office.

Pavón, R. C. C. 2002. Systemic intra-organizational industry demand analysis to support an organization's strategy making. PhD diss., Helsinki University of Technology, Institute of Strategy and International Business.

Perlow, L. A. 1999. The time famine: Toward a sociology of work time. *Administrative Science Quarterly* 44:57–81.

Pettigrew, A. M. 1992. On studying managerial elites. *Strategic Management Journal* 13:163–182.

Platje, A., H. Seidel, and S. Wadman. 1994. Project and portfolio planning cycle: Project based management for multiproject challenge. *International Journal of Project Management* 12(2):100–106.

Project Management Institute. 2000. *A guide to the Project Management Body of Knowledge.* Project Management Institute Standards Committee. Newtown Square, PA: Project Management Institute.

Porter, M. E. 1980. *Competitive strategy: Techniques for analyzing industries and competitors.* New York: Macmillan.

————. 1996. What is strategy. *Harvard Business Review.* (November–December).

Poskela J., M. Korpi-Filppula, V. Mattila, and I. Salkari. 2001. Project portfolio management practices of a global telecommunications operator. In *Project portfolio management: Strategic management through projects*, ed. K. A. Artto, M. Martinsuo, and T. Aalto. 81–102. Helsinki: Project Management Association.

Quinn, J. B. 1985. Managing innovation: Controlled chaos. *Harvard Business Review*, (May–June):73–84.

———. 1995. Strategies for change. In *The strategy process*. H. Mintzberg, J. B. Quinn, and S. Ghoshal. Prentice Hall London, Europe. Excerpted from J. B. Quinn. 1980. *Strategies for change: Logical incrementalism*. Copyright Richard D. Irwin, Inc.

Rouhiainen, P. 1997. Managing new product development: Project implementation in metal industry. PhD diss., Tampere University of Technology, Publications 207, Tampere, Finland.

Rämö, H., 2002. Doing things right and doing the right things: Time and timing in projects. *International Journal of Project Management* 20 (7, October): 569–574.

Saravirta, A. 2001. Project success through effective decisions: Case studies on project goal setting, success evaluation, and managerial decision making. PhD. diss., Acta Universitatis Lappeenrantaensis 121, Lappeenranta University of technology, Lappeenranta, Finland.

Schoonhoven, C. B., and M. Jelinek. 1996. Dynamic tension in innovative, high technology firms: Managing rapid technological change through organizational structure. In *Managing strategic innovation and change: A collection of readings*, ed. M. Tushman and P. Anderson. 233–254. New York: Oxford University Press.

Senge, P. M. 1990. The leader's new work: Building learning organizations. *Sloan Management Review* (Fall): 7–23.

Shenhar A. J., O. Levy, and D. Dvir. 1997. Mapping the dimensions of project success. *Project Management Journal* 28(2):5–13.

Shenhar, A. J., D. Dvir, T. Lechler, and M. Poli. 2002. *One size does not fit all: True for projects, true for frameworks*. 99–106. PMI Research Conference, Seattle, July 14–17. Newtown Square, PA: Project Management Institute.

Simon, H. A. 1957. *The new science of management decision*. New York: Macmillan.

Simons, R. 1995. *Levers of control: How managers use innovative control systems to drive strategic renewal*. Boston: Harvard Business School Press.

Steyn, H. 2002. Project management applications of the theory of constraints beyond critical chain scheduling. *International Journal of Project Management* 20 (1, January): 75–80.

Terwiesch, C., C. H. Loch, and M. Niederkofler. 1998. When product development performance makes a difference: A statistical analysis in the electronics industry. *Journal of Product Innovation Management* 15:3–15.

Turner, J. R., and A. Keegan. 1999. The management of operations in the project-based organization. In *Managing business by projects*, ed. K. A. Artto, K. Kähkönen, and K. Koskinen. Vol. 1, pp. 57–85, and Vol. 2. Helsinki: Project Management Association Finland and Nordnet

Turner, J. R. 1999. *The handbook of project-based management: Improving the processes for achieving strategic objectives*. 2nd ed. London: McGraw-Hill.

Turner, J. R., and A. Keegan. 2000. Processes for operational control in the project-based organization. In *Project management research at the turn of the millennium: Proceedings of the PMI Research Conference*, ed. D. P. Slevin, D. I. Cleland, and J. K. Pinto. 123–134. Newtown Square, PA: Project Management Institute.

Turner, J. R., A. Keegan, and L. Crawford. 2000. Learning by experience in the project-based organization. In *Project management research at the turn of the millennium: Proceedings of the PMI Research Conference*, ed. D. P. Slevin, D. I. Cleland, and J. K. Pinto. 445–456. Newtown Square, PA: Project Management Institute.

Wheelwright, S. C., and K. B. Clark. 1992. Creating project plans to focus product development. *Harvard Business Review* (March–April): 70–82.

Wolff, M. F., 1987. To innovate faster, try the skunk works. *Research Technology Management* (September–October: 7–8.

Yeo, K. T., and J. H. Ning. 2002. Integrating supply chain and critical chain concepts in engineer-procure-construct (EPC) projects. *International Journal of Project Management* 20 (4, May): 253–262.

Youker, R., 1999. The difference between different types of projects. *Proceedings of Project Management Institute Annual Seminars and Symposium*. Newtown Square, PA: Project Management Institute.

CHAPTER TWO

MOVING FROM CORPORATE STRATEGY TO PROJECT STRATEGY

Ashley Jamieson, Peter W. G. Morris

Developing and implementing corporate strategy is one of the most actively researched, taught, and talked about subjects in business today. Projects and project management are often quoted as important means of implementing strategy, but the way this implementation happens in practice is rarely the subject of detailed review. This chapter addresses this relationship head-on.

Many organizations move from corporate strategy to project strategy using highly structured and fully integrated sets of business management, strategic management, and project management processes, practices, and methods, deployed by highly skilled professional staff. This chapter looks at these means. The key findings from a series of studies of the way major organizations do this, together with a survey that we recently undertook of project management professionals' practice in this area are also included.

Developing Corporate Strategy

Corporate strategy is created as a means of thinking through and articulating how an organization's corporate goals and objectives will be pursued and achieved. This strategy is then typically cascaded through several strategic business units (SBUs) and then ends up being represented as collections—portfolios—of programs or projects. These become the vehicles for implementing the approved strategic initiatives.

Various processes are employed for doing this. Highly structured business management models are used by some of organizations; these identify the major "value delivery" processes and/or the key business processes that enable the organization to operate effectively. Much of traditional management writing tends only to cover the strategic management processes that formulate and implement strategy at the corporate level; there is a real dearth of writing about how corporate strategy gets translated into comprehensive program or project management strategies. The two sets of activities are, in practice, well interconnected and are the main means by which strategy is moved and aligned through corporate, business unit, and project levels. Where they often work less well, however, is in the feedback and adjustment process as events unfold and as program or project definition needs realigning (or vice versa: as project or program definition creates new situations that, in turn, impact the enterprise's strategic intent, thereby requiring the process to iterate!).

Developing effective strategy for major programs or projects from corporate and business strategies is a complex activity. It involves the use of key project management processes such as project definition and includes strategic elements from a wide range of project management practices, such as risk management, value management, and procurement management, and encompasses the entire project life cycle. (See the chapter by Thiry.) The roles, responsibilities, and accountabilities for those operating these processes and practices are usually identified within the process documentation, and their competency levels specified in competency frameworks or job descriptions. Recent research that we conducted shows that some companies have developed and implemented highly effective means of doing this. Many other organizations use less ambitious and effective approaches, however.

Positioning Project Management within a Business Management Context

To understand the way corporate strategy is translated into project strategy, it is important to start by considering the business management context and the position of project management within it, and how the business management functions perceive the project management function. While project management professionals and practitioners may think their function is central to the success of a company, it has little meaning unless it is clearly established and embedded within the organizational structure and business management processes of the enterprise. An indication of the position and the relative prominence given to project management can be gauged from the business management models used by organizations.

Watson (1994) gives McKinsey the credit for developing the generic business enterprise model shown in Figure 2.1. This model shows the structure of an organization in terms of processes rather than the traditional functional or matrix form and serves as a process framework for an organization.

The model also identifies what are considered to be the major processes of the value delivery system of a company and the key business processes that enable them, and implies a high level of connectivity between the two. Recent research findings by Morris and Jamieson (2004) show that many major organizations regard the management of projects as a key business process, and Figure 2.1 has been modified to reflect this fact.

FIGURE 2.1. GENERIC BUSINESS ENTERPRISE MODEL.

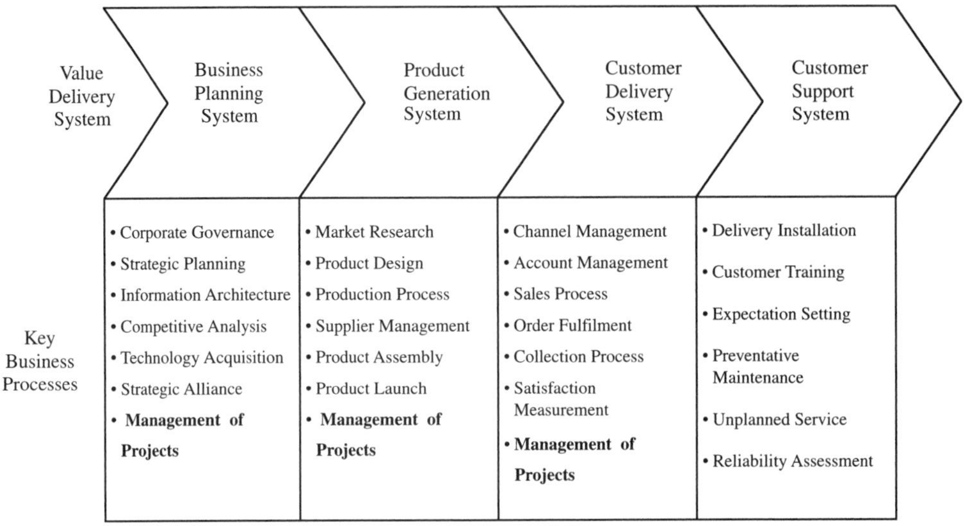

Value Delivery System	Business Planning System	Product Generation System	Customer Delivery System	Customer Support System
Key Business Processes	• Corporate Governance • Strategic Planning • Information Architecture • Competitive Analysis • Technology Acquisition • Strategic Alliance • **Management of Projects**	• Market Research • Product Design • Production Process • Supplier Management • Product Assembly • Product Launch • **Management of Projects**	• Channel Management • Account Management • Sales Process • Order Fulfilment • Collection Process • Satisfaction Measurement • **Management of Projects**	• Delivery Installation • Customer Training • Expectation Setting • Preventative Maintenance • Unplanned Service • Reliability Assessment

Source: Watson (1994); adapted by Morris and Jamieson (2004).

Creating Corporate and Business Strategies

The strategy management process is dealt with extensively by numerous authors. Most include the concepts and processes associated with strategy analysis, strategy creation (formulation), strategy evaluation, and strategy implementation. But few explicitly connect corporate and business unit strategy with project strategy or suggest that it should be taken into account at the strategy formulation stage or when determining the capability of an organization at the strategy implementation stage.

Strategy management is a dynamic process; and Mintzberg and Quinn (1996) show that "emergent" strategy is a key factor—that is, strategy that becomes evident as it, and events, emerge with time—in influencing realized strategy. Hill and Jones (2001) demonstrate how emergent strategy influences intended strategy through the components of the strategic management process, as, for example, those shown in Figure 2.2. This model indicates that strategy formulation flows from an organization's mission and goals through functional, business, and corporate levels. But, in common with much of the strategic management literature (Mintzberg, Ahlstrand, and Lampel, 1998, for example, point out that there are hundreds of different strategic planning models), Hill and Jones do not explicitly take account of the influence or impact of projects or project management activities on the creation and implementation of strategy, although it is clearly implied.

The findings presented in this chapter, however, show that most of the components of the strategic planning process, such as internal analysis, organizational structures, and con-

FIGURE 2.2. COMPONENTS OF THE STRATEGIC MANAGEMENT PROCESS.

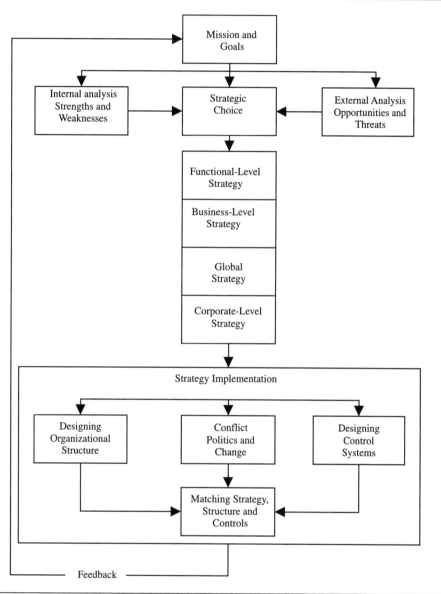

Source: Hill and Jones (2001).

trol systems, have strong links to project management processes and activities, and thereby strongly influence intended corporate and business strategies.

The availability of resources is a major factor in deciding the strategy of an organization. Figure 2.3 shows a schematic used by Mintzberg and Quinn (1996) to analyze strategy and serves to indicate some of the key activities involved in the formulation and implementation of strategy, one of which is determining the managerial resources of the organization. A fundamental responsibility of project management in organizations managing major projects is to manage and/or coordinate effectively whatever company resources are at its disposal. A company's skill at managing projects, and therefore its project management capability, is an essential managerial resource that can and does influence corporate and business strategies strongly—particularly in companies that see themselves in dynamic change situations where agility is important or that are driven by major projects (as in aerospace, construction, new product development, etc.). (See the chapters by Archer and Ghasemzadeh, Shenhar and Dvir.)

Hierarchy of Objectives and Strategies

A hierarchy of objectives and strategies can be formed as a result of using a strategy planning process; this can be a very effective means of structuring and managing strategy and communicating it to the organization. One such model, used by Cleland (1990), is Archibald's hierarchy of objectives, strategies, and projects, shown in Figure 2.4.

This model maps out in detail the structure and relationship of objectives and strategies at the policy, strategic, operational, and project levels. Specific objectives and strategies are developed at each level from higher-level ones and cascaded down, thereby ensuring alignment and continuity of strategy. Projects and their objectives, strategies, and plans are shown at the operational level.

Hierarchy of Strategy Plans

The corporate and business strategies created by organizations using a strategic planning process are incorporated into a hierarchy of strategy plans. Kerzner (2000) provides an example of a typical hierarchy, as shown in Figure 2.5.

This model shows typical strategic plans cascading corporate strategy to SBU level from a single corporate strategic plan and supporting plans cascading business strategy from each SBU. Another example of a hierarchy of strategic plans is the Stanford Research Institute's (SRI's) "system of plans," which is shown in Mintzberg, Ahlstrand, and Lampel (1998).

If this, roughly, is how corporate strategy is created from an organization's mission, goals, and objectives and how this strategy is used to create business unit objectives and strategies, the next stage is to consider how business unit strategy is moved through portfolios and programs into projects.

FIGURE 2.3. STRATEGY AS A PATTERN OF INTERRELATED DECISIONS.

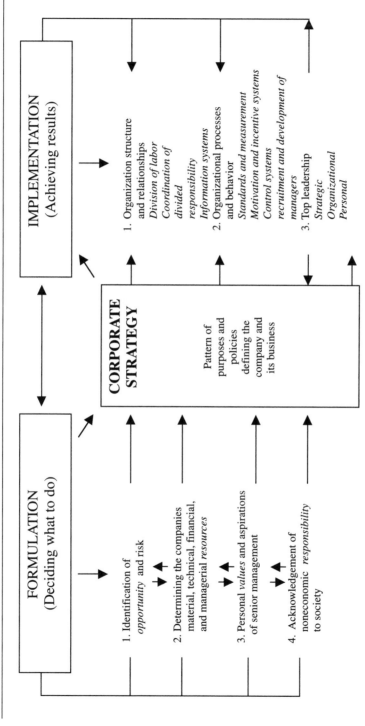

Source: Mintzberg and Quinn (1996).
© Andrews, K. R. (1999) Custom Ed. *The Concept of Corporate Strategy.* Reprinted by permission of McGraw-Hill Companies.

FIGURE 2.4. HIERARCHY OF OBJECTIVES, STRATEGIES, AND PROJECTS.

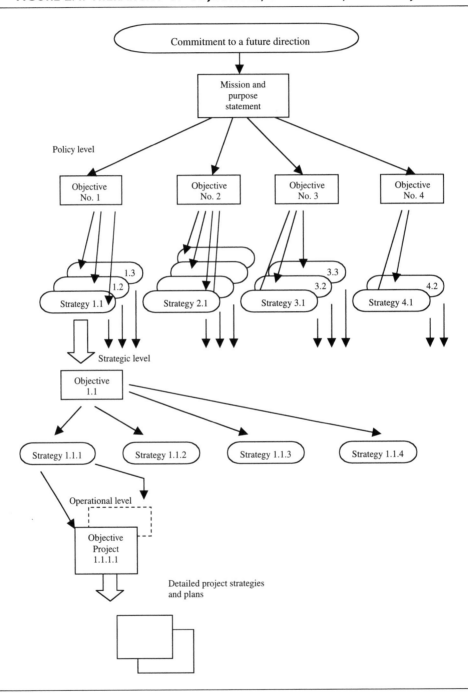

FIGURE 2.5. HIERARCHY OF STRATEGIC PLANS.

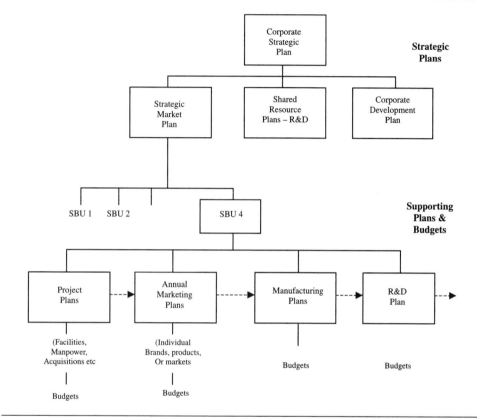

Source: Kerzner (2000).

Moving Business Strategy through Portfolios, Programs, and Projects

Turner (2000) points out that the majority of projects take place as part of a portfolio of several projects. Program management is the way of coordinating projects that have a shared business aim (see the chapter by Thiry). Both portfolio management and program management focus on prioritizing resources and optimizing the business benefit (Kerzner, 2000; Morris and Jamieson, 2004) Portfolio management, as Archer and Ghasemzadeh outline in their chapter, tends to be about the selection and prioritization of projects or programs. Program management is about the day-to-day management of programs (products, platforms, brands, or multiple projects) to deliver business value (OGC, 1999).

Turner (1999) cites Youker to illustrate how organizations undertake programs and projects to achieve their development objectives. Morris and Jamieson (2004) have adapted this model to include business strategy and portfolios, as shown in Figure 2.6, and to indicate that a portfolio may comprise of groups of programs and/or projects.

Portfolios

Project portfolio management is predominantly about choosing the right project, whereas project management is about doing the project right. It has also been described as the activity of aligning resource demand with resource availability, to achieve a set of strategic goals (see the chapter by Archer and Ghasemzadeh). Crawford (2001) believes that making the conceptual leap from the tools-and-techniques-focused variety of project management to portfolio management (and indeed program management), with its broader focus on business strategy and enterprise-wide integration, is a special challenge and one that many now face with little in the way of standards, best practices, or other generally accepted knowledge to guide them.

Knutson (2001) provides a generic project portfolio management process model, schematized in Figure 2.7, which takes strategy issues into account. The solicitation stage of the process ensures that a potential program or project has a credible strategy that is aligned with the organization's objectives and strategy; and that a business case for the project, containing these details, is developed. The relative value of the project and its overall synergy

FIGURE 2.6. LINKING CORPORATE AND PROJECT STRATEGY.

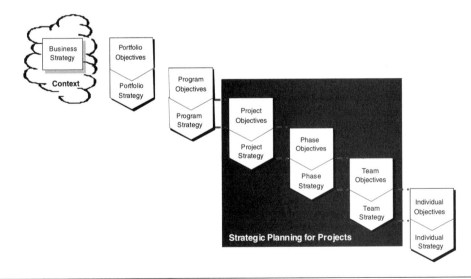

Source: The Handbook of Project-Based Management, 2nd ed. J. R. Turner © 1999 Reproduced by kind permission of the Open University Press, McGraw-Hill Publishing Company.

FIGURE 2.7. A GENERIC PORTFOLIO MANAGEMENT PROCESS.

Source: J. Knutson (2001) and schematized by Morris and Jamieson (2004).

with the organization's strategy is evaluated during the selection process. If a project is selected, the next step is its prioritization. A scoring system is used to determine the priority of the project with respect to other projects, during the prioritization stage (see Archer and Ghasemzadeh's chapter); and subsequently, depending upon availability, resources are allocated to the project to allow it to proceed.

Archer and Ghasemzadeh (1999) also provide a general framework for project portfolio selection that demonstrates the need for strategy to be set at corporate level and then filtered down to a project level.

Programs

Murray-Webster and Thiry (2000) suggest that the discipline of program management is emerging as a fundamental method of ensuring that an organization gains the maximum benefit from the integration of project management activities. They also suggest that program management involves more iteration and strategic reflection than the more "single-shot" project management, and they describe programs as "the missing link"—the means of effectively bridging the gap between a strategy, subjected to emergent change, and projects. (See also Thiry's chapter in this book.)

The UK Office of Government Commerce (OGC) (1999) considers the alignment between strategy and projects to be one of the main benefits of program management. The program management process they use comprises the following stages:

- Identifying, defining and establishing a program
- Managing the portfolio
- Delivering benefits
- Closing the program

They perceive the environment of program management to be that shown in Figure 2.8, which indicates programs emanating from business strategies and initiatives, and an iterative hierarchy of programs, projects, and business operations cascading from them.

The objectives and strategies for the programs are created and aligned with the objectives and strategies of the organization; and the objectives and strategies for individual projects are created and aligned with their respective programs. We will consider in detail how project strategy is created within projects.

FIGURE 2.8. THE PROGRAM MANAGEMENT ENVIRONMENT.

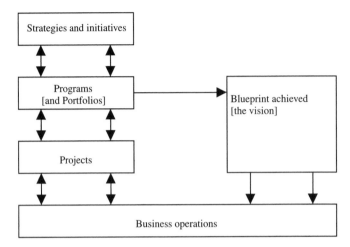

Source: UK Office of Government Commerce (OGC) (1999). Adapted by Morris and Jamieson (2004).

Projects

To move business strategy into project strategy most effectively, whether a project interfaces directly with business units or indirectly through portfolios and programs, there needs to be coherent project management processes that integrate seamlessly with the strategic management processes. We (the authors) developed a schematic to show the way business strategy is translated into project strategy at the front-end of a project, predicated on sections of the **PMI PMBOK** (2000) and the **APM BoK** (2000). This model, shown in Figure 2.9, comprises two stages and a number of key project management activities or processes within each stage. The stages are as follows:

- Translating business strategy stage, including:
  - Project definition (PMBOK)
  - Project scope management (PMBOK
  - Requirements management (APM BoK)
  - Strategic framework (APM BoK)
- Creating project strategy stage, including:
  - Project management planning and integration processes (PMBOK)
  - Project plans development process (PMBOK)
  - Generic project management knowledge and competencies (APM BoK)
  - Elements of project strategy

Translating Business Strategy in Projects. The project definition process is an essential part of developing project strategy from business strategy. Turner (1999) builds on a model for the strategic management of projects developed by Morris (1987). This model indicates that

FIGURE 2.9. CREATING PROJECT STRATEGY FROM CORPORATE AND BUSINESS STRATEGIES.

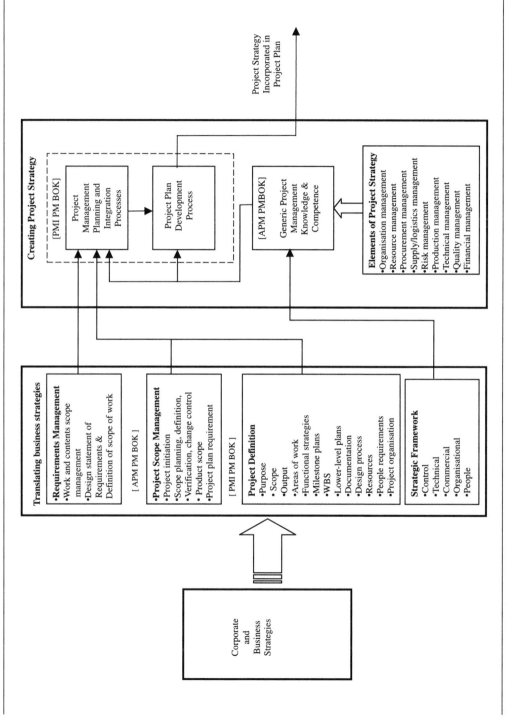

Source: Morris and Jamieson (2004).

projects are subjected to seven forces and the two that relate to strategy are identified as attitudes and definition.

Turner also advocates developing a comprehensive definition of a project at the start of the project, and is achieved by:

- setting the project objectives;
- defining the scope through a strategic, or milestone plan;
- setting the functional strategies for assessing technical risk;
- carefully managing the design process; and
- managing resources and the context.

Through the project definition process:

- the vision of the project is created;
- the purpose of the project is defined;
- the project plans are aligned with the business plans; and
- the basis of cooperation for the project is agreed upon.

The project scope management processes described in the **PMI PMBOK** (2000) identify strong links with corporate and business strategies. The initiation process (the first process of the **PMBOK** project scope management) links a project to the ongoing work in the performing organization and recognizes that a project can be authorized as a result of market demand or a business need. The inputs to the initiation process include the strategic plan of the performing organization, which is also used in project selection decisions and is seen as providing the link through which the project strategy is aligned to the organization's strategy. The output of the initiation process reflects the organization's strategy and is fed as an input to the second process of project scope management.

The second process is "scope planning," the output of which is the project scope statement containing the project justification, product, deliverables, and objectives. All of these are seen as key elements of project strategy; and are inputs to "scope definition" process. Thus, the scope of the project is driven by the key elements of project strategy that are directly linked to business strategy. Figure 2.9 lists what is generally accepted in the **PMI PMBOK** (2000), the **APM BoK** (2000), and Turner (1999) to be the key elements constituting the definition or scope of a project and what is usually contained in the project definition document for a project.

The **APM BoK** (2000) addresses work content and scope management as a single topic and places requirements management within the technical context of a project.

Requirements management covers the process of defining the user/customer technical requirements and building the system requirements for a project. The technical requirements of a project frequently affect project strategy significantly, and for this reason, it is identified as a key project management process.

Creating Project Strategy in Projects. Figure 2.9 shows the project requirements, project scope, and project definition from the "translating business strategies" stage being inputs to the project management planning and integration processes in the "creating project strategy" stage, and used in the project plan development process to create project strategy. The project plan development process also takes the outputs from the planning processes from other project management knowledge areas (not shown in Figure 2.9), such as resource plans, risk plans, procurement plans, and so on, and subsidiary plans such as scope management and schedule management plans, and incorporates them in a consistent, coherent project plan document or set of documents. A summary of the contents of these key documents and management plans is incorporated into the project plan, which, in the PMBOK schema, encapsulates the project strategy.

Strategic Framework and Project Management Practices

The APM BoK identifies a broad range of topics essential to the effective management of projects. These topics are grouped into a number of sections, one of which is labeled "Strategic." This section covers the topics that manage the strategic framework of a project. Such a framework is an important factor in the process of creating project strategy, and for this reason it is incorporated in Figure 2.9. The strategic framework provides the overall integrative framework for managing projects efficiently and effectively and potentially includes strategic elements from almost all the topics contained in the Control, Technical, Commercial, Organizational, and People sections of the APM BoK. Figure 2.9 identifies a number these elements, all of which are also commonly referred to as project management practices.

A strategic framework containing these elements can be developed using generic project management knowledge and competence, predicated on the APM BoK, and applied to the project integration management processes identified in the PMBOK, thereby enhancing the quality of the inputs to the processes from which project strategy is created, and ultimately the quality of the project strategy. Some may perceive there to be a risk of overlap between the APM's strategic framework and project management practices, and the planning processes of project management knowledge areas described in the PMBOK, and used in Figure 2.9. This is not the case. The idea is that by using the PMBOK project management planning and integration process model, within the context of a much broader generic strategic (APM) framework, a more effective project strategy can be created.

A Structured Approach to Creating and Moving Strategy within Projects

Figure 2.9 demonstrates the large number of factors involved in creating project strategy at the front end of a project. This highlights the need for organizations to have an effective way to manage the whole process of project strategy creation, which covers not only the front end of a project but the entire project life cycle. Our research (Morris and Jamieson, 2004) shows that companies have developed structured approaches for creating and managing project strategy that cover the entire project life cycle and are integrated with the business strategy development processes. Figure 2.10 is an example of an actual structured approach of a major manufacturing company. It comprises the following:

- A business process model
- The managing major projects process
- The project management process

The project management process comprises five stages covering the entire project life cycle. The key tasks undertaken during each of the process phases are stipulated in the supporting documentation and reflect many of the front-end elements shown in Figure 2.9. Defining project strategy is one of the key tasks undertaken in all of the phases and is done in accordance with a list of key topics by those working on the project. This approach enables project strategy to be managed throughout the life cycle of the project, by all those involved, in a manner that is specific, comprehensive and dynamic, and highly visible.

Competencies, Roles, Responsibilities, and Accountabilities for Moving Strategy

Moving strategy by means of sophisticated processes, such as those shown in Figure 2.9, requires an extensive range of competencies, highly skilled staff, and a clear definition of their roles, responsibilities, and accountabilities. Armstrong (1999) points out that the descriptions of competencies may be called competency frameworks, competency maps, competency profiles, or competency clusters—and that competency frameworks define the competency requirements that cover all the key jobs in the organization or all the jobs in a job family. An example of a generic approach in a project/technical area, is the UK Institution of Civil Engineers' competency framework structure (2000). Marchington and Wilkinson (2002) have observed that a competency framework provides a set of performance criteria at organization and individual levels, and that it identifies the expected outcomes of achieving those criteria. Whiddett and Hollyforde (1999) provide further information on competency frameworks.

Behavioral competencies describe how we behave while performing our work tasks. Armstrong (1999) reports that a survey of 126 organizations shows the most common behaviors sought by the organizations, one of which is strategic capability. Crawford (2000) reveals a number of knowledge, skills, and personal attributes of project managers, including that of strategic direction. Our research (Morris and Jamieson, 2004) shows the core behavioral competences for project directors and project managers in one global organization, as shown in Figure 2.11. It is clear that a high percentage of these competencies, to a greater

FIGURE 2.10. A STRUCTURED APPROACH TO CREATING AND MOVING PROJECT STRATEGY.

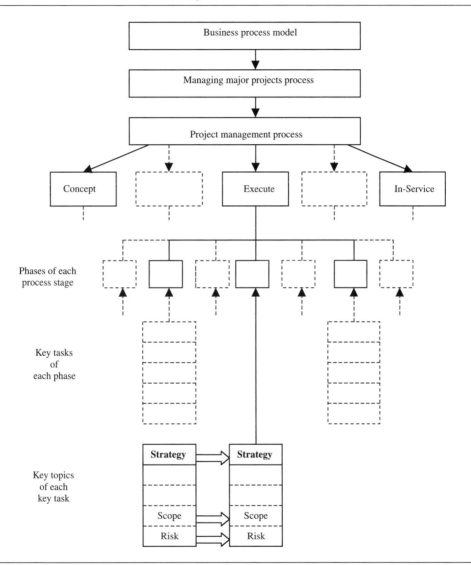

Source: Morris and Jamieson (2004).

FIGURE 2.11. COMPETENCIES FOR PROJECT DIRECTORS AND PROJECT MANAGERS.

Core Competencies	Project Director	Project Manager
Managing vision and purpose	X	
Business acumen	X	
Customer focus	X	X
Priority setting	X	X
Directing others	X	X
Leading from the front	X	X
Drive for results	X	
Dealing with ambiguity	X	
Composure	X	
Comfort around higher management	X	X
Negotiating	X	X
Building effective teams	X	X
Conflict management		X
Timely decision making		X
Motivating others		X
Organizing		X

Source: Morris and Jamieson (2004). This material is used by permission of the Project Management Institute.

or lesser extent, are required to move strategy from corporate to projects, particularly at the project director level. A number of the competencies for a project manager are the same as those for a *project* director—but not program director (see below)—with two notable exceptions: managing vision and purpose and business acumen.

In addition to the behavioral competencies, this company has also defined project management functional competencies—that is, knowledge and skills, based on the APM's BOK (2000). These functional competencies, grouped into several categories, are those required to develop the elements of a strategic framework, shown in Figure 2.9, and subsequently used to create project strategy, using processes similar or equivalent to those identified in

Figure 2.9. The categories of strategic management competencies, and the processes to which they are applicable, are as follows:

- *Strategic.* Including strategy/project management plan and value management
- *Control.* Including work content and scope management
- *Technical.* Including requirements management
- *Commercial.* Including business case, financial management, and procurement
- *Organizational.* Including life cycle design and management, and organizational structure and roles

These competencies are also related to the job requirements for project directors, project managers, and other professional project management staff. The levels of competency for a project director and a project manager, for the elements mentioned previously, in most instances are the same, but in one or two cases, such as requirements management, the project manager is required to have a higher level of competency.

A second example from our research (Morris and Jamieson, 2004) reveals how a family of job descriptions focuses on strategy. The role and purpose of a global program director are expressed in terms of leading and coordinating the resources required to implement a strategic change program for the organization. This involves *inter alia* using project management disciplines, ensuring the initiatives are aligned to organizational strategy and values, and the delivery of the operational vision on behalf of the sponsor. A program normally involves multiple projects and working across lines of business and function. The emphasis on leading and coordinating resources to implement strategic change endorses the point made previously—namely, that it is a fundamental responsibility of project management to effectively manage and/or coordinate whatever company resources are at its disposal, using well-structured project management processes, practices, and methods, and professional project management staff.

A global program director's responsibilities, in the company studied in the research, include the following strategy related areas:

- Global strategic direction of the program and its alignment with business unit vision direction and priorities;
- Business case management;
- Release management of change program operational implementation;
- Definition management; aligning the solution to the business unit vision and ensuring the solution remains aligned;
- Management of the implementation of the solution, and its scope, resource, and schedule requirements; and
- Development of program plans to reflect all progress, change, and problematic issues.

To effectively discharge these responsibilities, a global program director in this company typically possesses the following knowledge, skills, and experience:

- Business and commercial management
- Task management
- Planning and organization
- Project management
- Time management
- People/relationship management, including leadership

Additionally, as with the previous example, the company also specifies behavioral competencies for the role holder, of which the following are relevant to project strategy:

- Strategic thinking
- Conceptual thinking
- Innovativeness
- Analytical thinking

The knowledge, skills, behaviors, and experience required for project managers in the company fall into similar categories as those for a global program director, but the levels of each are within the context of a project and not a program, and accordingly are at a commensurately lower level. As with the previous example, these knowledge, skills, experience, and behaviors are integrated with the project management process and described in the associated documentation.

Companies document the roles, responsibilities, and accountabilities for the staff involved in managing projects using many different formats. Figure 2.12 provides an example of the approach taken by one company (the example could be expanded to show the many staff involved in managing projects). It is essential that the roles and responsibilities are defined and rolled out in a way that is user-friendly and are bought into by those using the processes.

FIGURE 2.12. ROLES, RESPONSIBILITIES, AND ACCOUNTABILITY MATRIX.

Tasks / person accountable or responsible	Project Board	Business Manager	Project Manager	-	-	-	-	-
Understand and assess the brief								
Covers 4 subactivities, e.g., Define project success criteria		R	A/R					
Review team composition								
Covering 1 activity			A/R					
Gather business case information								
Covers 5 activities.			A					
Finalise project definition document								
Covers 5 activities.		R	A/R					

Source: Morris and Jamieson (2004). This material is used by permission of the Project Management Institute.

The matrix covers one subprocess, associated with creating project strategy of a major projects process that comprises numerous subprocesses. The person or bodies involved in the subprocess are identified; in this case there are three, but normally all those involved are identified, which frequently amounts to a much larger number. All the activities of the subprocess are also identified, and either an A or R (there are other indicators) on the matrix indicates who has the accountability and/or the responsibility for them. (the company defines accountability as having full responsibility for the task, and responsibility is limited to completing the task.) The matrix approach is a very quick and effective means of identifying the roles, responsibilities, and accountabilities of all those involved in managing major projects, although it may be perceived by some as being very prescriptive.

Four Studies of Moving from Corporate to Project Strategy

In our research we studied four companies: a global aerospace company, a division of a global pharmaceutical company, a group within a global financial services company, and an international airport owner and operator. The resulting case studies provide valuable evidence and insight into the way corporate strategy is created and moved through SBUs, portfolios, and into programs and projects by means of, and through the activities and support of, processes, practices, and people. A number of the key findings of the case studies are summarized in the following.

Business Models

Two of the companies used business models that are equivalent to the McKinsey model outlined previously. These top-level frameworks were used to manage the activities of all the business units of the companies concerned and included strategy management, portfolio management, and program and project management processes.

Cascading Corporate Strategies into Projects and Strategy Plans

The companies created corporate objectives, goals, and strategies using processes that are typically like those strategic management processes described by Mintzberg, Hill, and Thompson. Like the model shown in Turner (1999), these objectives, goals, and strategies were cascaded to the SBUs or equivalent organizational entities, which in turn, and in conjunction with corporate strategy planners, developed their own objectives, goals, and strategies, in some instances using additional processes that were fully integrated with the business strategy processes. The SBUs subsequently developed objectives, goals, and strategies with and for their respective program and project teams, again, in some instances using fully interconnecting business and project management processes. In all four cases, the program and/or project teams developed strategies that aligned with the SBU and corporate strategy using project strategy or similar processes. The outputs of the processes containing the objectives, goals, and strategies included strategy plans, business plans, deployment plans, and project plans, the hierarchy of which, in most cases, was similar to Archibald's hierarchy of objectives, strategies, and projects, and SRI's system of plans. The most comprehensive case is shown in Figure 2.10.

Portfolio Management

The importance of project portfolio management was recognized by all the companies, and one had developed a dedicated project portfolio management practice. Within the companies, portfolio management was used primarily to select and prioritize programs and projects, and not to manage programs or projects. Corporate and business units assembled a strategic portfolio of programs and projects, or measured the strategic contribution of a program or project, using strategic management and project management processes, tools, and techniques. Company management boards or committees of senior managers adopted or rejected projects based on this information. (This is very similar to the pattern described by Artto and Dietrich in their chapter.)

Program Management

Program management was also practiced by the majority of companies, primarily within the context of managing a large group of high-value projects with a common aim and/or of delivering regular benefits over a protracted period of time. Program management and project management activities were carried out using the same set of common processes, variously called integrated program management, program management, or even project management; and accordingly, the development of program strategy and its alignment with corporate and business strategy was achieved in a similar way to that for projects.

Business Cases and Project Strategy

The creation of business cases was a key element of the business and project management interface within all the companies. An outline project strategy was developed during this activity and was aligned with corporate and business strategies. Subsequently, business strategy, in most of the companies, was translated into a comprehensive project strategy using project management processes, similar to those used in Figure 2.9 (see also Cleland's chapter in this book). Project strategy was generally in the form of a diversity of management plans and project plans rather than a single comprehensive document.

A Structured Approach to Creating and Managing Project Strategy

Two of the companies used a very structured approach to create and manage project strategy. One had institutionalized a project strategy management practice that was equivalent to, for example, risk management or technical management. The other had identified specific project-strategy-related issues for each phase and stage of the project management process and the project life cycle, as shown previously in Figure 2.10. Both companies assigned roles and responsibilities for the execution of the processes. The other companies used a less structured approach and developed management plans for their projects, but they tended not to summarize the plans nor develop a single project strategy statement from them. They also tended not to use the term project strategy in their project management processes. (There is a research issue left open here—namely, whether it would be beneficial to manage project strategy as a more formal, single document and process.)

Two out of the four companies manage project strategy for effectively the entire project life cycle and not just at the front end of a project.

Processes and Procedures

Of the processes identified in the preceding text, the ones that were most consistently used were those in which the structure and content were described at a practical level—for instance, flowcharts with inputs and outputs for key processes—and those who were accountable and responsible for carrying the process activities were identified. Conversely, when the procedures for these processes were described in too much detail, staff tended not to use them. The best examples of the deployment of the business models and associated processes were those that were fully documented and incorporated within a company's quality management system, and that were Web-based and available online throughout the organization. (Again, see Artto and Dietrich's chapter.) The companies that had not implemented such sophisticated systems or extensively integrated business and project management processes nevertheless linked the activities of their business units and projects to ensure alignment of strategy.

Roles, Responsibilities, and Accountabilities

All of the companies specified the roles, responsibilities, and accountabilities of all those involved in the business management and project management processes within the process documentation, some using comprehensive sets of tables and matrices that were linked directly to the processes. These tables and matrices covered in detail all the phases and stages of the project management process and project life cycle, including those for creating and maintaining project strategy, and identified who does what and when at any point along the process. In one company, a family of project management job descriptions was used in conjunction with the tables and matrices. These identified the job holder's roles, responsibilities, and accountabilities for specific project management process activities and outputs; they provided an unusually high degree of integration between the process and the individual or team.

Competencies and Frameworks

The companies also employed a number of other methods to identify and specify the skills, knowledge, behaviors, and experience required to manage projects and project strategy. These included, for example, company-wide competency frameworks that defined the competency requirements for all the key jobs and comprised families of job descriptions, including those for project management staff; core behavioral competencies for senior project management staff, such as managing vision and strategy; and project management functional competencies covering knowledge and experience of strategy-related areas like scope management.

Survey Data on How Companies Move Strategy from Corporate to Projects

The case study data briefly summarized in the preceding text provides a rich qualitative context in which to explore whether companies do in fact move from corporate to program and project strategy as we have hypothesized. But the data sample is obviously small. To provide a bigger data sample, we carried out a survey of the way members in a number of PMI chapters in European countries moved from corporate to program and project strategy (Morris and Jamieson, 2004).

A series of 32 questions were developed and used to examine the processes, practices, and people issues involved in moving strategy from the corporate level to projects, covering business management, strategic management, portfolio management, program management, and project management. The questionnaire was sent out randomly to PMI chapter members to obtain their views, based on their experience and knowledge. Seventy-five responses (about 50 percent from the UK) were received from people, at various levels of seniority, in small, medium, and large enterprises in a diverse range of business sectors, such as aerospace, automotive, IT, telecommunications, pharmaceuticals, retail, transportation, publishing, and academia and consultancy.

The findings of the survey, summarized in Exhibits 2.1 to 2.5, are the result of the analysis of the answers to questionnaire. The exhibits cover the following areas:

1. Business management
2. Program management and portfolio management
3. Project management and project strategy
4. Value management
5. Project management competencies

Most items in each Exhibit have a population % figure, which is a percentage of the total number of organizations represented in the survey. For example, in Exhibit 2.1, Item 1, 67 percent of the organizations in the survey indicated that they used a generic business model. Or in Exhibit 2.2, Item 3, 85 percent of organizations surveyed use programs to implement change.

The first major area the survey explored was the extent to which processes were used within organizations and to what degree was the continuity of strategy achieved through them.

EXHIBIT 2.1. SURVEY FINDINGS—BUSINESS MANAGEMENT.

1. The extent to which a business model was used:

Model used	Generic*	Equivalent
Population %	67	23

*Including project management processes

2. 80% of the organisations indicated they were process-oriented organisations as follows:

Level of process-oriented	Extensively	Adequately	Inadequately
Population %	40	40	17

3. Those extensively or adequately process-oriented indicated they had processes for moving corporate goals and objectives to project strategy as follows:

Level of processes	Extensive	Adequate
Population %	50	33

4. The level of interconnection between the corporate, business, and project management processes for the organizations with extensive processes was as follows:

Level of interconnection	Extensive	Adequate	Inadequate
Population %	40	50	10

5. Organizations having extensive processes/subprocesses for moving corporate goals and objectives to project strategy consider continuity of strategy is achieved as follows:

Level of continuity achieved	Fully	Well	Inadequate or poor
Population %	20	55	25

6. Hierarchy of objectives and strategies developed and deployed for structuring strategy:

Hierarchy span	Corporate through to project	SBU to project
Population %	53	68

How organizations perceive and use program management and portfolio management and the extent to which they moved strategy through programs to projects was the second major area investigated.

EXHIBIT 2.2. SURVEY FINDINGS—PROGRAM MANAGEMENT AND PORTFOLIO MANAGEMENT.

	Population %
1. Program management was defined as the management of a portfolio of projects sharing a business objective of strategic importance, probably utilizing shared resources.	95
2. Programs were considered to comprise:	
Groups of projects	20
Number of separate projects	20
Combination of both	70
3. Programs were used to implement change:	85

EXHIBIT 2.2. (*Continued*)

4. Some form of portfolio management was implemented:	50
5. Portfolio management was considered as:	
Selecting the right project quantitatively	50
Maintaining a balanced portfolio	60
Managing projects grouped around a common theme	66
6. Hierarchy of objectives and strategies were developed and deployed at program and project level	75
Of which, the levels deployed were as follows:	
Program	90
Program and project	60
Program, project, and project team	45
Program, project, project team, and individual	28
7. Program management was implemented	90
Of which, program management included the following:	
A. [i] Managing an integrated set of projects to achieve a common theme, aim, or working off a common platform;	55
[ii] Integrated project teams	
[iii] Managing resources in an integrated manner	
B. [i] and [ii]	10
C. [i] and [iii]	15
8. Program management implied the management of business benefits	75
Of which:	
It was normal practice to formally identify a benefits process within the overall program management process.	70
Those who do not incorporate a benefits process believed they should.	60
Nonfinancial measures were used to track benefits in programs.	70
9. Program management implied the aggregation of risks	75
Of which:	
It is normal practice to formally identify risk aggregation as part of the overall risk management activity.	80
Those who do not identify risk aggregation believed it should be incorporated into program management.	60

The third area investigated, project management, followed on naturally from program management. The survey focused on identifying the key strategy inputs and some of the project management activities, which were employed to create project strategy.

EXHIBIT 2.3. SURVEY FINDINGS—PROJECT MANAGEMENT AND PROJECT STRATEGY.

	Population % Almost all
1. Organizations had extensive or partially integrated project management processes to help manage project strategy, which contained:	
Project strategy management	85
Requirements management, project strategy, project definition and project scope management	75
Requirements management, project definition and project scope management.	85
2. Organizations had specific strategy inputs to integrated project management processes	
Which included:	most
Corporate strategy	75
Corporate strategy and business strategies	65
Corporate, business, and portfolio strategies	50
Corporate, business, portfolio, and program strategies	45
Portfolio and program strategies only	55
Program strategy	75
3. The integrated project management processes delivered the following outputs:	
A project or program plan and strategy plan	50
Other project management plans	75
A project or program plan, strategy plan and other plans	45
4. Organizations with integrated project management processes managed project strategy dynamically	65
5. The roles and responsibilities for developing, implementing and updating project strategy were specified in:	
Project management procedures	60
Project plans	55
6. Project plans were formally reviewed at project "gates"	85
Those who did not and thought they should	85
7. Peer groups formally reviewed project plans	75
Those who did not and thought it would be sensible to do	65
8. It was clear who approved and signed off project strategy	75
9. In broad terms, a project sponsor was the individual or group within the performing organisation who provides the financial resources, in cash or in kind, for the project; and as the owner of the business case, represents the funder's interests.	90
The relationship between the project sponsor and the project management team was normally defined in project plans.	70
10. Strategy was expected to be upgraded and reviewed:	
During the development of the project	65
Systematically as projects develop from concept to execution	55
Of which: it was systematically undertaken at project review gates	85

The survey also explored a two other areas:

- Project value management and its link to project strategy
- To what extent do companies define competencies to manage program and project strategy, and incorporate them in competency frameworks and job descriptions.

EXHIBIT 2.4. SURVEY FINDINGS—VALUE MANAGEMENT.

	Population %
1. A process was used for optimizing the value of proposed project/program strategy	55
Of which:	
Value was expressed as benefit over resources used	80
The process was formalized as value management	55
Of which value management workshops were held at strategic stages in the life of the project.	40
Those not using a process for optimizing the value of project/program strategy believed they should.	55
2. Value engineering was practiced on programs and projects	25
Of which:	
Value engineering (optimizing the value of the technical configuration) was distinguished from value management.	80
Those not practicing value engineering on programs and projects thought they should.	56
3. The value optimization process was integrated with risk management	75
Those that did not thought it should be done.	40

EXHIBIT 2.5. SURVEY FINDINGS—PROJECT MANAGEMENT COMPETENCIES.

	Population %
1. Project management skills and knowledge competencies required to manage programs or projects were formally defined.	80
Of which included:	
Those required to develop program and project strategy	75
Linking the competencies to personal appraisal and development systems	80
Linking personal objectives to project objectives	65
2. Those that did not formally define the project management skills and knowledge competencies incorporated the management of project strategy in job descriptions or job specifications.	50
3. Organization-wide behavioral competency frameworks were used	60
Those that did not use them considered they should.	45
4. Competency support programs for program and project managers were provided	70
Of which covered support for project strategy development.	66

Summary

Project and program management is widely used as a means of implementing corporate strategy. Normatively, we can expect strategies to be aligned and moved from the corporate level through programs and projects in a systematic and hierarchical manner that provides cohesion, visibility, and an effective means of communication. There is a cascade of moving from corporate planning at the enterprise level through portfolio management into programs and projects. Within this framework, project strategy is managed dynamically.

Enterprise-wide business models play an important part in effecting this transformation. Many organizations have project management as a core process.

Programs are important vehicles for implementing corporate strategy and for implementing change. Most companies consider that program management implies the management of business benefits (as well as the ideas of product, brand, or platform management).

Project strategy management is widely recognized as an important project management practice that systematically relates project definition and development to corporate goals and strategies. Structured project management approaches are now being used by organizations, covering the entire project life cycle, with project strategy development, review, and optimization occurring at specific points. Value management is quite widely used in optimizing the strategy, often in combination with risk management.

Project management resources and capabilities are key factors in creating, deploying, and maintaining enterprise, portfolio, program, and project strategies. The project management roles, responsibilities, and accountabilities required for this are generally well defined. A high percentage of those organizations surveyed define the personal project management competencies required to develop project strategy.

Project strategy management is an underexplored and insufficiently described subject in the business and project literature but is in fact a relatively well-trodden area within industry and commercial practice, and deserves more recognition.

References and Further Reading

Andrews, K. R. 1999. *The concept of corporate strategy*. Custom ed. New York: McGraw-Hill Companies, Inc.

Archer, N. P., and F. Ghasemzadeh, F. 1999. An integrated framework for project portfolio selection. *International Journal of Project Management* 17:207–216.

Armstrong, M. 1999. *A handbook of human resource management practice*. 7th ed. London: Kogan Page Ltd.

Association of Project Management. 2000. *Project management body of knowledge*. 4th ed. High Wycombe, UK: Association of Project Management.

Cleland, D. I. 1990. *Strategic design and implementation*. New York: McGraw-Hill Companies, Inc.

Crawford, J. L. 2001. Portfolio management: Overview and best practices. In Project Management for Business Professionals. J. Knutson (ed). New York: Wiley.

Crawford, L. 2000. Profiling the competent project manager. *Proceedings of PMI Research Conference*. Newtown Square, PA: Project Management Institute.

Hill, C. W. L., and G. R. Jones. 2001. *Strategic management: An integrated approach*. 5th ed. Boston: Houghton Mifflin Co.

Kerzner, H. 2000. *Applied project management: best practices on implementation.* New York: Wiley.

Knutson, J. 2001. *Succeeding in project-driven organizations: People, processes and politics.* New York: Wiley.

Marchington, M., and A. Wilkinson. 2002. *People management and development: Human resource management at work.* 2nd ed. London: Chartered Institute of Personnel and Development.

Mintzberg, H., B. Ahlstrand, and J. Lampel. 1998. *Strategy safari.* London: Pearson Education.

Mintzberg, H., and J. B. Quinn. 1996. *The strategy process: Concepts, contexts and cases.* 3rd ed. Upper Saddle River, NJ: Prentice Hall.

Morris, P. W. G. 1997. *The management of projects.* 2nd ed. London: Thomas Telford.

Morris, P. W. G., and G. H. Hough. 1987. *The anatomy of major projects.* Chichester, UK: Wiley.

Morris, P. W. G., and H. A. Jamieson. 2004. *Translating corporate strategy into project strategy.* Newtown Square, PA: Project Management Institute.

Murray-Webster, R., and M. Thiry. 2000. Managing programmes of projects. In *The Gower Handbook of Project Management.* 3rd ed., ed. R. Turner and S. Simister. (Aldershot, UK: Gower.

Project Management Institute. 2000. *A guide to the Project Management Body of Knowledge.* Newtown Square, PA: Project Management Institute.

Thompson, J. L. 2001. Strategic management. 4th ed. London: Thomson Learning.

Turner, R. J. 1999. The handbook of project-based management. Maidenhead, UK: McGraw-Hill. Reproduced by kind permission of the Open University Press/McGraw-Hill Publishing Company.

Turner, R., and S. Simister, eds. 2000. *The Gower handbook of project management.* 3rd ed. Aldershot, UK: Gower.

The Institution of Civil Engineers. 2000. Management development in the construction industry: Guidelines for the construction professional. London: The Institution of Civil Engineers.

The Office of Government Commerce. 1999. *Managing successful programmes.* London: The Stationery Office. Crown copyright material is reproduced with the permission of the Controller of HMSO and Queen's Printer for Scotland.

Watson, G. H. 1994. Business systems engineering: Managing breakthrough changes for productivity and profit. Chichester, UK: Wiley.

CHAPTER THREE

STRATEGIC MANAGEMENT: THE PROJECT LINKAGES[1]

David I. Cleland

"There is nothing permanent except change."

HERACLITUS OF GREECE, 513 B.C.

The purpose of this chapter is to present an overview of what is involved in the development of project strategy. Project strategy does not stand alone; rather, it is an integral part of the overall strategic plan for the management of change in the enterprise. A general philosophy of what project strategy is, as well as how and where project strategy falls in the overall scheme of the strategic management of the enterprise, is presented.

This chapter also provides an overall perspective—a philosophy—to guide the design and development of project strategy. A *philosophy* is defined as a system of thought based on some logical relationship between concepts and principles that explains certain phenomena, and supplies a basis for rational solutions of related problems (Davis, 1951). A philosophy, taken in its most basic sense, is simply a way of thinking about a field of endeavor. A sound philosophy of project strategy is a precondition of starting the project planning process.

The dictionary defines *strategy* as "The essence of art or military command as applied to the overall planning and conduct of large-scale combat operations; a plan of action resulting from the practice of strategy, the art or skill of using stratagems, especially in politics and business (*Webster's II New College Dictionary*, 1999).

The details of project strategy, such as scheduling, networking techniques, scope planning, risk assessment, resource planning, cost estimating, life cycle planning, project specifications, and so forth, will not be presented in this chapter. Rather, what will be presented is an overall description of project strategy, including how it fits into the overall scheme of

---

[1]In the development of this chapter I have drawn material from David I. Cleland & Lewis R. Ireland, *Project Management: Strategic Design & Implementation*, 4th Edit., McGraw-Hill, New York, NY 2002.

enterprise strategy, and some general guidelines on the "work packages" that should be developed in the preparation of project strategy.

We tend to live, worship, work, socialize, and so forth, based on some philosophy—some way of thinking about our world. Such a philosophy may be nebulous and fleeting, or it may be well defined and provide the performance and behavior standards of our life.

The key question that the reader needs to ask is this: What philosophy do I have regarding the development of project strategy? Can the philosophy that is presented in this chapter help me—and the people with whom I work—in the development of project strategy?

Background

Project management is currently being used worldwide in a wide range of applications, both in developed and developing countries. Theses applications include the following::

- *Industrial.* For new or improved products, services, or organizational processes
- *Social.* To support new or enhanced programs for society
- *Economic.* For stimulation of local, regional, national, and international economies
- *Technological.* To advance the state-of-the-art of organizational products, services, or processes
- *Legal.* For the development of new or modified laws and their application
- *Political.* Governmental initiatives to support local, regional, and national strategies
- *Military.* For campaigns to support military strategies
- *Discovery.* To find new territories and worlds

Pinto has stated: "Many of the products that are being created today in a variety of industries, from children's toys to automobiles, are becoming more technically complex to develop, manufacture, and use. Technologically driven innovation presents a tremendous challenge for organizations in the areas of engineering, design, production, and marketing. As a result, many organizations are relying on project teams composed of cross-functional groups to create and move to market these products in as efficient a time frame as possible" (Pinto, 1998).

Project teams are being applied to a wide variety of uses within the individual organization. These uses are shown in Table 3.1 (Cleland, 1996). It should be noted that four of these teams are ongoing rather than ad hoc, as would be expected of a traditional project team. However, setting up the ongoing teams is often handled through traditional projects. The continuous challenge of coping with the inevitable changes facing contemporary organizations will likely expand the use of these teams in the future.

A broad definition of a *project* is that it is a combination of organizational resources being pulled together to create something that did not previously exist and that will, when completed, provide a performance capability to support strategic management initiatives in the enterprise. Four key considerations are always involved in a project:

TABLE 3.1. CLASSIFICATION OF TEAMS.

Type	Output/Contribution	Time Frame
Reengineering teams	Handle business process changes	Ad hoc
Crisis management teams	Manage organizational crises	Ad hoc
Product and process development teams	Handle concurrent product and process development	Ad hoc
Self-directed production teams	Manage and execute production work	Ongoing
Task forces and problem-solving teams	Evaluate/resolve organizational problems/opportunities	Ad hoc
Benchmarking teams	Evaluate competitors/best-in-industry performance	Ongoing
Facilities construction project teams	Design/develop/construct facilities/equipment	Ad hoc
Quality teams	Develop/implement total quality initiatives	Ongoing
General-purpose project teams	Develop/implement new initiatives in the enterprise	Ad hoc
Audit teams	Evaluate organizational efficiency and effectiveness	Ad hoc
Plural executive teams	Integrate senior-level management decisions	Ongoing
New business development teams	Develop new business ventures	Ad hoc

Source: David I. Cleland, *The Strategic Management of Teams* (New York: Wiley, 1996, p. 10).

1. What will be the expected cost of designing and creating the expected project results?
2. How much time is required to develop, produce, and place the project results in an operational environment?
3. What capability will the project results provide the project owner?
4. How will the project results fit into the operational or strategic capability of the project owner's enterprise?

Project Evaluation and Selection

In this chapter it has been assumed that the decisions concerning which projects to develop to support strategic management initiatives have been made by the responsible executives of the enterprise. Indeed, the selection of projects to support strategic management initiatives is a most important decision—certainly an important decision in the development of project strategy.

The proper choice of how enterprise resources will be used to support the strategic management is crucial to the long-term survival and growth of the enterprise. Many of the existing texts on project management suggest procedures and processes for how projects can best be selected. A few criteria to use in determining which projects to select to support strategic and operational initiatives in the enterprise are suggested. These criteria are presented in the form of questions.

- Will there be a customer for the expected results of the project?
- Will the project results provide the enterprise a competitive edge?
- Will the project results make a distinctive contribution to existing products, services, or organizational processes?
- Will the enterprise be able to handle the risk and uncertainty likely to come forth as the project is undertaken?
- What is the probability of the project being completed on time, within budget, and at the same time satisfy its technical performance objectives?
- Will the project results provide value to the expected customer?
- Will the project provide a satisfactory return on investment to the customer's organization?
- Will the project have a high probability of supporting the enterprise's strategic initiatives?

These are key questions to be considered during the project selection process. In addition, these questions should be examined for each project during its life cycle, particularly during major reviews of the project's status.

Relationship to Strategic Management

Strategic management is the management of the organization as if its future mattered. Strategic management has two interrelated elements: (1) strategic planning and (2) strategic implementation.

In the design and execution of strategic management initiatives, an early assessment of expected real and potential environmental changes is required. This assessment should consider the general environmental conditions that currently exist and what should be expected for the future. In addition, the current and expected strengths and weaknesses of the competition should be evaluated. Once a meaningful database has been established regarding the present and forthcoming environmental conditions, the strengths and weaknesses of the organization should be compared to what is expected in the competitive future in which the enterprise will exist.

The assessment of current and expected environmental change is usually done in the following areas:

- *Political.* What political conditions might have an impact on the development of the project? For example, a Department of Defense (DoD) defense contractor would carefully watch the political developments in Washington, D.C., as such developments might impact an existing or expected DoD market. The current debate under way regarding whether or not the United States military force should buy or lease military transport aircraft is being watched carefully by aircraft manufacturers and lessors.

- *Social.* What social and cultural considerations might have an impact on the development and use of the project results in an emerging country?
- *Economic.* Will there be sufficient resources available to develop and sustain the project results in its operational environment?
- *Technological.* Will the project be supported by the required state-of-the-art resources in its development? For example, a project to build a new manufacturing plant would include a consideration of the current state-of-the-art in design and production processes.
- *Legal.* What legislation might exist, or can be expected to develop, regarding the use of the project results? Enterprises that develop projects in the construction industry need to be mindful of what environmental impact requirements need to be satisfied in the use of the project.
- *Competitive.* What are the strengths, weaknesses, and probable strategies of competitive firms performing in the same marketplace? Benchmarking of competitive firms is particularly important in the defense and construction industries.

As the assessment of the possibilities and probabilities of the expected future facing the project results is carried out, key decisions can be made through the project planning process that will provide a strategy for the project results..

Strategic management is inextricably interwoven into the entire fabric of the management process; it is not something separate and distinct from the process of management of the organization. Strategic management keynotes the shift of organization management from operations to long-range strategy. Stated another way, strategic management should not be distinguished from the rest of the organization management process, in particular the project management process where the strategic management of change is carried out.

In earlier times the managerial emphasis was primarily on operations being sufficiently efficient to maximize the return on investment in operational resources. Then came the need to develop organization strategies to deal with the growing turbulence and rapidly changing environments facing the enterprise. Today, there is a continuing emphasis on effective strategic management and efficient operational management, but an increasing emphasis on the use of project management to better facilitate the management of both operational and strategic change.

Organizational Changes

Strategic management provides a focus for the development and providing of support for the future *product, service,* and *organizational processes* of the enterprise. It is in these areas that competition becomes most critical for the enterprise through providing a capability for the following:

- *New or improved products* offered by the organization
- *New or improved services* offered by the organization

- *New or improved organizational processes* used to support the organization's products and services. This would include marketing, manufacturing, financial, R&D, and other organizational functions required to maintain the enterprise functions as entities capable of competitive performance in the marketplace.

As an ongoing entity, the competitive enterprise needs to maintain a balance in its utilization of resources. The key is to maintain a balance between operational competence, strategic effectiveness, and functional excellence. Figure 3.1 portrays this balance (Cleland, 1996).

Operational competence concerns the ability to use resources to provide customers with quality products and/or services that are delivered on time and provide value for customers. Operational competence requires the ability to use resources in a cost-effective manner to produce and deliver products and services that exceed what competitors are offering.

The final measure of operational competence is profit or, in the case of a not-for-profit entity, greater value to the customers than was required to develop, produce, and deliver the products and services to the customer. To maintain operational competence, the enterprise depends on the strategic effectiveness with which the enterprise is managed.

Strategic effectiveness is the ability to assess what may be possible and probable in the enterprise's future products and services, as well as the organizational processes required to

FIGURE 3.1. BALANCE MUST BE ACHIEVED AND MAINTAINED AMONG STRATEGIC MANAGEMENT CHALLENGES.

Strategic Effectiveness

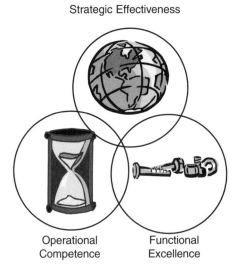

Operational
Competence

Functional
Excellence

Source: David I. Cleland / Lewis R. Ireland, *Project Management: Strategic Design and Implementation,* 4th ed. (New York: McGraw-Hill, 2002, p. 8). Reproduced with permission of the McGraw-Hill Companies.

support future purposes. Strategic effectiveness is concerned with doing the "right things" to prepare the organization for its future. Project teams are crucial in preparing the enterprise for its future, thereby contributing to the strategic effectiveness of the enterprise.

Functional excellence is the use of state-of-the-art resources in the disciplines that support the enterprise's organizational processes. Functional excellence includes ongoing improvement in employees' capability as well as other organizational resources so that the effective and efficient use of resources is sustained for the organization. Project teams can be used to develop employee training programs, improve the use of resources, and design and develop new initiatives in the application of functional resources to the enterprise.

The bottom line of strategic management is to maintain balance among strategic effectiveness, operational competence, and functional excellence. Project management can provide critical support to the maintenance of this balance.

Context of Project Strategy

The development of project strategy has to be done in the context of the strategic management of the enterprise. Projects do not stand alone in the enterprise. Rather, they are part of the overall organizational initiatives. Thus, project strategy has to be developed to support larger organizational plans. In Figure 3.2, projects are portrayed in this larger context of the enterprise—essentially as building blocks in the "choice elements" of the enterprise. In the material that follows, these choice elements are briefly described. A more detailed explanation can be seen in the source cited in Figure 3.2.

FIGURE 3.2. CHOICE ELEMENTS OF STRATEGIC MANAGEMENT.

Source: David I. Cleland/Lewis R. Ireland, *Project Management: Strategic Design and Implementation*, 4th ed. (New York: McGraw-Hill, 2002, p. 8).

The choice elements are as follows:

- *Vision.* The development of intelligent and relevant foresight of probable future opportunities. One company sees its vision to be a "world-class competitor and we keep it that way—we have programs and projects in place to do just that." Another company states its vision as "People working together as a global enterprise for aerospace leadership" (Boeing, 2002). Another company included in its vision statement: "We will enhance our competitiveness by being first in the development of advanced technology that supports our world-class products and services." A statement of the vision of an enterprise is a mental image of that organization's current and future reason for existing.
- *Mission.* A broad, enduring intent that an organizational entity pursues—essentially the "business" that the organization pursues. One drug manufacturer stated its mission as "The development of model drug absorption systems for therapeutic companies that provide distinctive benefits for the physician and patient."
- *Objectives.* The desired future destination of the enterprise stated in quantitative and/or qualitative terms. A computer company describes one of its objectives as "Leading the state-of-the-art of technology in our product lines."
- *Goals.* Specific time-sensitive milestones to be accomplished using organizational resources. Attainment of a goal signifies that progress has been made toward attaining an organizational objective. One company stated that "We will, by the end of 2002, complete the transition from a predominantly R&D services company to an industrial manufacturer." Goals are supported by projects within the organization.
- *Strategies.* The design of the means through the use of resources to accomplish organizational purposes. Strategies include designation of the use of resources to design and implement organizational programs, projects, operational plans, and organizational design arrangements. Sometimes a strategy for an enterprise is stated in general terms. For example, one major aerospace company described their overall strategy as "...to excel in all principal aerospace markets to reduce our dependence on the cyclical commercial airplane market."
- *Facilitative services.* Those plans for the development of policies, procedures, protocols, and systems to support organizational resources. For example, many companies have published a project management guidebook to provide guidance on how projects will be managed within the enterprise.

Every failure in the development and implementation of resources directed to these choice elements will impact the other choice elements. A failure in the development of credible project plans can cause problems in the execution of these plans and will ultimately impact other choice elements, in particular the goals of the enterprise. From an organizational perspective, completion of a project means that an organizational goal has been completed. Indeed, the breakdown of the choice elements for the enterprise where the relative position of the vision, mission, objectives, goals, programs, and projects could be considered a conceptual work breakdown philosophy for the strategic management of the organization. For example, if a project (an organization goal) is not being developed and produced, then

- the objective of the organization will be adversely affected;
- the organization mission can be compromised;
- failure to obtain a full realization of the vision can occur; and
- the overall performance of the organization in its marketplace will be impacted.

Truly "everything is related to everything else" in the management of an organization as well as in the management of the appropriate supporting choice elements in the organization.

In the material that follows, the role of project planning is presented.

The Importance of Project Planning

Writers in project management have recognized the importance of project planning:

> During the early 1960s, after hundreds of projects had been completed, it became apparent that many projects successfully achieved their basic project objectives, whereas some failed to achieve budget, schedule, and performance objectives originally established. The history of many of these projects was carefully reviewed to identify conditions and events common to successful projects, vis-à-vis those conditions and events that occurred frequently on less successful projects. A common identifiable element on most successful projects was the quality and depth of early planning by the project management group. Execution of the plan, bolstered by strong project management control over identifiable phases of the project, was another major reason why the project was successful (Duke, 1977).

The author of what is believed to be the first book on project management stated: "No other aspect of project management is so essential to success as planning. Most of the troubles that confront project management on the rocky road to completion of the project are traceable directly to faulty planning—unrealistic, incomplete, too-broad, or just plain lack of plans" (Baumgartner, 1963). Several hundred books on project management have appeared since Baumgartner published his leading contribution to the project management discipline. No other books have stated the importance of project planning any better than this first book in the field.

Basically, project planning is a process for achieving success in the future of the project and of the organization. It is a plan of action for getting the best return from the resources that are going to be used on the project during its life cycle. The project plan is an expected arrangement for dealing with the ever-changing environment facing the project and the enterprise. Project planning starts with the development of work breakdown structure (WBS).

The Work Breakdown Structure (WBS)

The most basic consideration in the development of project strategy is the WBS. The WBS divides the overall project into work units that can be assigned either to the organization or to an outside agency. The underlying philosophy of the WBS is to divide the project into work packages that are assigned for which accountability can be expected. Each work package becomes a performance control unit; it is negotiated and assigned to a specific organizational professional or manager. The person to whom the work package is assigned becomes the "work package manager." The process of developing the WBS is to establish a scheme for dividing the project into major groups and then dividing these groups into tasks, subtasks, and so forth. Projects should be planned, organized, monitored, and controlled around the lowest level of the WBS. If a small work package is not on schedule or is running over its budget or has failed to provide the expected performance objective, the question can be raised concerning the impact on the overall project.

With an aircraft project the WBS might look like the model that is shown in Figure 3.3.

A WBS can also be portrayed as shown in Figure 3.4.

The WBS provides for the following:

- Summarizing all products, services, and processes that make up the project
- Displaying how the work packages are related to each other, to other activities within the organization, including support, as well as to outside agencies such as contractors and other stakeholders
- Providing the basis for establishing the authority-responsibility patterns in the organization, usually reflected in some form of matrix structure

FIGURE 3.3. WORK BREAKDOWN STRUCTURE CODING SCHEME FOR AIRCRAFT (EXAMPLE).

X-33 Aircraft

1. **System**
    - 1.1   Airframe
    - 1.2   Tail Section
    - 1.3   Wings
    - 1.4   Engine
    - 1.5   Avionics
2. **Documentation**
    - 2.1   Operator's manual
    - 2.2   Repair manual
3. **Test and Demonstration**
    - 3.1   Static system test on ground
    - 3.2   Dynamic air test
        - 3.2.1   Initial flight for aerodynamics
        - 3.2.2   Initial flight maneuver test
        - 3.2.3   Endurance flight test
4. **Logistics**
    - 4.1   Maintenance tools
    - 4.2   Repair parts (spares)

Source: David I. Cleland/Lewis R. Ireland, *Project Management: Strategic Design and Implementation*, 4th ed. (New York: McGraw-Hill, 2002, p. 319).

FIGURE 3.4. WBS IN A GRAPHIC DIAGRAM (EXAMPLE).

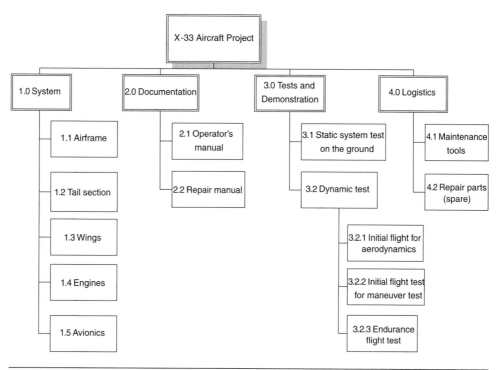

Source: David I. Cleland / Lewis R. Ireland, *Project Management: Strategic Design and Implementation*, 4th ed. (New York: McGraw-Hill, 2002, p. 320).

- Estimating cost
- Performing risk analysis
- Scheduling the work packages
- Building the project management information system
- Monitoring, evaluating, and controlling the application and use of resources on the project
- Providing a reference point for getting people committed and motivated to support the project

Work packages are the goals to be accomplished on the project. There are a few key criteria that should be applied to the WBS work packages:

- Are the work packages clear and understandable?
- Are they specific?

- Are they time-based and capable of being scheduled in the assignment of work on the project?
- Are the work packages measurable?

If an adequate and accurate WBS has not been established for the project, the project cannot be effectively and efficiently managed.

Project planning can be carried out through a representative model of the work packages. Figure 3.5 depicts these work packages. The subsequent discussion provides basic descriptions of these packages.

Project Planning Work Packages

A project plan can be broken down into work packages. The following is a general guide to these work packages:

- *Establish the strategic and/or operational fit of the project.* Ensure that the project is truly a building block in the design and execution of organizational strategies and that it provides

FIGURE 3.5. PROJECT PLANNING WORK PACKAGES.

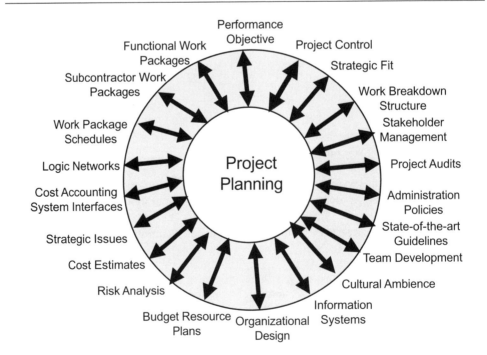

the project owner with an operational capability not currently existing or improves an existing capability.

- *Create a WBS.* Develop a product-oriented family tree division of hardware, software, services, and other tasks to organize, define, and graphically display the product to be produced, as well as the work to be accomplished to achieve the specified results.
- *Determine who the project stakeholders are and plan for the management of these stakeholders.* Determine how these stakeholders might change through the life cycle of the project.
- *Plan for the nature and timing of the project audits.* Determine the type of audit best suited to get an independent evaluation of where the project stands at critical junctures.
- *Design project administration policies, procedures, and methodologies.* Administrative considerations are often overlooked. Take care of them during early project planning, and do not leave them to chance.
- *Integrate contemporaneous state-of-the-art project management philosophies, concepts, and techniques.* The art and science of project management continue to evolve. Take care to keep project management approaches up-to-date.
- *Develop the project team.* Establish a strategy for creating and maintaining effective project team operations.
- *Assess the organizational cultural ambience.* Project management works best where a supportive culture exists. Project documentation, management style, training, and attitudes all work together to make up the culture in which project management is found. Determine what project management training would be required. What cultural fine-tuning is required?
- *Provide for the project management information system.* An information system is essential to monitor, evaluate, and control the use of resources on the project.
- *Select the organizational design.* Provide the basis for getting the project team organized, including delineation of authority, responsibility, and accountability. The linear responsibility chart (LRC) is a useful tool to determine individual and collective roles in a project-oriented matrix organization.
- *Develop the project budgets, funding plans, and other resource plans.* Establish how the project funds should be utilized, and develop the necessary information to monitor and control the use of funds on the project.
- *Perform risk analysis.* Establish the degree or probability of suffering a setback in the project's schedule, cost, or technical performance parameters.
- *Estimate the project costs.* Determine what it will cost to design, develop, and manufacture (construct) the project, including an assessment of the probability of staying within the estimated costs.
- *Identify the strategic issues that the project is likely to face.* Develop a strategy for how to deal with these issues.
- *Ensure the development of organizational cost accounting system interfaces.* Since the project management information system is tied in closely with cost accounting, establish the appropriate interfaces with the function.
- *Develop the logic networks and relationships of the project work packages.* Determine how the project parts can fit together in a logical relationship.
- *Develop the master and work package schedules.* Use the appropriate scheduling techniques to determine the time dimension of the project through a collaborative effort of the project team.

- *Identify project work packages that will be subcontracted.* Develop procurement specifications and other desired contractual terms for the delivery of the goods and services to be provided by outside vendors.
- *Make provisions for the assignment of the functional work packages.* Decide which work packages will be done in-house, obtain the commitment of the responsible functional managers, and plan for the allocation of appropriate funds through the organizational work authorization system.
- *Develop the project technical performance objective.* Describe the project end product(s), services, and/or processes that satisfies a customer's needs in terms of capability, capacity, quality, quantity, reliability, efficiency, and such performance standards.
- *Develop project control concepts, processes, and techniques.* How will the project's status be judged? On what basis? How often? By whom? How? Ask and answer these questions prospectively during the planning phase.

The Project Plan

Of course, the size and organization of the project plan depends on the project. The *key essentials* of a project plan, regardless of the size of the project should include the following:

- A summary of the project that states briefly how the project is to be developed and the processes and techniques that will be used. A clear statement of the expected results of the project should be provided so that these results can be identified and compared to the standards used on the project during its life cycles.
- A list of tangible goals of the project, usually expressed in terms of a work element or work package without ambiguity so that it can be easily determined that a goal has been achieved.
- A brief outline of how organizational resources will be used to accomplish the project ends.
- A Gantt chart and activity network that shows the sequence and relationship of work packages to provide a roadmap of how project work is to be done.
- Budgets and schedules for all the work packages of the project.
- A description of how authority and responsibility is delegated to the project team members, as well as other to project stakeholders.
- Policy guidelines on how the project work will be monitored, evaluated, and controlled.
- A list of the key project participants and their assignments relative to the project WBS.

The results of the development of project plans should be reflected in project and organizational documentation.

Project Strategy Documentation

Every project plan that is prepared requires documentation that provides a written record of the content of the resources committed and the action initiatives required to start the project and bring it to a successful conclusion through its life cycle. During the planning and implementation of a project, the required documentation should include the following elements:

- Citation of resources required
- A schema for the allocation of resources
- A project plan
- Individual and collective roles of project team
- Assigned project responsibility and authority
- Work breakdown structure
- Cost estimates
- Schedules
- Anticipated contents of project guidebook
- Summary of project planning work packages
- Quotes or bids
- Financial plan
- User requirements
- Marketing plan
- Preliminary construction/manufacturing plan
- Project evaluation, monitoring, and control strategies
- Protocol for progress payments
- Project team performance review strategies
- Customer reporting strategies
- Configuration baselines
- Communication plan
- A project risk assessment plan
- Logistic support plans
- Stakeholder support strategies

The initial documentation should include summary material about the project during its life cycle. For example, the documentation required during the closeout of the project could include such things as:

- Final report to project owner
- Technical drawings
- As-built drawings
- User guidebooks
- Project history to include "lessons learned"
- Closeout of resources used on project

Understanding Linkages

An effective and clear link between the strategic management and project management of the enterprise depends on several considerations:

- Recognition that strategic management involves a continued examination of how enterprise resources can be used in the future.
- An understanding of the mutuality between strategic management and project management.
- Acceptance of a reference focus for the use of choice elements in allocating organizational resources in strategic management.
- Recognition that project results are building blocks in the design of new or improved products, services, or organizational processes.
- Acceptance that projects provide the most reasonable means for dealing with environmental changes facing the enterprise.
- The existence of policy and procedures documentation that clearly links strategic management and project management—and provides a protocol for how such linkages can be maintained and strengthened.

Project Planning Principles

This chapter concludes with the citation of a few principles of project planning. A principle is a basic truth, law, or assumption; a fixed or predetermined policy or mode of action; the essence or vital philosophy to deal with phenomena.

Henry Fayol, truly one of the founding fathers of modern management, put forth his General Principles of Management in his book (Fayol, 1949). He noted that

> . . . I shall adopt the term principles whilst dissociating it from any suggestion of rigidity, for there is nothing rigid or absolute in management affairs . . . Therefore principles are flexible and capable of adaptation to every need; it is a matter of knowing how to make use of them, which is a difficult art requiring intelligence, experience decision and proportion.

It is in the spirit inherent in Fayol's description of principles that the following principles of project planning are presented:

- Planning for projects is a key responsibility of project managers.
- The project planning process should be carried out with those stakeholders that have an interest in the project.
- A project plan will typically require some change during the life cycle of the project as new information becomes available or basic project requirements change.
- Project planning is a process that pulls information from the organization as well as from other key stakeholders.
- The quality of project planning relates directly to the probability of success on the project.
- All plans will require revision to meet emerging changes and new information.

Summary

This chapter presented an overview of a philosophy of project planning. Such a philosophy is a way of thinking about how project planning should be carried out to support project and enterprise purposes. Project planning must support strategic management protocol and processes of the enterprise. Projects are building blocks in the design and execution of enterprise strategies that provide for the principal means for dealing with the change facing the enterprise.

References

Barkley, B. T., and J. H. Saylor. 1993. *Customer-driven project management: A new paradigm in total quality implementation*. Boston: McGraw-Hill.

Baumgartner, J. S. 1963. *Project management*. Homewood, IL: Richard D. Irwin.

Cleland, D. I. 1996. *The strategic management of teams*. New York: Wiley.

Cleland, D. I., and L. R. Ireland. 2000. *Project manager's portable handbook*. New York: McGraw-Hill.

———. 2002. *Project management: Strategic design and implementation*. 4th ed. New York: McGraw-Hill.

Davis, R. C. 1951. *The fundamentals of top management*. New York: Harper & Brothers.

Duke, R. K. et al. 1977. Project management at Fluor Utah, Inc. *Project Management Quarterly* 3:33.

Fayol, H. 1949. *General and industrial management*. London: Pitman.

Hamilton, A. 1997. *Management by projects: Achieving success in a changing world*. Dublin: Oak Tree Press.

Knutson, J., ed. 2001. *Project management for business professionals: A comprehensive guide*. New York: Wiley.

Lewis, J. P. 2000. *Project planning, scheduling and control: A hands-on guide to bringing projects in on time and on budget. 3rd ed*. New York: McGraw-Hill.

Meredith, J. R. and S. J. Mantel. 1985. *Project management: A managerial approach*. New York: Wiley.

Pinto, J. K., ed. 1998. *Project management handbook*. San Francisco: Jossey-Bass Publishers.

Randolph, W. A., and B. Z. Posner. 1988. *Effective project planning and management: Getting the job done*. Englewood Cliffs, NJ: Prentice Hall.

Steiner, G. A. 1979. *Strategic planning: What every manager must know: A step-by-step guide*. New York: Free Press.

Turner, J. R., ed. 1999. *Handbook of project-based management*. 2nd ed. London: McGraw-Hill.

CHAPTER FOUR

MODELS OF PROJECT ORIENTATION IN MULTIPROJECT ORGANIZATIONS

Joseph Lampel, Pushkar P. Jha

More than 30 years of research on project management have produced a relatively coherent body of principles for effective project planning and execution. Over the same period, there have been consistent reports of the failure of managers and organizations to follow these prescriptions. The obvious response to this failure has been to look for "best practices" that can bridge the gap between prescriptions and outcomes. The rationale behind this effort is to comparatively analyze the conduct and performance of different projects, and then translate effective managerial practices into a set of robust practices that are widely accepted.

While projects may fail because of factors that are internal to the project itself, such as poor leadership, conflict, poor coordination, and so forth, organizational factors that are external to the projects also play a role in the failure of prescriptions to produce expected outcomes. More specifically, researchers hypothesize that problems in the relationship between organizations and projects will hamper the implementation of project management prescriptions. Thus, to understand why project management prescriptions often fail, it is necessary to identify these problems. Research that deals with new product innovation (Wheelwright and Clark, 1992; Lewis and Welsh, 2002) suggests that projects usually fail because they are insufficiently resourced, insufficiently empowered, and/or, generally lack wider organizational support.

When these problems are examined in some depth, the normative conclusion that follows is that organizations that take the trouble to do these things properly are more likely to successfully plan and execute projects than organizations that do not. The ideal organization, from this point of view, is one in which structures, empowerment, and resources are aligned as far as possible with the needs of projects. In practice, most projects take place in organizations that depart from this ideal. And by the same token, the further they depart from this ideal, the more likely are the project management specialists to encounter problems when translating prescriptions into practice.

While this point is often raised in the literature (Raz et al, 2002; Shenhar et al., 2001), what is clearly lacking is a construct that allows going beyond simple observations of the shortcomings of organizations that do not properly resource, support, and empower projects. This is a task made difficult by the very nature of projects, as organizations often have projects that serve different clients, meet different needs, require different resources, and develop along different trajectories (Lampel, 2001).

In this chapter, we define the degree to which organizations resource, empower, and support projects as "project orientation". The term project orientation has been used in the literature to examined a multitude of interconnected concepts, from project leadership to team and project management competencies (Gareis and Hueman, 2000; Gareis, 2002) and management design for project coordination (Gareis, 1992). The term is used here to define an integrated construct that is likely to be of considerable utility to the project management specialist. The concept of project orientation allows project management specialists to discuss and measure the constraints in which they operate with greater clarity and precision. This may not only be useful for communication among managers that are directly involved in planning and implementing projects but is also useful in discussions with support functions and top management.

The construct of project orientation, however, has wider utility and can be of importance when it comes to managing the overall strategy of the firm. The starting point for considering the use of the construct is the interrelationship between project selection and management on the one hand and organizational strategy on the other (Cleland, 1994, p. 81). Attaining a "fit" between strategy and project management activities is an important aspect of the search for overall sustainable advantage (Porter, 1996). According to Porter (1996), the search for fit operates on three levels. First-order fit is the consistency of different activities with overall strategy. Second-order fit occurs when activities reinforce each other. And third-order fit is the optimization of efforts across activities.

Attaining fit on any of these levels, let alone on all three, requires organizations to choose between different combinations of structures, systems, and processes (Mintzberg, Ahlstrand, and Lampel, 1998, pp. 302–347). The choice is often a function of organization type (Miller and Mintzberg, 1983). For example, small law firms are likely to approach the choice differently from large, diversified consumer goods firms. In the case of projects, we would argue that the search for fit between strategy and project management activities is likewise dependent on the type of organization. Research suggests (Wheelwright and Clark, 1992) that when it comes to projects, organizations generally fall into one of the following types:

- *Project-based.* Organizations with centralized operations tuned to support projects. Here, a formal body (for example, a project office) may lead project planning. This is typical of organizations with large-scale projects and with a project portfolio characterized by a large number of projects for clients' external to the organization. Organizations whose projects are characterized by extensive post project support in client handover also tend to fall in this category.

- *Project-led.* Organizations where operations provide support for projects. Here core operations in functions or areas such as manufacturing, information systems, human resources and product development, to name several, tend to have a significant say in project initiation and design. The term project-led typically describes an organization that may or may not have a project portfolio aligned to service an external client base but is very likely to be characterized by a reasonable density of internal change/process improvement projects. Complex alliancing requirements in project initiation and design stages may also be a feature of projects in such an organization.
- *Core-operations-led.* Organizations where projects are support mechanisms to augment/improve the efficiency of core operations. Projects are likely to be internal and could be low down in top leadership involvement beyond the initiation and conceptualization stage. This is in comparison to organizations that can be categorized as project-based and project-led.

In general, project orientation tends to be high in project-based organizations, moderate in project-led organisations, and relatively low in core-operations-led organizations. Strategically, however, organizations tend to move within this typology. Project-based organizations may change their portfolio of projects and clients. Project-led organizations may become focused on one type of projects. And finally, core-operations-led organizations may decide that their future depends on a specific type of transformational projects.

In all these cases, organizations may experience a mismatch between their strategy and their project orientation. Put differently, their strategy may demand a different project orientation than the one that has evolved over time. To become more effective strategically, they would need an audit of their project orientation; first, in order to determine their project orientation, and second, in order to decide to what extent it should change.

Models of Project Orientation

In the sections that follow, two models of project orientation are presented. The first is the project perspective of project orientation and the second the organizational perspective. The interfacing environment between the projects and the organization is viewed from both sides of this interface: the project side and the organization side, respectively. The primary basis for these models are insights drawn from the first phase of exploratory interviews with seven partner organizations from different sectors in the EPSRC research project on "Organisational Learning and Business Performance in Project Based Organisation" (PROBOL; 2001–2003). The exploratory interviews attempted to understand the project portfolio of these organizations and how they made sense of project experiences to configure organizational strategy for projects.

The aim of the study is to explore and validate project orientation as a construct. We have followed the standard approach of using principal component analysis to evaluate construct validity (Carmines and Zeller, 1979). An outline of the data and analysis is provided after the discussion on the two models. The interpretations drawn from the analysis are at present exploratory. They are outlined in the form of some key relationships that may dominate the interfacing environment that are defined as project orientation.

Model 1: Project Perspective of Project Orientation

Our reading of the literature (Bugetz, 1992; Lawrence and Johnson, 1997; DeFillippi and Arthur, 1998; Bugetz, 1992; Rose et al, 2000; Nobelius, 2001; Sydow and Staber, 2002) suggests that seen from the perspective of projects, three project dimensions shape the interaction between projects and organizations. These are, project scoping, project programming, and project autonomy. Figure 4.1 uses these dimensions to visualize a space inhabited by the project portfolio of the organization. This is the "project to organisation" perspective of project orientation as it has evolved over time.

Project scoping consists of the basic definition of the project plus the extent to which the concept is fully mapped during the early phase of the project, as opposed to allowing these details to emerge subsequently. The contrast is between fixing the scope in some detail and curtailing exploration very early in the life -cycle of the project, as opposed to simple sketching of the scope and allowing considerable room for exploration as further information and learning is gained.

Project programming consists in the extent to which the project is tightly constrained by budgets, schedules, and targets, as opposed to being loosely constrained by these factors. A tightly programmed project is often a project in which every effort is made to foresee

FIGURE 4.1. PROJECT PERSPECTIVE ON PROJECT ORIENTATION.

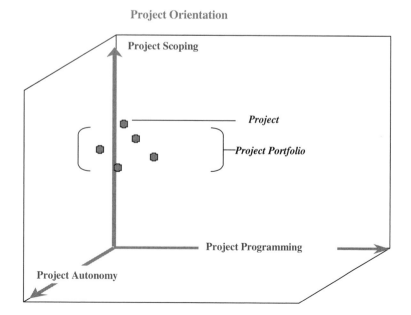

contingencies, break down the process into discrete steps (i.e., gating the process), and insert strong incentives for compliance with schedules and targets. A loosely programmed project is one in which schedules are simple and approximate, and incentives are tied to achievement of generally formulated rather than highly specified targets.

Project autonomy refers to the degree to which the project is allowed to evolve without constant report and input from the organization. Here we have two extremes. At one extreme we have projects that (a) constantly report progress to other parts of the organization (in particular, to higher organizational levels) and (b) accept input and request for changes. At the other extreme we have projects that are (a) given high latitude to move forward with little reporting and (b) do not expect or easily accept input and suggestions for changes.

These three dimensions constrain and define project evolution within the confines of the organization. They are in effect evolutionary trajectories that are shaped by the interface with the rest of the organization. With this in mind, project orientation from the perspective of projects is determined by the extent to which these dimensions are set by the needs of the projects, as opposed to the needs of the organization as a whole. Such needs do not privilege specific polarity along these dimensions.

In some industries it is better if projects are fully scoped in advance, while in others basic scoping with plenty of room for exploration tends to produce better results. In some industries tight programming is regarded as best practice, whereas in other industries loose programming is often seen as preferable. The same can be said of project autonomy. Here not only will industry practice vary, but company practice will likewise vary, since autonomy, as you will see, is as much a political as a strategic issue.

A note of caution should be sounded at this point. From a statistical point of view, the dimensions are probably not independent. In general, fully scoped projects often go together with tightly programmed projects; and tightly programmed projects often go together with projects that have only low autonomy. Having said this, it is also worth emphasizing that this statistical observation is not necessarily strategic.

Organizations do have a choice when it comes to how they design, program, and control their projects. In some industries and organizations, one finds fully scoped projects with loose programming and projects with high autonomy that are tightly programmed. As Table 4.1 shows, the key issue is the evolutionary trajectories that are defined by these dimensions. Some may be rare, but they are all in principle possible.

Model 2: Organizational Perspective of Project Orientation

Organizations are systems of power and resources. From the point of view of projects, most decisions regarding power and resources should devolve to the level of the project. From the point of view of organizations, this is not always possible or indeed desirable. There is frequently a tug-of-war here between what is best for the project and what is best for the organization. The tension may be at odds with prescriptive models of project management that ordain that projects should be aligned with the strategy of the organization as a whole (Cleland, 1994), but it is a tension that persists for a number of reasons.

First, organizations often have multiple projects with different clients and different needs. Second, projects overlap in terms of planning and execution. Today's strategy may give rise to new projects, but the organization may still be structured to serve projects that

TABLE 4.1. TYPES OF PROJECTS.

Illustrative Projects Defined by Some Lead Characteristics	Scoping	Programming	Resource Allocation	*Autonomy* Low or High (Interpreted by a combination of:) Monitoring Light (low frequency, loosely structured) or heavy (high frequency and very structured)
1. Client-specified solution where similar projects for the client have been carried out successfully in the past	Rigid	Detailed	Up front	It could be either light or heavy.
2. Problem-solving R&D project with high sanction from the top	Rigid	It could be either detailed or sketchy.	Up front	It could be either light or heavy.
3. New-product development targeted at specific market needs	Rigid	It could be either sketchy or detailed.	Either up front or gradual	It could be either light or heavy.
4. Exploratory R&D project for Product technology/core technology improvements anticipated to provide competitive edge and championed from the top	Flexible	Sketchy	Up front	It could be either light or heavy.
5. Internal change project where solution specifications are provided by external R&D	Rigid	Detailed	Gradual	It could be either light or heavy.
6. Internal project where solution specifications are to be developed in-house for a given need; also, utility projects that involve critical resources and are intended "to keep the lights on" (Shenhar, et al., 2002)	Flexible	Detailed	Gradual	Heavy
7. Client-specified "problem only" projects where detailed specifications for output are missing, e.g., software development projects and consultancy projects	Rigid	Sketchy	Gradual	Heavy
8. New-product development characterized by a fast-changing market	It could be either rigid or flexible.	Sketchy	Gradual	Heavy

were conceived some time back. Third, firm strategy is often interpreted and enacted differently in different part of the organisation (Gann and Salter, 2003). Finally, projects are not always aligned with the strategy because strategy itself is frequently an umbrella that permits a range of options rather than a clearly and tightly defined set of goals (Mintzberg and Waters, 1985). This is also highlighted in numerous industry specific case studies (Drake, 1999; Westney, 2003) in recent times.

All of the preceding leads to a perspective of project orientation that is centered on the organization, as distinct from model 1, which is centered on the project. From an organizational point of view, the following issues matter most when it comes to making key project decisions: First, what resources does the project need to perform effectively and how should these resources be allocated? And second, how much political backing does the project need from top management (Green, 1995) and support systems to perform effectively?

These two issues are clearly related. For example, resource allocation often requires top management support. And accessing support systems is often a matter of resource allocation. But beyond this, the ability of projects to gain resource and political support at the front end and subsequently their ability to gain backing and more resources from both top management and support systems is a measure of the organization's project orientation.

In this respect, organizations are likely to be different. In some organizations, projects receive proportionately limited attention and support. They are regarded as one area of action among others. In other organizations, projects are central to the strategy of the company and thus are more likely to top the agenda and have first claim to resources.

As Figure 4.2 suggests, the distinction between these two types can be perceived by an alignment typifying high project orientation in the latter case against low project orientation in the former case. This alignment evolves over time and may need to be redefined by deliberate action. This may be through key transformational projects when the organization feels the need to develop this emerged pattern to fit its interpretation of business needs from projects. Some organizations may also engage in experimental projects where they test the performance of deliberate action used to change this alignment, before setting in motion organization-wide change in the way projects are done.

Project Orientation Variables

Exploratory research during the first phase of the PROBOL research and the theorization for the models of project orientation have been used to generate a list of variables that manifest themselves in the interfacing environment between the projects and the organization. Based on these variables, an attempt to investigate the key relationships in the project organisation interface has been undertaken. A list of these variables, which forms the basis of the data collection instrument, is as follows:

- Treatment of past project performance
- Top leadership's support for change
- Basic characteristics of project systems, including volatility of the technology involved
- Baseline for project management performance (time, cost, and scope)

FIGURE 4.2. DEGREES OF PROJECT ORIENTATION.

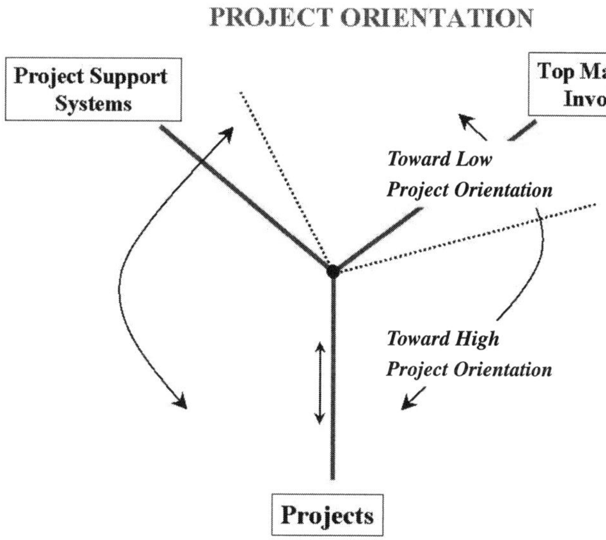

PROJECT ORIENTATION

- Top leadership involvement in project planning including role in giving visibility to projects
- Experience and role clarity of multiproject leadership
- Role span of the project manager
- Capacity building for project roles
- Project alliancing decisions (e.g., contractor alliancing and inter-organizational alliancing.)
- Support department's involvement in project control
- Senior management's involvement in project control
- Approach toward project teams in terms of physical location and performance assessment
- Project resourcing modes
- Incentive for gaining project experience
- Nature of communities of practice (and networking of knowledge management structures) for projects

Project Orientation Data: A Profile of the Sample

Using a structured instrument, quantitative data has been collected from 54 respondents over six of the seven organizations in the second phase of the PROBOL research. All seven organisations had contributed to the qualitative data generated from exploratory interviews in the first phase of the research.

The six partner organizations that compose the data in the second phase of the research are drawn from the sectors of information technology, pharmaceuticals and healthcare, poverty alleviation, banking and finance, construction, aerospace and energy.

Nearly three-fourths of the respondent sample was composed of project practitioners and one-fourth was composed of personnel who are working with project support departments. Some project leadership and top leadership personnel are also included in this stratification. The work experience range for the entire sample was from 1 to 36 years of experience with the organization. This provides a comprehensive "perception map" of how the aforesaid variables influence project orientation. It has also allowed for an identification of the key driving relationships for each organization. These key drivers may significantly aid organizational efforts to mold the evolved project orientation profile to match the strategic needs of the organization from projects.

Instrument Testing

The structuring of the variables and the design of indicators to measure them was based on inputs from the literature survey and insights about partner organizations from the first-phase exploratory data collection under the research. Inter-rater reliability, based on rank order correlation across respondents for each organization, was at acceptable levels. This implies that different respondents with their respective profiles understood the indicators as measuring the same phenomenon in their respective organisations.

Multiple indicators were designed to measure each variable for reasons of validity. A total of 57 indicators make up the questionnaire that measures these 15 variables. The responses were generated using a Likert scale [A scale where the respondent specifies a level of agreement or disagreement with statements that express a favorable or unfavourable attitude towards the concept under study/being measured]. For content validity purposes, questions were worded both negatively and positively (based on what constituted high or low project orientation) to make sure that the right phenomenon were being measured and each question represented an aspect of the variables under investigation.

Content validity was supported by interim data analysis that investigated relationships between the indicators. These relationships were presented to the respective organizations and were recognized by them as phenomena that characterize the interfacing environment between the project and the organization, thus establishing that the expected values for project orientation were being measured.

Data and Analysis

The 57 questionnaire items were aggregated over the 15 variables under consideration. The main part of the analysis has been carried out using principal component analysis to reduce dimensionality because of a large number of variables as enumerated earlier.

Simulation was carried out for reasons of data limitation. Standard deviation in the original or base sample was used as a factor to increase the data set for statistically significant analysis. The simulation was carried out for the sample as a whole and also for individual organization data sets to achieve a significant ratio between variables and number of samples.

Open-ended responses and post-instrument-administration group discussions were also used to further augment the understanding of project orientation. The interpretations were discussed in partner organization group sessions and received confirmation as a reliable interpretation of the phenomenon of project orientation.

Limitations of sample size and the extensive range of factors involved in configuring an organization's project orientation make the results and implications of the study less definitive than we would like them to be. Nevertheless, the findings provide useful insights to illustrate the concept of project orientation.

Findings and Implications

Principal component analysis was used on the sample as a whole and also to individually process patterns for each organization. These findings support the introduction of the third element, top management involvement at the project-organization interface level as also enumerated in the discussion leading to the second model (Figure 4.2). This was because variables related to this aspect (top leadership and multi-project leadership) load strongly on the principal components.

The two models of project orientation have been related to six of the seven organizations that contributed data toward the second phase of the research. These organisations are profiled in the following:

Organization 1: Characterized by external-client-focused projects, fast-changing technology, and projects with small- to medium-length[1] life cycles.

Organization 2: Characterized by internal projects, product development, infrastructure, and/or change focus.

Organization 3: Characterized by long life cycle and fast-changing technology; projects move in clusters using centralized resources, product development, and external client focus.

Organization 4: Characterized by large-scale external-client-focused projects, moderate to high project life cycle, widespread in physical location of project sites, and an ongoing restructuring to improve the interface between project roles and top management roles.

Organization 5: Characterized by internal change and process efficiency improvement projects, relatively smaller project life cycle.

Organization 6: Characterized by clusters of highly interconnected projects within divisions, which almost always exist as part of larger programs.

---

[1] Length of lifecycle is discussed as relative over the organisations in the sample.

Table 4.2 links the preceding organization types and the project orientation models. The implications for each of the organizations in the sample lie in the key relationships enumerated here. As an illustration, these may be understood as levers that can be effectively used to affect project orientation and thus, by extension, also affect organizational objectives essential to the projects.

The table is essentially a matrix that has several possible manifestations of organizational perspective of project orientation in columns and manifestations of the project perspective in rows. For categorization purposes here, high, moderate, and low qualifiers are used. The six partner organizations in the PROBOL research that contributed to the data in the second phase are placed in the respective intersection of the two models as in this matrix. The categorization of organizations is based on exploratory interviews and background research on the organizations to understand the organizations in the sample and the projects they do. Principal component analysis outputs are used to express relationships that dominate the interfacing environment between the organization and the projects.

Conclusions

Morris (2003) has argued that we should make a distinction between project management and the management of projects. The first deals primarily with practices and principles that are internal to the discipline of project management as it was traditionally conceived. The second looks to a wider perspective of the relationship between projects and a more holistic view of their context—whether their context is the organization or the larger economic and social environment. The construct of project orientation belongs to the management of projects, rather than to project management. It is put forward as a core construct that defines the relationship of the organization and its projects and perspectives from both sides of this interface (i.e., the project-to-organization perspective and the organization-to-project perspective, respectively) are employed.

The results presented from ongoing research (PROBOL, 2001–2003) that partners from seven global firms in different business areas provide an illustration of the utility of the construct of project orientation. There are statistical limitations because of sample size in the second-phase quantitative data used in the research (that was drawn from six of the seven organizations). The range of variables that impact the project organization interface also contributed to limitations in analysis. However, use of simulation techniques has allowed us to partly overcome the limitations of a small sample. This has been further aided by use of associated qualitative data from discussions with respondents from the partner organizations and by sharing findings with the partner organisations to get feedback on the accuracy of the diagnosis and constructs, which has been affirmative.

Projects have an ambiguous relationship to the corporate environment in which they evolve (Garbher, 2002). On the one hand, projects often underpin the organization's main business and are a direct expression of its strategy. On the other hand, however, they often stretch and change both operations and strategy. This tug-of-war between projects and organizations is inevitable and, when properly managed, can be constructive. Often, however, the tension translates into friction and failure—in part because such tensions trigger

TABLE 4.2. PROJECT ORIENTATION PERSPECTIVES.

Organizational Perspective / Project Perspective	Low Project Orientation	Moderate Project Orientation	High Project Orientation
Low spread in the project portfolio cluster	**Organization 5.** A strengthening of project manager role is instrumental in relaxation in project standards of cost, time, and scope.		**Organization 4.** Top leadership support for projects critically influences variation in core technology in projects and independence of individual project planning in relation to the larger portfolio. Clearer career progression paths from project to project leadership levels are vital for coherent project strategy for the organization
Moderate spread in the project portfolio cluster	**Organization 6.** Strengthening of the project manager's role encourages seeking project experience. Top leadership support is critical to knowledge sharing across projects and developing communities of practice.	**Organization 1.** Incentive for gaining project experience depends how project performance is treated (buried or magnified). Clear role definition for multiproject leadership may help reduce support department's control on project execution.	**Organization 3.** Senior–middle management involvement may be critical in influencing top management support and sponsorship for projects. Clear role definition for multiproject leadership may help reduce support department's control on project execution.
High spread in the project portfolio cluster	**Organization 2.** Incentive for gaining project experience depends upon a greater boundary spanning in the project manager's role. Top leadership support is critical to knowledge sharing across projects and developing communities of practice.		

power politics but also because there is an insufficient understanding of the causes and dynamics of the tension between projects and organizations in the first place. Project orientation is a useful way of looking at the relationship between projects and the corporate environment. It opens the way to deeper understanding and better management of the factors that shape the interface between project management at the project level and the management of the organization at the higher, strategic level.

References

Burgetz, B. A. 1992. Project design: The critical step to successful systems. *CMA Magazine* 66(4):10.

Carmines, E. G., and R. A. Zeller. 1979. *Reliability and validity assessment.* Beverly Hills: Sage Publications.

Cleland, D. I. 1994. *Project management: Strategic design and implementation.* 2nd ed. p. 81. New York: McGraw-Hill.

Drake, D. L. 1999. Projects must fit into company strategies. *Business News New Jersey* 12(1):22–23.

DeFillipi, R., and M. Arthur. 1998. Paradox in project based enterprise: The case of filmmaking. *California Management Review* 40(2):125–139.

Gann, D. M., and A. J. Salter. 2003. Project baronies: Growth and governance in project-based firm. 19th EGOS Colloquium, Copenhagen.

Gareis, R. 1992. Management of network of projects. *AACE transactions.* www.wu-wien.ac.at/pmg/pos/docs/pub_portfolio_management.pdf.

Gareis, R., and M. Hueman. 2000. Project management competencies in the project-oriented organisation. In *The Gower Handbook of Project Management.* ed. J. R. Turner and S. J. Simister. 709–721. Aldershot, UK: Gower Publishing.

Grabher, G. 2002. Cool projects, boring institutions: Temporary collaboration in social context. *Regional Studies* 36(3):205–214.

Green, S. G. 1995. Top management support of R& D projects: A strategic leadership perspective. *IEEE Transactions on Engineering Management* 42(3):223–232.

Lampel, J. 2001. The core competencies of effective project execution: The challenge of diversity. *International Journal of Project Management* 19(8):471–483.

Lawrence, B., and B. Johnson. 1997. The project scoping gamble. *IEEE Transactions* 14(3):107–109.

Miller, D., and H. Mintzberg. 1983. The case for configuration. In *Beyond Method*, ed. G. Morgan. Beverly-Hills: Sage.

Mintzberg, H., and J. Waters. 1985. Of strategies, deliberate and emergent. *Strategic Management Journal* 6:257–272.

Mintzberg, H., B., Ahlstrand, and J. Lampel. 1998. *Strategy safari.* London: Prentice Hall.

Morris, P. W. G. 2003. Irrelevance of project management as a professional discipline. IPMA 17th World Congress on Project Management, Moscow.

Nobelius, D. 2001. Empowering project scope decisions: Introducing. R&D content graphs. *R&D Management* 31(3):265–274.

Porter, M. E. 1996. What is strategy? *Harvard Business Review.* (November–December): 61–78.

PROBOL 2000–2003. UK government-funded (EPSRC) research project on Organisational learning and business performance in project based organisations. www24.brinkster.com/probol.

Raz, T., A. J. Shenhar, and D. Dvir. 2002. Risk management, project success, and technological uncertainty. *R&D Management* 32(2):101–109.

Rose, K. H., J. K. Pinto, J. W. Trailer, and K. Rose. 2000. Essentials of Project Control. *Project Management Journal* 31(2):60–61.

Shenhar, A. J., D. Dvir, T. Lechler, and M. Poli. 2002. One size does not fit All: True for projects, true for frameworks. Project Management Institute Research Conference. Seattle.

Shenhar, A. J., D. Dvir, D., O. Levy, and A. C. Maltz. 2001. Project success: A multidimensional strategic concept. *Long Range Planning* 34:699–725.

Sydow, J., and U. Staber. 2002. The institutional embeddedness of project networks: The case of content production in German television. *Regional Studies* 36(3):215–227.

Turner, R., R. Peyami. 1996. Organizing for change: A Versatile Approach. *In The Project Manager as a Change Agent*, Composed and edited by J. R. Turner, K. V. Grude, and L. Thurloway, 62–75. New York: McGraw-Hill.

Westney, R. 2003. Setting the strategy for successful projects. *Offshore* 63(5):133–134.

Wheelwright, S. C., and K. B. Clark. 1992. *Revolutionizing Product Development: Quantum Leaps in Speed, Efficiency, and Quality*. 175–196. New York: Free Press.

CHAPTER FIVE

PROJECT PORTFOLIO SELECTION AND MANAGEMENT

Norm Archer, Fereidoun Ghasemzadeh

A project portfolio is a group of projects to be carried out under the sponsorship of a particular organization. These projects must compete for scarce resources (labor, finances, time, etc.), since there are usually not enough resources to carry out every proposed project. Project portfolio selection is the periodic activity involved in selecting a portfolio from the set of available project proposals and from projects currently under way. Compared to the managerial and operational decisions that are usually involved in managing individual projects, portfolio selection is a strategic decision. To ensure a maximum return on selected projects, the selection process must be linked to the business strategy of the organization.

The purpose of this chapter is to outline portfolio selection and management as a strategic process, using known techniques and tools in a logical and organized manner. We begin with a literature review, and then discuss the characteristics of projects affecting their selection and outline some portfolio selection methodologies. Choosing among these methodologies and organizing a logical portfolio selection process requires a framework that can be followed easily, and we describe one such framework in detail. Finally, we review some current issues in portfolio management.

Literature Review

Project portfolio management is an essential concept in many industries, and applications include new product development (NPD), construction, pharmaceuticals, process development, product maintenance, fundamental research, and so on. These tend to differ along certain dimensions, and we will demonstrate some of their differences by examples.

The majority of innovation-based manufacturing industries generate products obtained predominantly via innovative R&D processes. NPD portfolios are more likely to be successful if they include a limited number of carefully selected, positioned, and balanced projects (Wheelwright and Clark, 1992; Cooper, Edgett et al., 2000). A portfolio is balanced if there is a suitable distribution of projects on dimensions such as technology and market risk, completion time, and return on investment. If there are too many projects in the pipeline, there is a high degree of conflict over existing resources, slowing project progress and reducing successful completion rates. This may result in missed market opportunities and limited investment returns.

On the other hand, discovery-based research activities dominate in the development of new pharmaceuticals. This involves a search and a trial-and-error test or screening process, whereas NPD involves a concept, proof of feasibility, a product design, and an evaluation. In the pharmaceutical industry, the risk is so high in the initial (discovery) phase that literally hundreds of selected compounds or molecules may be investigated in order to improve the probability of finding a few that are worthy of the next phase of development (Prabhu, 1999).

In the construction industry, technology risk is relatively low, but there are other risks resulting from uncertainties in weather, political decisions, and labor availability. Another feature of portfolios in the construction industry is that most projects are outsourced, and this industry is characterized by a lack of long-term relationships between supplier and client. Welling (Welling and Kamann, 2001) suggests that cooperation can be improved when the same individuals, representing the general contractor and subcontractors, respectively, deal with each other in a series of projects, rather than dealing with different individuals for each project. This suggests an increasing emphasis on long-term cooperation, networking, and strategic alliances.

Most recent advances in portfolio selection and management have occurred in the new product development area. A comprehensive review of this field has been completed by Cooper, Edgett, and Kleinschmidt (Cooper, Edgett et al., 2001). In addition, this team did an extensive survey of NPD industry practice of portfolio management (Cooper, Edgett et al., 1999) in 205 U.S. companies. From their study, they were able to cluster companies into four groups, with "benchmark" businesses as the top performers. This group's new product portfolios consistently scored the best in terms of performance. These companies tended to choose high-value projects and their portfolios were aligned with the company's business's strategy, with the right balance and the right number of projects. Benchmark businesses employed a much more formal, explicit method in managing their portfolios. They relied on clear, well-defined portfolio procedures, were consistent in applying their methods to all projects, and there was strong management support for portfolio selection and management.

Risk and outsourcing are currently having the most active impact on changes and advances in portfolio selection and management. For these reasons, the following two subsections are devoted to these particular topics.

Risk

There are multiple sources of risk, including technology, market, schedule, cost, legal, political and so on, depending on the field of application. Sophisticated methodologies to assess risk amount and impact prior to project commitment, through risk identification, quantification, response development, and control, have been identified as one of the areas in which major advances have recently occurred (Pinto, 2002). A key criterion for successfully applying risk evaluation in portfolio selection is that risk assessment and quantification be uniformly applied across all projects and teams in order to distinguish among projects that have acceptable and unacceptable levels of risk. For construction portfolios, since there is less uncertainty in the expected costs, schedule, and performance, financial approaches are the accepted norm for project selection. However, there is often a significant uncertainty in such factors as weather, labor availability, and political environment. Disagreement frequently occurs among parties to lump-sum construction projects, because of a lack of advance agreement in apportioning risk between client and contractor (Hartman, Snelgrove et al., 1998).

Assessing and managing risk in NPD programs is discussed in detail by Githens (Githens 2002). Although risk is one of the most important characteristics of projects, it is often difficult to evaluate. The frequency with which project risk analysis has been applied in industry (Raz, Shenhar, et al., 2002) has been disappointing, although when it is used, the chances of success are much improved.

The risks of failure of individual research and development projects can be very high when truly new technology is involved in new product development. A recent survey indicates that it takes an average of 6.6 ideas to produce one successful new product, while on average 50 to 60 percent of NPD projects fail (Griffin, 1997). Companies that rely heavily on internal R&D for new products need to be able to take risks without compromising the profitability of the company. It is critical to link company strategy to portfolio development when company strategy involves both a high degree of innovation and a high rate of growth (Wadlow, 1999).

Risk-value project selection procedures that seek to minimize risk for a certain specified value tend to bias against higher risks, both for technical risk and commercial risk uncertainties. Another option is to maximize value for a certain level of risk. If innovation and growth are major strategic goals, project selection must first select on the basis of growth without unduly compromising other important objectives. However, formal product portfolio selection may bias against the inclusion of more highly innovative projects in a portfolio. For example, Roseneau (Roseneau, 1990) notes a reduction in the degree of innovativeness of new products through internal competition, with the attractive short lead times associated with low-risk, less innovative projects. Speed to market and high degrees of innovation in new products tend to be conflicting objectives.

Typical models used in analyzing risk include Monte Carlo simulation, decision theory and Bayesian statistical theory (Hess, 1993; Riggs, Brown et al., 1994; Martino, 1995), and decision theory combined with influence diagram approaches (Rzasa, Faulkner et al.,1990). One new approach to portfolio selection in NPD is risk strategy analysis (Wadlow, 1999). This focuses on risk as a key factor in the development of new technological products. A risk strategy, in this context, describes a specific extent and type of involvement in research in terms of the different levels of technical and commercial risk associated with the mix of projects making up the portfolio. Accordingly, a risk strategy in terms of technology and commercial risk is analyzed in terms of statistical risk of failure as opposed to probable outcome. Given a number of basic project descriptors, this approach can be used to design as well as analyze risk strategies while screening projects for portfolio selection, based on the firm's strategic business objectives. While there is uncertainty in almost every project parameter, an interesting application of cost uncertainty in portfolio analysis is provided by Stamelos (Stamelos and Angelis, 2001).

The unique nature of projects, and their range in objectives, size, complexity, and variety of technological content, is almost limitless and not confined by industry boundaries. Shenhar et al. (Shenhar, 2001) have derived a general classification system for project management commonalities, which can be helpful in establishing portfolio risk profiles, since an optimal choice of projects for a portfolio may need to span these dimensions. Although market risk also reflects the potential acceptance of a new product, two dimensions they suggest are technological uncertainty and contribution to business. Table 5.1 reflects these two dimensions, although some of the examples given in the table might in fact extend across more than one of the given categories on one or both dimensions. But there are other uncertainties in addition to technological uncertainty, particularly in the construction industry.

The applications given in Table 5.1 reflect classifications of (a) construction projects that tend to have uncertainties in areas other than technology, (b) derivatives or enhancements of existing products or processes, (c) platform projects or pharma (pharmaceutical) line extensions that entail new combinations and extensions of existing technologies, (d) positioning or breakthrough projects that incorporate revolutionary new technologies or manufacturing processes, possibly enhancing the company's ability to develop new markets, (e) scouting or probing offerings to early adopters in new markets, and (f) stepping-stone R&D and pharma R&D discovery, which are long-term, high-risk projects in areas where both markets and technologies are relatively undeveloped (Shenhar and Wideman, 1997; Wheelwright and Clark, 1992; MacMillan and McGrath, 2002; Bunch and Schacht, 2002).

Outsourcing

The construction industry almost invariably outsources projects, so the selection and management of portfolios of construction projects by general contractors is an established dis-

TABLE 5.1. PROJECT CHARACTERISTICS.

Contribution to Business	Technological Uncertainty			
	Established	Mostly Established	Advanced	Highly Advanced
Immediate Short Term	Construction			
		Derivatives; Enhancement		
Medium Term			Pharma line extensions; Platform development	Scouting
Long Term			Positioning; Breakthrough	Pharma R&D discovery; Stepping-stone R&D

cipline. On the other hand, the practice of outsourcing technology development in other industries (Kimzey and Kurokawa, 2002) is playing an increasingly important role in achieving competitive advantage, providing access to larger technology resource pools, increasing ability to develop products that could not be developed using internal resources, shortening product cycle times, and reducing development costs. However, technology outsourcing tends to be on a tactical case-by-case basis rather than forming a part of corporate strategy.

A typical application of outsourcing is the pharmaceutical industry, where firms face high R&D risks and costs as well as intense new product competition. Here, larger companies need new blockbuster products that smaller biotechnology companies have to offer, but lack the pharmaceutical company research funding and marketing muscle. This often leads to alliances that combine the core strengths of both types of firm (Bunch and Schacht, 2002). Smaller pharma firms often try to mitigate their risks and leverage limited R&D resources by contracting out upstream (laboratory scale) research for a portfolio of R&D projects to smaller technology institutions. These firms can then concentrate limited R&D resources on downstream (commercial scale) R&D that utilizes the limited stream of successful upstream research outputs received from their collaborators (Prabhu, 1999).

Although portfolio selection may have increased flexibility through outsourcing, it must be evaluated carefully for the potential gains. There are many differences in managing external projects or subcontracts versus managing internal projects. For example, the Royal Bank of Canada has won awards over the years in the quality of its IT portfolio development and operations. It consistently handles development operations internally, with hundreds of projects across some 12 product groups under development in any given year (Melymuka 1999). Outsourcing their IT portfolio to possibly inferior developers would unnecessarily increase costs and risks.

Characteristics of Projects Affecting Portfolio Choice

The most important characteristics of projects that affect their selection depend upon the industry sector. Common across all fields are project scope, the total investment, labor, required expertise, materials, facilities, estimated return on investment, and timing. Alignment with the firm's business strategy is also important in each case. In the NPD and pharmaceutical sectors, technological and market risk becomes a critical issue, and whether the proposal involves an incremental or platform project (Tatikonda, 1999) is a major consideration in the selection decision.

The availability of outsourcing resources plays a crucial role in construction, and it is becoming of more importance in NPD and pharmaceuticals. The division of risk between client and contractor is an increasingly important topic in construction. If this is not decided in advance, it can lead to costly and lengthy legal disputes over responsibility when time, cost, or quality shortfalls are in question.

At the portfolio level, constraints exist that affect project selection. Reaching a proper balance of risk, timing, diversity, total investment, and return is critical to the overall portfolio selection. In addition, sharing resources (labor, facilities, materials) among projects, project priorities, and interdependencies all play a role in the selection process.

The ground needs to be well prepared in advance of the project selection process, to improve the likelihood that the best kinds of projects are being proposed. As project descriptions are generated, there must be a continuous dialogue between management and project proponents, covering issues such as (a) management's strategic intent, (b) implementation issues, (c) related projects that are competing for resources, and (d) corporate technology practices and rules (Bordley, 1998).

Clearly, in order that an informed judgment be made in selecting the portfolio, reliable project data are required to apply models such as those discussed in the following section. In the construction industry, for example, estimating costs and timing is based on experience with many similar projects. This may also be the case in NPD for maintenance projects, but for new and even incremental product development, estimates are needed for market, revenue, pricing, manufacturing or operations, risk, and resource requirements (Cooper, Edgett et al., 2001, Chapter 8). These data tend to be highly uncertain, particularly at the time the selection decision must be made.

Portfolio Selection Models and Methodologies

The use of specific project characteristics for portfolio selection is situation-dependent. For example, a product development organization may use market research, economic return, and risk analysis to develop project characteristics that can be useful in selection exercises. Or a government agency may use economic and cost benefit measures. Measures used may be qualitative or quantitative, but regardless of the techniques used to derive them, common measures must be used during portfolio selection so projects can be compared equitably.

An important consideration is that while there are many possible methodologies that can be used in selecting a portfolio, there is no consensus on which are the most effective. As a consequence, each organization tends to choose, for the project classes being considered, the methodologies that suit its culture and that allow it to consider the project attributes it believes are the most important. Also, the methodologies most useful in developing a portfolio for one class of projects may not be the best for another (e.g., good estimates of quantitative values such as costs and time may be readily available for certain construction projects, but qualitative judgment is more likely to be used for development of advanced new products). A major difficulty with portfolio selection is the number of possible combinations of projects that can be selected, with resources and schedules for each project affecting those available for the remainder of the portfolio. This number of possible solutions grows geometrically as the number of projects increases. Many companies use more than one approach in the selection process. For example, the NPD portfolio selection process utilizes an average of 2.4 techniques in the typical firm (Cooper, Edgett et al., 1999). The following is an outline of major classifications of methodologies used in portfolio selection.

Economic Return

These techniques typically require financial estimates of investment and income flows over the time frame of the project, often based on experience with similar projects. They are the best choice in construction portfolios, where most of the costs and schedules can be estimated with reasonable accuracy. Results from the calculations can be used in ranking or displaying information for decision making. Techniques include net present value (NPV), discounted cash flow (DCF), internal rate of return (IRR), return on original investment (ROI), return on average investment (RAI), payback period (PBP), and expected value (EV) (Martino, 1995; Souder, 1984). The latter allows a consideration of expected risk at various project stages, usually based on either IRR or NPV. A variation is the productivity index approach, popularized by the Strategic Decisions Group (SDG) (Matheson, Matheson et al., 1994). It tries to maximize the economic value of a portfolio for given resource constraints. NPV and DCF approaches tend to penalize risky projects because they do not include a provision for early project termination. Options pricing theory has recently become more popular for financial calculations because it takes this possibility into account (Rzasa, Faulkner et al., 1990). Economic techniques are the most widely used in establishing NPD portfolios but at the same time tend to yield poor outcomes (Cooper, Edgett et al., 1999).

Market Research

Market research can be used to collect data for forecasting the demand for new products or services, based on concepts or prototypes presented to potential customers, to gauge the potential market. Techniques used include consumer panels, focus groups, perceptual maps, and preference mapping, among many others (Wind, Mahajan et al., 1981).

Portfolio Matrices

Portfolio matrices ("bubble diagrams") are popular for displaying parameter values on three or four project dimensions. For example, probability of success can be plotted against net present value for projects in the portfolio, with the size of the icons used on the chart representing expected return on investment. A wide variety of bubble diagrams used to plan new product portfolios in a number of companies is demonstrated by Cooper et al. (Cooper, Edgett et al., 2001, Chapter 4). Although bubble diagrams are popular for graphical representations and comparisons, they have little theoretical or empirical support, and they may lead decision makers to overlook profit maximization (Armstrong and Brodie, 1994). They should therefore be used in conjunction with other tools, and primarily to illustrate relative characteristics of projects and the outcomes of balancing processes.

Comparative Approaches

Included in this classification are Q-sort (Souder, 1984), pair wise comparison (Martino, 1995), the Analytic Hierarchy Process (AHP) (Saaty, 1990), and Data Envelopment Analysis (DEA) (Linton, Walsh et al., 2002). Q-sort is the most adaptable of these in achieving group consensus. In these methods, first the weights of different objectives are determined, then alternatives are compared on the basis of their contributions to these objectives, and finally a set of project benefit measures is computed. The *Expert Choice*® commercial software package can be used to support the use of AHP in portfolio selection. Once the projects have been arranged on a comparative scale, decision makers can use these measures for guidance in selecting projects. An advantage of these techniques is that quantitative, qualitative, and judgment criteria can be considered. A major disadvantage is the large number of comparisons involved, making them difficult to use for analyzing large portfolios. Also, anytime a project is added or deleted from the list, the process must be repeated.

Scoring Models

Scoring models use a relatively small number of decision criteria, such as cost, workforce availability, probability of technical success, and so on, to specify project desirability (Martino, 1995). The merit of each project is determined with respect to each criterion. Scores are then combined (when different weights are used for each criterion, the technique is called "weighted factor scoring") to yield an overall benefit measure for each project. A major advantage is that projects can be added or deleted without recalculating the merit of other projects. Scoring models are probably the easiest to use of all the models, and a major advantage is that they can combine scores for both quantitative and qualitative measures. They are the third most preferred in NPD, after economic and strategic approaches (Cooper, Edgett et al., 1999).

Optimization Models

Optimization models select from the list of candidate projects a set that provides maximum benefit (e.g., maximum net present value). These models are generally based on some form of mathematical programming, to support the optimization process and to include project interactions such as resource dependencies and constraints, technical and market interactions, or program considerations (Martino, 1995). Some of these models also support sensitivity analysis, but most do not seem to be used extensively in practice. Probable reasons for disuse include the need to collect large amounts of input data, the inability of most such models to include risk considerations, and model complexity. Optimization models may also be used with other approaches that calculate project benefit values. For example, 0-1 integer linear programming can be used in conjunction with AHP to handle qualitative measures and multiple objectives, while applying resource utilization, project interaction, and other constraints (Ghasemzadeh, Archer et al., 1999).

Portfolio Decision Support Systems

Decision support systems in portfolio selection are typically based on a mathematical optimization approach that begins by selecting a portfolio from a set of candidate projects, providing a maximum overall benefit (e.g., net present value based on financial measures, benefit based on AHP calculations, or results from scoring models) (Ghasemzadeh, Archer et al., 1999; Ghasemzadeh and Archer, 2000; Dickinson, Thornton et al., 2001). If a mathematical programming approach is used to support the optimization process, other considerations may also be included, such as resource dependencies and constraints, precedence dependencies among the projects, risk, timing, technical and market interactions, or program considerations. Unless there are only a few projects to be considered, it is virtually impossible to take such constraints into account without the assistance of such models. For example, almost all other portfolio selection models assume that all projects begin at the same time, whereas it is simple to use mathematical programming to schedule individual projects in a way that minimizes competition for scarce resources. Solving the model optimally gives a beginning point. This solution, when combined with an interactive display of results, gives decision makers the ability to make adjustments to the portfolio, based on nonquantifiable judgments such as balancing risk. This approach can be effective if decision makers receive feedback on the resulting consequences, in terms of optimality changes and effects on resources. However, decision support approaches such as this are rarely used, partly because of the perception that a large amount of detailed financial data is required.

Cooper and his co-workers (Cooper, Edgett, et al., 1999) recently investigated the relative popularity of various portfolio selection methods in the NPD field. They found that financial methodologies were the most popular and dominated in portfolio decisions. However, the businesses that achieved the best results from their portfolios placed less emphasis on financial approaches and more on alignment of their portfolios with corporate strategy, and they tended to use more multiple approaches than other firms. Strategic methods that aligned the portfolio with business strategy, along with scoring approaches, yielded the best portfolios; financial methods yielded poorer results.

Process Model for Portfolio Selection

Portfolio selection is usually a committee process, where objective criteria such as predicted rate of return and expected project cost are mingled with subjective criteria relating to the needs of the different organizations represented on the project selection committee. All committee members should have access to information with which project intercomparisons are made, as well as information on the project portfolio as a whole. Portfolio selection can be managed most effectively in the context of an integrated framework that decomposes the process into a flexible and logical series of activities, involving full participation by the portfolio selection committee. Such an approach can take advantage of the best characteristics of a combination of existing methods, well grounded in theory. Because of the variance in interests and experience of responsible decision makers, it is important to include a flexible choice of techniques and interactive system support for those involved. The following discussion outlines such an integrated framework, suited to either distinct manual use of methodologies or to potential decision support system (DSS) application for portfolio selection.

Stages in the Project Portfolio Selection Framework

The process of portfolio selection can be highly complex unless approached logically and carefully. With process simplification in mind, the portfolio selection process should be organized into a series of stages, allowing decision makers to move logically toward an integrated consideration of projects most likely to be selected, based on sound theoretical models. However, each step should have a sound theoretical basis in modeling and should generate suitable data to feed the following step. Users need access to data underlying the models, with "drill-down" capability to develop confidence in the data being used and the decisions being made. At the same time, users should not be overloaded with unneeded data; it should be available only when needed and requested. Users also need training in the use of techniques that specify project parameters to be used in making decisions. An overall balance must be achieved between the need to simplify and the need to generate well-founded and logical solutions. A series of discrete stages for portfolio selection is depicted in Figure 5.1, where *process* stages are represented by the five outlined boxes linked horizontally in the center of the diagram. The ovals in the diagram represent *preprocess* activities, including methodology selection and strategy development. *Post-process* stages (below and to the right of the portfolio selection process) are also shown for completeness, since these may result in data generation and project evaluation during project development, affecting future portfolio selection activities and decisions.

Table 5.2 summarizes the stages in the framework, the associated activities, and some of the potential methodologies previously mentioned, for each stage. Each of the three phases and the included stages are considered in detail in the following discussion.

FIGURE 5.1. PROJECT PORTFOLIO SELECTION PROCESS.

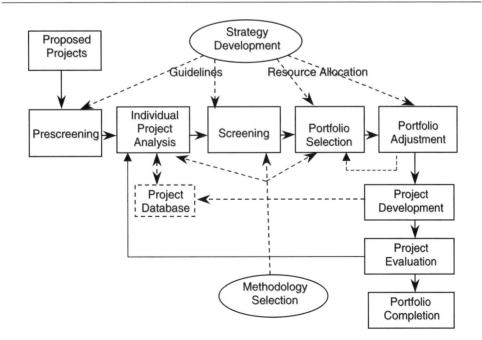

Source: Adapted from Archer and Ghasemzadeh (1999).

TABLE 5.2. ACTIVITIES AND METHODOLOGIES IN THE PORTFOLIO SELECTION FRAMEWORK.

Process Stage	Selection Stage	Activity	Potential Methodologies
Pre-process	Methodology selection, strategy development	Choice of modeling techniques, development of strategic focus, budgeting, resource constraints	Business strategy correlation and allocation, cluster analysis, etc.
Portfolio selection process	Prescreening	Rejection of projects that do not meet portfolio criteria	Manually applied criteria; strategic focus, champion, feasibility study availability.
	Individual project analysis	Calculation of common parameters for each project	Decision trees, risk est., NPV, ROI, resource req'ts., etc.
	Screening	Rejecting non-viable projects	Ad hoc techniques
	Portfolio selection	Integrated consideration of project attributes, resource constraints, interactions	AHP, constrained option, scoring models, sensitivity analysis
	Portfolio adjustment	User-directed adjustments	Matrix displays Sensitivity analysis
Post-process	Final portfolio	Project development	Project management techniques, data collection

Source: Adapted from Archer and Ghasemzadeh 1999.

Preprocess Phase

Methodology Selection. Methodology selection is a strategic process that clearly should be done in advance of any other activities in portfolio selection. It need only be done once for all time, with minor adjustments from time to time if other methodology choices appear to be better matches for the task at hand before each cycle of portfolio selection. Methodology selection should be flexible, based on decision maker understanding of the candidate methodologies or willingness to learn new approaches. Each stage involves choices, which are at the discretion of users, in order to gain maximum acceptance and cooperation of decision makers with the portfolio selection process. Choosing and implementing techniques suitable to the project class at hand, the organization's culture, problem-solving style, and project environment may also depend upon previous experience. It is also critical that common measures (e.g., NPV, scoring attributes, valuation of risk, etc.) are chosen so they can be calculated separately for each project under consideration, allowing an equitable comparison of the projects.

Strategy Development. Strategic decisions concerning portfolio focus and overall budget considerations should be made in a broader context that takes into account both external and internal business factors, *before* the project portfolio is selected. The strategic implications of portfolio selection are complex and varied (Cooper, Edgett, et al., 2001, Chapter 5) and involve considerations of many factors, including the marketplace and the company's strengths and weaknesses. These considerations can be used to build a broad perspective of strategic direction and focus, and specific initiatives for competitive advantage. This strategy can be used to develop a focused objective for a project portfolio and the level of resources needed for its support. Firms engaged in contract competition often use consensus-driven approaches to decide budget allocations for research and development. Consensus is derived at meetings among key personnel, based on informal arguments and assumptions regarding customer priorities, selection criteria, competitive position, and political priorities (Vepsalainen and Lauro, 1988). All may bear on the probability of successfully completing the projects being considered.

The front-end planning process is often done poorly (Khurana and Rosenthal, 1997). However, project portfolio matrices (Hax and Majluf, 1996) and graphs (Cooper, Edgett, et al., 2001) are useful tools for evaluating the strategic positioning of the firm, where various criteria for a firm's position are shown on one or more displays on two or three descriptive dimensions. These displays can be used by decision makers to evaluate their current position, and where they would like their firm to be in the future.

Overall resource allocation to different project categories also involves high-level decisions that must be made before the portfolio selection process, since this in turn determines the resource levels available for projects to be undertaken. Rules regarding the admissibility of projects into the portfolio need to be decided in advance of the selection process.

Process Phase

Prescreening. Prescreening precedes portfolio calculations. It is based on guidelines developed in the strategy development stage and ensures that any project being considered for the portfolio fits the strategic focus of the portfolio. Projects should be classified in advance of portfolio selection, according to criteria that can override other considerations. Frequently there are criteria that override all other considerations and mandate selection of specific projects, including sacred cow (mandated by influential stakeholders), operating necessity, and competitive necessity. Chien (Chien, 2002) also suggests that projects should be classified as independent, interrelated, or synergistic in advance of the selection process.

The upfront work to prepare for a project's evaluation is critical to its acceptability (Bordley, 1998). Essential requirements before the project passes this stage should include a feasibility analysis and estimates of parameters needed to evaluate each project, as well as a project champion who will be a source of further information. Elimination of projects not ready for serious consideration at this time helps to simplify following activities by reducing the number of projects to be considered. Mandatory projects are also identified, since they will be included in the remainder of the portfolio selection process. Mandatory projects are projects agreed upon for inclusion, including improvements to existing products no longer competitive, projects without which the organization could not function adequately, and so on.

Individual Project Analysis. Projects are analyzed individually after prescreening. Here, a common set of parameters required for equitable comparisons in following stages is calculated separately for each project, based on estimates available from feasibility studies and/

or from a database of previously completed projects. For example, project risk, net present worth, return on investment, and so on can be calculated at this point, including estimated uncertainty in each of the parameter estimates. Scoring, benefit contribution, risk analysis, market research, or checklists may also be used. Note that current projects that have reached certain milestones may also be reevaluated at this time, but estimates related to such projects will tend to have less uncertainty than those projects that are proposed but not yet under way. The output from this stage is a common set of parameter estimates for each project. For example, if the method to be used were a combination of net present value combined with risk analysis, data required would include estimates of costs and returns at each development stage of a product or service, including the risks. Uncertainty could be in the form of likely ranges for the uncertain parameters. Other data needed could include qualitative variables such as policy or political measures. Quantitative output could be each project's expected net value, risk, and resource requirements over the project's time frame, including their calculated uncertainties.

Screening. The screening stage is shown in Figure 5.1 following the individual project analysis stage. Here, project attributes from the previous stage are examined in advance of the actual selection process, to eliminate any projects or interrelated families of projects that do not meet preset criteria such as estimated rate of return, except for those projects that are mandatory or required to support other projects still being considered. The number of projects that may be proposed for the portfolio may be quite large, and the complexity of the decision process and the amount of time required to choose the portfolio increases geometrically with the number of projects to be considered. In addition, the likelihood of making sound business choices may be compromised if large numbers of projects must be considered unnecessarily. Screening may be used to eliminate projects that do not match the strategic focus of the firm, do not yet have sufficient information upon which to base a logical decision, do not meet a marginal requirement such as minimum internal rate of return, and so on. Care should be taken to avoid setting thresholds that are too arbitrary, to prevent the elimination of projects that may otherwise be very promising.

Another consideration at this point is the optimal number of research projects to be developed for NPD portfolios (Lieb, 1998). Lieb developed a model that views research and development as a two-stage process, where the task of research is to reduce the uncertainty for eventual development. Research projects require both technical and business evaluation, including marketing research. The number of research projects undertaken that maximizes the likelihood of development success depends on the cost-effectiveness of the research effort and the ability of the organization to support the development effort. Lieb concludes that the optimal fraction of research projects to be undertaken is critically dependent on the relative average research project cost and effectiveness compared to development.

Optimal Portfolio Selection. Optimization is performed in the second-to-last stage, as a beginning point for final portfolio selection. Here, interactions among the various projects are considered, including interdependencies, competition for resources, and timing, with the value of each project determined from a common set of parameters that were estimated for each project in the previous stage. AHP, scoring models, and portfolio matrices are popular among decision makers for portfolio selection, because they allow users to consider a broad range of quantitative and qualitative characteristics as well as multiple objectives. However, none of these techniques consider multiple resource constraints and project interdependence. AHP, pairwise comparison, and Q-sort also become cumbersome and unwieldy for larger numbers of projects. We suggest a two-step process for the portfolio selection stage.

In the first step, the relative total benefit is determined for each project. A comparative approach such as Q-sort, pairwise comparison, or AHP may be used in this step for smaller sets of projects, allowing qualitative as well as quantitative measures to be considered. This may involve extensive work by committee members in comparing potential project pairs. For large sets of projects, scoring models are more suitable, as these do not involve comparison of large numbers of project pairs. The result of either of these approaches would be to establish the relative worth of the projects on some scale that could combine both quantitative and qualitative attributes.

In the second step of this stage, all project interactions, resource limitations, and other constraints should be included in an initial optimization of the overall portfolio, based on the relative worth established for each proposed project. At this point, relationships among projects, and other constraints must be considered (Verma and Sinha, 2002). For example, (a) projects compete for scarce resources, including funding, labor, and facilities; (b) one project may be dependent upon the completion of another (e.g., construction of a new subdivision must await the construction of access roads); (c) projects may be mutually exclusive (e.g., a choice must be made among alternative solutions to the same requirements); and so on. Other constraints on project inclusion may include (a) a project must be completed prior to a particular date, (b) a project is mandatory (e.g., maintenance work), and so on.

If all the project measures could be expressed quantitatively, the first step in this stage could be omitted, since optimization could be performed directly in a mathematical program in the second step. In the unusual case where interdependence and timing constraints were not important, and there is only one resource that is binding, it might be tempting in the second step to simply select the highest-valued projects until available resources are used up. However, this does not necessarily select an optimal portfolio (combinations of certain projects may produce a higher total benefit than individual projects with higher individual benefits). The relative worth of each project should be input to a computerized process, which can be a 0-1 integer linear programming model that applies resource, scheduling, interdependence, and other constraints to maximize total benefit (Ghasemzadeh, Archer, et al., 1999).

Portfolio Adjustment. This is the final stage, using as input the initial optimal portfolio from the preceding stage. The end result is to be a portfolio that meets the objectives of the organization optimally or near-optimally, but the approach must have provisions for final judgmental adjustments, which may be difficult to model. This is a strategic decision, and the relevant information must be presented so it allows decision makers to evaluate the portfolio without being overloaded with unnecessary information. The portfolio adjustment stage needs to provide an overall view, where the characteristics of projects of critical importance in an optimized portfolio (e.g., risk, net present value, time-to-complete, etc.) can be represented, using matrix-type displays, along with the impact of any suggested changes on resources or selected projects (Cooper, Edgett, et al., 2001, Chapter 4). It is important to use only a limited number of such displays to avoid confusion (cognitive overload) while the final decisions are being made. Decision makers should be able to make changes at this stage, and if these changes are substantially different from the optimal portfolio developed in the previous stage, it may be necessary to recycle back to recalculate portfolio parameters such as project schedules and time-dependent resource requirements. In addition, sensitivity analysis should be available to predict and provide feedback to decision makers on the impact of their suggested changes (addition or deletion of projects) on resources and portfolio optimality.

There are a variety of ways in which portfolio balance can be attained; this requires the right mix of project size and/or duration, and a risk profile that is suited to the company environment. Balance on project size is important, because the commitment of a high proportion of resources to a few large projects can be catastrophic if more than one fails. And too many long-term projects, no matter how promising they are, may cause cash flow problems. In achieving portfolio balance on the risk dimension, the greater the risks taken, the greater the potential rewards should be. This should not preclude the greatest care in choosing the portfolio so that risky projects are balanced with less risky projects to avoid jeopardizing the overall company strategy of maximizing profitability and ensuring long-term survival. The average risk of projects should be such that decision makers view the portfolio as manageable, without dominance of high-risk projects. At the same time, there is danger in reducing risk that there will be a bias against selecting breakthrough projects that may provide a large long-term profit to the company. In NPD portfolio selection, an overall objective may be to maximize value while minimizing risk. In NPD portfolios, balancing the portfolio is second in importance only to having the right number of projects in the pipeline at one time (Cooper, Edgett et al., 2001).

Post-Process Phase

Project Development and Project Evaluation. Development and evaluation are post-process activities that can generate data from experience that are highly useful to learning and project evaluation, for future portfolio selection exercises. This could involve both the evaluation of existing projects that may be candidates for termination or generation of data for future use in estimating parameters for contemplated projects that are similar to already completed projects. Current projects that have reached major milestones or gates can be reevaluated at the same time as new projects being considered for selection. This allows a combined portfolio to be generated within available resource constraints at regular intervals because of (a) project completion or abandonment, (b) new project proposals, (c) changes in strategic focus, (d) revisions to available resources, and (e) changes in the environment.

The framework we have outlined can be used for project portfolio selection in an environment that is partially supported by computerized modeling and databases, since users are given the flexibility of choosing their own techniques or models at each stage. However, many of the stages in the framework (see Figure 5.1) can be integrated into a decision support system, including a model management module to handle models of the many different types that may be chosen. To be effective as a decision support system, the system requires a common interface, a database through which data may be interchanged among the models, and a user interface that allows user control and overrides of model calculations (Archer and Ghasemzadeh, 1999).

Portfolio Management

A program manager is the organizational leader charged with executing a portfolio of projects. In managing a portfolio, the problems are not centered on just managing individual projects, but they extend to a variety of complexities, including multiple, cross-functional, global, overlapping, interdependent projects; resource allocation; politics; sponsorship; and culture. Addressing these issues requires centralized support, increasingly provided in larger organizations through a project office. Global competency standards and best practices for project organizations such as this have been provided by Toney (Toney, 2002; see also the chapter by Lynn Crawford). Project offices can fulfill portfolio needs by giving support and gathering data that can be used for monitoring project progress and providing estimates for future portfolio selection activities. They provide standard tools and methodologies, as well as coaching. They collect information by means of project management tools that give a current view of the company-wide IT project portfolio, including staff deployment (Melymuka, 1999).

After some time in operation, a project office can provide a historical view of project estimates compared with actual performance, enabling management to see where projections part with reality. Some staffs see their project offices as centers of excellence for project management; see, for example, the chapter by Powell & Young. Theoretically, a project office forces the business to face the reality of limits on staff, time, and budget, and requires it to prioritize projects. It manages the portfolio selection process and assists in evaluating ongoing projects. It gives project managers the documentation they need to demand adequate staffing, funding, and reasonable deadlines. This results in better planning, portfolio selection, coordination and execution, and more projects completed on time and within budget. Virtual project management also becomes an issue when there is more than one site involved in projects within the portfolio, and communications arrangements can be supported through the project office (face-to-face visits, e-mail, video, telephone conferencing). Virtual projects are almost certainly an issue when projects are outsourced. When portfolios are developed internally, management can direct its attention to understanding and managing the relationships among the resources and project managers involved. When projects are outsourced, the firm has to cooperate with one or more external organizations in managing the portfolio, and it is confronted with the task of managing not just a portfolio of projects, but a portfolio of new and evolving relationships.

Summary

In this chapter we provided a review of project portfolio selection and management, along with a suggested logical approach to the selection process. In recent years, there has been a great deal of interest in this topic as project approaches have become more widely used but project failure rate experience in poorly managed operations has continued to be unacceptable. Improved portfolio selection and management practice can greatly enhance the likelihood of success, contributing to the rate of innovation in new products or controlling other problems in industries such as construction where disputes over performance or quality have long been a problem.

There are some areas of portfolio selection and management that require further research. For example, because of the increased emphasis on outsourcing, the management of outsourced and high-technology risk portfolios needs more study. Collaborative commerce (c-commerce), a major area of e-commerce application in which a Web-based system is used for communication, design, planning, information sharing, and information discovery (Turban, McLean et al., 2002), has significant potential for further research. This includes extensions of online project management and partner relationship management (PRM) tools and techniques for portfolios that include outsourced projects. In spite of new technological support for these types of cooperation, a problem arising from this trend is that knowledge sharing across organizational boundaries becomes more difficult due to legal, cultural, and virtual management issues

References

Archer, N. P., and F. Ghasemzadeh. 1999. An integrated framework for project portfolio selection. *International Journal of Project Management* 17(4):207–216.

Armstrong, J. S., and R. J. Brodie. 1994. Effects of portfolio planning methods on decision making: Experimental results. *International Journal of Research in Marketing* 11:73–84.

Bordley, R. F. 1998. R&D project selection versus R&D project generation. *IEEE Transactions on Engineering Management* 45(4):407–413.

Bunch, P. R., and A. L. Schacht. 2002. Modeling resource requirements for pharmaceutical R&D. *Research Technology Management* 45(1):48–56.

Chien, C.-F. 2002. A portfolio evaluation framework for selecting R&D projects. *R & D Management*.

Cooper, R. G., S. J. Edgett, and E. J. Kleinschmidt. 1999. New product portfolio management: Practices and performance. *Journal of Product Innovation Management* 16:333–351.

———. 2000. "New problems, new solutions: Making portfolio management more effective." *Research Technology Management* 43(2):18–33.

———. 2001. *Portfolio Management for New Products*. Cambridge, MA: Perseus Publishing.

Dickinson, M. W., A. C. Thornton, and S. Graves. 2001. Technology portfolio management: Optimizing interdependent projects over multiple time periods. *IEEE Transactions on Engineering Management* 48(4):518–527.

Ghasemzadeh, F., and N. Archer. 2000. Project portfolio selection through decision support. *Decision Support Systems* 29:73–88.

Ghasemzadeh, F., N. Archer, and P. Iyogun. 1999. A zero-one model for project portfolio selection and scheduling. *Journal of the Operational Research Society* 50:745–755.

Githens, G. D. 2002. How to assess and manage risk in NPD programs: A team-based risk approach. In *The PDMA Book for New Product Development*, ed. P. Belliveau, A. Griffin, and S. Somermeyer, 187–214. New York, Wiley.

Griffin, A. 1997. PDMA research on new product development practices: Updating trends and benchmarking best practices. *Journal of Product Innovation Management* 14:429–458.

Hartman, F. T., P. Snelgrove, and R. Ashrafi. 1998. Appropriate risk allocation in lump-sum contracts—Who should take the risk? *Cost Engineering* 40(7):21–26.

Hax, A., and N. S. Majluf. 1996. *The Strategy Concept and Process: A Pragmatic Approach*. Upper Saddle River, NJ: Prentice Hall.

Hess, S. W. 1993. Swinging on the branch of a tree: Project selection applications. *Interfaces* 23(6):5–12.

Khurana, A., and S. R. Rosenthal. 1997. Integrating the fuzzy front end of new product development. *Sloan Management Review* 38:103–120.

Kimzey, C. H., and S. Kurokawa. 2002. Technology outsourcing in the U.S. and Japan. *Research Technology Management* 45(4):36–42.

Lieb, E. B. 1998. How many R&D projects to develop? *IEEE Transactions on Engineering Management* 45(1):73–77.

Linton, J. D., S. T. Walsh, and J. Morabito. 2002. Analysis, ranking, and selection of R&D projects in a portfolio. *R & D Management* 32(2):139–149.

MacMillan, I. C., and R. G. McGrath. 2002. Crafting R&D project portfolios. *Research Technology Management* 45(5):48–59.

Martino, J. P. 1995. *Research and Development Project Selection*. New York: Wiley.

Matheson, D., J. E. Matheson, and M. M. Menke. 1994. Making excellent R&D decisions. *Research Technology Management* 37(6):21–24.

Melymuka, K. 1999. The project office: A path to better performance. *Computerworld*.

Pinto, J. K. 2002. Project management 2002. *Research Technology Management* 45(2):22–37.

Prabhu, G. N. 1999. Managing research collaborations as a portfolio of contracts: A risk reduction strategy by pharmaceutical firms. *International Journal of Technology Management* 18(3,4):207–231.

Raz, T., A. J. Shenhar, and D. Dvir. 2002. Risk management, project success, and technological uncertainty. *R & D Management* 32(2):101–109.

Riggs, J. L., S. B. Brown, and R. P. Trueblood. 1994. Integration of technical, cost, and schedule risks in project management. *Computers and Operations Research* 21(5):521–533.

Roseneau, M. D. J. 1990. *Faster New Product Development: Getting the Right Product to Market Quickly*. New York: AMACOM.

Rzasa, P. V., T. W. Faulkner, and N. L. Sousa. 1990. Analyzing R&D portfolios at Eastman Kodak. *Research Technology Management* 33(1):27–32.

Saaty, T. L. 1990. *The Analytic Hierarchy Process*. Pittsburgh: RWS Publications.

Shenhar, A. J. 2001. Contingent management in temporary, dynamic organizations: The comparative analysis of projects. *Journal of High Technology Management Research* 12:239–271.

Shenhar, A. J., and R. M. Wideman. 1997. Toward a fundamental differentiation between project types. *Innovation in Technology Management: The Key to Global Leadership—PICMET '97 conference*. Portland, Oregon, July.

Souder, W. E. 1984. *Project Selection and Economic Appraisal*. New York: Van Nostrand Reinhold.

Stamelos, I. and L. Angelis. 2001. Managing uncertainty in project portfolio cost estimation. *Information and Software Technology* 43:759–768.

Tatikonda, M. V. 1999. An empirical study of platform and derivative product development projects. *Journal of Product Innovation Management* 16:3–26.

Toney, F. 2002. *The Superior Project Organization*. New York: Marcel Dekker.

Turban, E., E. McLean, and J. Wetherbe. 2002. *Information Technology for Management: Transforming Business in the Digital Economy*. New York, Wiley.

Vepsalainen, A. P. J., and G. L. Lauro. 1988. Analysis of R&D portfolio strategies for contract competition. *IEEE Transactions on Engineering Management* 35(3):181–186.

Verma, D., and K. K. Sinha. 2002. Toward a theory of project interdependencies in high tech R&D environments. *Journal of Operations Management* 20(5):451–468.

Wadlow, D. 1999. The role of risk in the design, evaluation and management of corporate R&D project portfolios for new products. *Sensors Research Consulting* 32.

Welling, D. T., and D.-J. F. Kamann. 2001. Vertical cooperation in the construction industry: Size does matter. *Journal of Supply Chain Management* 37(4):28–34.

Wheelwright, S. C., and K. B. Clark. 1992. Creating project plans to focus product development. *Harvard Business Review* 70(2):70–82.

Wind, Y., V. Mahajan, and R. N. Cardozo. 1981. *New Product Forecasting.* Lexington, MA: Lexington Books.

CHAPTER SIX

PROGRAM MANAGEMENT: A STRATEGIC DECISION MANAGEMENT PROCESS

Michel Thiry

Recent large-scale studies have demonstrated that 30 percent of projects are canceled before the end (Standish, 1996; KPMG, 1997); or that large, long-term projects—more than three years—are "significantly less predictable" in terms of time and scope (Cooke-Davies, 2002). These studies and others have sorely exposed the failure of project management to respond to emergent inputs, as well as the lack of integration between strategic intents and the results generated by projects. Recently, a number of authors (Frame, 2002; Thomas et al., 2000; Kendall, 2001) have advocated the evolution of the role of project manager toward a more business-focused "function," or a stronger focus on business benefits for projects (Cooke-Davies, 2002; Morris, 1997). This is probably part of the answer, but it will not be sufficient if not supported by a robust framework. The business world is moving toward a systemic process-based framework for the management of strategic decisions (Kaplan and Norton, 2000; EFQM, 1999).

This chapter develops a view in which program management (PgM) may effectively link strategic decision making with its successful implementation through projects. It first discusses the need to combine a performance (project-based) and a learning (value-based) process in the management of programs. The chapter then explains the need for a complete strategic decision management process, which includes decision making and decision implementation, and suggests an integrated learning-performance model for program management. The chapter then outlines in detail a program management life cycle, based on that model and grounded in a long experience of managing stakeholders' needs and expectations through value, project, and program management in a variety of industries.

Program Management Paradigms

Program management is an emerging discipline that has its roots in project management; however, as the need for better strategic focus is recognized, a project paradigm based purely on performance efficiency becomes insufficient. In their research on selling project management to executives, Thomas et al. (2000) have argued that "successful messages reflect the buyer's needs as the buyer understands them." Program management needs to reflect the concepts and rhetoric of strategic long-term management, rather than the tactical short-term view of project management, in order to gain executive management support and truly be able to support strategic decision management. Projects, as currently defined (PMI, 2000; APM, 2000), are time-framed and require clear objectives; they are based on a performance paradigm (Thiry, 2002a) embedded in an "uncertainty-reduction" process (Winch et al., 1998). This does not sell well to executives.

A number of textbooks and papers (CCTA, 1999; Reiss, 1996; Gareis, 2000; DSMC, 2000; NASA, 2001) have suggested program processes that, albeit their different designation, are in most instances just transpositions of the project paradigm into program management. Although it is now generally agreed that programs need to produce business-level benefits and are a link between strategy and projects, few management concepts and little rhetoric have made their way into the program management literature and practice. In the report of an ongoing survey of organizations practicing program management, Reiss and Rayner (2001; 2002) point out that most respondents still consider "organization," "issues and risk," "planning," and "accounts and finance" as key to the success of programs, whereas "achievement of benefits," "stakeholder management," "communications," and "configuration management," which were identified by the developers of the model as key program components, seem less important. Generally speaking, organization, risk, planning, and cost management are key elements of project management. Benefits, stakeholders, communications, and configuration management are about the management of stakeholders' needs and expectations and emergent change. All are linked to value management, which is often not part of the current PMI-based project management paradigm (PMI, 2000), although it has been included in the Association for Project Management's Body of Knowledge (APM, 2000).

Johnson and Scholes (1997) have written that strategic management is "ambiguous and complex; fundamental and organization-wide; and has long-term implications." Wijnen and Kor (2000) write that a program strives for the achievement of a number of, sometimes conflicting, aims and has a broader corporate goal than projects, which aim to achieve single predetermined results. Görög and Smith (1999) argue that strategic management is based on continuous reformulation and is a form of continuous adjustment, whereas projects concentrate on achieving one single particular result within set time and cost constraints. Partington (2000) argues that programs require integration across strategic levels, controlled flexibility, team-based structures, and especially, an organizational learning perspective, which is able to accept paradox and uncertainty. Murray-Webster and Thiry (2000) advocate a vision that includes mechanisms to identify and manage emergent change; they use an idea developed by Hurst (1995) to promote the concept of a "learning loop," which completes the project "performance loop."

Programs are often ongoing or long-term and are subjected to both uncertainty and ambiguity. They require a *strategic decision management* paradigm, taking into account a strategic perspective, organizational effectiveness, a systems view, and a learning (ambiguity reduction) approach. Whereas projects are essentially "deliberate" (planned) strategies, programs combine both deliberate strategies and "emergent" (unplanned) strategies.

As outlined in the preceding text, strategic decision making and change situations often mean that multiple stakeholders, with conflicting needs and expectations, are competing with each other, which creates ambiguity. Effective strategic decision making requires an "ambiguity reduction" process, based on a learning paradigm, to take place before any attempt is made at uncertainty reduction (see Figure 6.1); otherwise, it will lead to results that are not necessarily in line with stakeholders' needs. This process can effectively be supported by value management (VM), which uses a range of "soft" methodologies and techniques like sensemaking, stakeholder analysis, functional analysis, ideation, soft systems analysis, and others (as more fully described in the chapter on value management).

In a complex, continually changing environment, a program management paradigm must integrate both a learning-based value loop and a performance-based project loop to address both ambiguity reduction and uncertainty reduction to form a full strategic decision management framework. You will also see later that the learning loop must be part of both

FIGURE 6.1. THE UNCERTAINTY-AMBIGUITY RELATIONSHIP IN CHANGE SITUATIONS.

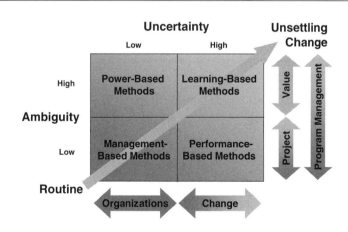

Source: © Michel Thiry (2000–2002).

project reviews and program appraisal in order to achieve strategic benefits and stakeholders' satisfaction at delivery.

Definitions

Following the description of program paradigms, it is important to define some basic terms: strategy, program management (PgM); portfolio management, value management (VM), and project management (PM). Although these definitions might not gain general consensus, they are sufficiently documented to gain acceptance by most scholars and practitioners.

Strategy

In a program management context, *strategy* is essentially the organization's response to external or internal pressures to change. For the purpose of this chapter, strategies are grouped in two major types—*deliberate* strategies, which are planned, and *emergent* strategies, which are responses to unplanned inputs (Mintzberg and Waters, 1994)—and include also the concept of *configuration* strategy (Mintzberg et al., 1998), a combined use of the two.

A deliberate strategy is submitted to a formal process of analysis, design, and planning; its implementation is subjected to a formal baseline control process. An emergent strategy is an unplanned response to an emergent input originating from the environment or from within the organization. Such strategies are usually based on a leader's intuitive decision (based on vision and/or experience), negotiation, a collective process, or, more simply, a knee-jerk reaction.

Project Management

There are currently two paradigms in the project management community; one, based on the PMI's PMBoK Guide (PMI, 2000), views project management essentially as a delivery process based on a performance paradigm; the other, based on the view developed by members of IPMA (International Project Management Association), views project management as a broader concept that includes preinitiation phases and project definition and, in that sense, is closely linked to program management. In any case, there is one point, once expected outcomes have been clearly defined, when project management becomes a performance process intended to deliver with the highest possible efficiency—best scope-quality versus lowest cost-time—though the *management of projects* concept is obviously broader (Morris 1997).

Value Management

In the context of this chapter, value management is based on a learning paradigm aimed at *ambiguity reduction* (see the chapter on value management in this book). Value management is a group decision-making process expected to develop a shared understanding of a complex situation and an agreement on options to resolve it. It uses creative thinking concepts.

Portfolio and Program Management

Program and portfolio management are both emerging disciplines and are often confused. The two are different, and it is therefore important to describe how they can be clearly distinguished. In Europe, researchers and practitioners have looked into this distinction for the last few years. In the United States, there seems to be less use of the word portfolio management, and program management is often associated with very large projects (DSMC, 2000; NASA, 1998) or the management of large, multifunctional projects as a portfolio of independent projects through a project office (CBP, 2002). A number of authors still associate program management with multiproject coordination or portfolio management, which is often associated with resource management (Patrick, 1999) or account/client management; this view is supported by so-called program management computer software, which is essentially designed to support resource management, planning, and cost/time control across many projects.

A sound, widely acceptable definition for both portfolio and portfolio management is this: "A project portfolio is a collection of projects to be managed concurrently under a single management umbrella where each project may be related or independent of the others" (Martinsuo and Dietrich, 2002). Portfolio management consists of "the management of a multi-project organization and its projects in a manner that enables the linking of the projects to business objectives" (Artto et al., 2002). Writing in the context of decision making, Spradlin (1997) links portfolio management to the need to choose among a number of options where resources are insufficient to fund all options. Portfolio management is a management approach where projects could be related or not to a common objective.

There are a number of definitions of *program*; all these definitions have elements in common as well as differences. The main common points are that programs usually cover a group of projects, that their management must be coordinated, and that they create a synergy, which will generate greater benefits than projects could do individually. The main differences are that the elements must have a common objective (APM, 2000), that they also cover ongoing operations (PMI, 2000), and that their impact is at the organizational level and concerns change (CCTA, 1999). A widely acceptable definition of program should integrate those three elements. The definition that follows does that with the concept of purposefulness that is related to setting common objectives; the word "*actions*," which refers to ongoing operations as well as to projects; and the concept of strategic or tactical benefits, which are always measured at the organizational level.

Programs are "a collection of change actions (projects and operational activities) purposefully grouped together to realize strategic and/or tactical benefits" (Murray-Webster and Thiry, 2000). Program management consists of the "purposeful and integrated direction and coordination of a group of actions, their interface and consequences for strategic effectiveness and/or tactical efficiency" (Thiry, 2002a); it is a management process that addresses both decision-making and decision implementation.

In summary, portfolio management is mostly concerned with the ongoing prioritization and management of a group of existing projects or programs, to efficiently support a project-oriented organization. Program management, on the other hand, is mainly a purposeful strategic decision management process, grounded in change and aimed at the effectiveness of solutions. It can include both projects and non-project actions. Figure 6.2 shows how program management draws on three main areas of management to combine processes belonging to recognized fields of knowledge and practice.

The Link between Programs and Strategy

Authors and researchers in various fields have argued that one of the major problems concerning the failure of decision implementation lies within inaccurate understanding of

FIGURE 6.2. PROGRAM MANAGEMENT AS AN INTEGRATING PROCESS.

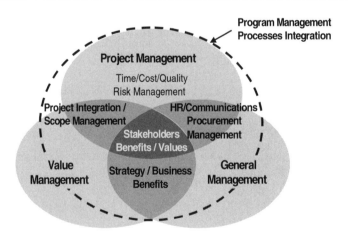

Source: © Michel Thiry (2001).

stakeholders' needs and expectations or the inability to respond to their changing needs (Carver and Scheier, 1990; Kubr, 1996; Standish, 1996; Kirk, 2000; Hartman and Ashrafi, 2002; see also the chapter by Winch in this book). The need to change, which triggers a strategic decision, is usually caused by an unsatisfactory condition. Quinn (1978) talks about the "change deficit" as a measure of the need for an organization to put in place change measures that go beyond incremental change or simple evolution. The change deficit represents the difference between the rate of change of an organization and its environment: competitors, market, and so on. Generally, internal pressures to change are addressed in an incremental way (Quinn, 1978), but it may sometimes be necessary for an organization to implement more fundamental changes to increase or maintain its effectiveness. An organization that falls back in terms of change deficit will quickly lose its competitive edge and market share. The problem with large, heavily structured organizations is that they cannot adjust quickly and are more adapted to incremental change than fundamental change. In a fast-changing environment, organizations have to find a way to be more flexible in making and implementing strategic decisions.

In this context, program and project management are effective means of modifying an unsatisfactory situation or pursuing an already ongoing change process. These strategic change processes generally aim to improve organizational effectiveness or competitiveness or to respond to a change deficit.

Authors have suggested an abundance of criteria to measure an organization's *effectiveness*, but it is generally accepted that effectiveness can be defined as the match between results and stakeholders' satisfaction, and that satisfaction is dependant on clearly agreed and understood objectives (EFQM, 2000; Kaplan and Norton, 1996; Porter, 1998). Goals and objectives must be clearly outlined as part of a strategic decision-making process, as satisfaction will be measured by the rate of progress from the actual situation toward the desired outcome (Carver and Scheier, 1990).

A number of authors (Guba and Lincoln, 1989; Kubr, 1996; Stake, 1975) also argue that when situations are complex and fluid, control needs to be more qualitative, based on negotiation between the stakeholders and with criteria evolving in time. This is the case for strategic decision implementation.

These concepts require that a strategic decision management framework be systemic and flexible, based at the same time on deliberate and emergent strategies.

Program Management Process

Because of their long-term characteristic, programs will be subjected to contextual change and therefore must include not only deliberate strategies (typically projects) but also elements of emergent strategies. In emergent strategies the focus must be on expected benefits and the process must "display consistent patterns of actions over time" (Mintzberg and Waters,

1994). This requires a learning approach where results are regularly appraised against benefits during implementation and changes are managed against these stated benefits.

Recent research has shown that there is a need for a decision management process in which anticipated results are directly linked to the justification for the decision (expected benefits) and the means to support their delivery (resources) (Thiry, 2002b; Hartman and Ashrafi, 2002). It has also been observed that there is a lack of communication between the strategic and tactical levels of management (Hatch, 1997; Neal, 1995; Pellegrinelli and Bowman; 1994, Thomas et al., 2000). This points to the need for a "strategic decision management process," to link strategic analysis and choice with strategy implementation. Some authors have suggested program management could rise to this challenge (Murray-Webster and Thiry, 2000; Partington, 2000; Wijnen and Kor, 2000).

A Program Management Framework

The program environment is complex: There are multiple stakeholders with differing and often conflicting needs, emergent inputs are always affecting the process, and both uncertainty and ambiguity can be high at the same time. Processes that are applicable to project management cannot always be readily applied to program management, as programs often have a finality that cannot be defined clearly and require processes that are both cyclic and aimed at reducing ambiguity.

Hurst (1995) described organizational change as a cyclic "Change Eco-Cycle" composed of a renewal loop (emergent strategy), which follows a crisis (emergent input), and a conventional loop (deliberate strategy), which follows a choice (deliberate input). Mintzberg et al. (1998) have argued that organizations have to adapt to both periods of stability and periods of transformation, and they identified this pattern as the *configuration* school of strategy. Since program management is linked to organizational change and can be seen as a strategic decision management process, a consistent program management framework will include both a *performance loop*, associated with a "deliberate" strategy process, based on a precise plan consistent with project management, and a *learning loop*, which can be likened to a largely responsive "emergent" strategy process and related to value management. This process is shown in Figure 6.3.

The learning loop is in fact a decision-making process; its first step is to make sense of the emergent input that justifies a change. The second step, ideation, is based on a lateral thinking concept (de Bono, 1990) and consists of identifying as many alternatives as possible to increase the quality of the decision. The third step, elaboration, consists of evaluating and combining alternatives to develop them into viable options, which can then be assessed and prioritized.

The performance loop is the decision implementation process; it can only be effective if a clear decision has been made and objectives clearly defined. It is only through the iteration of a number of learning-performance cycles that program management can be the framework to successful implementation of strategic decisions. The model in Figure 6.4

FIGURE 6.3. THE INTEGRATED PROGRAM MANAGEMENT CYCLE MODEL.

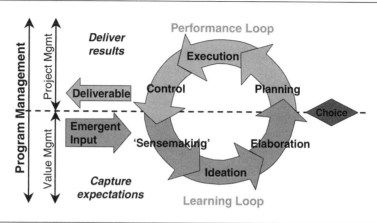

Source: © Michel Thiry (1999–2002).

FIGURE 6.4. STRATEGIC DECISION MANAGEMENT MODEL.

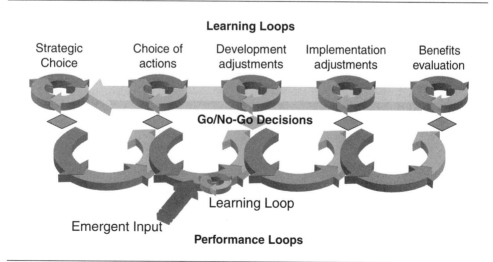

shows how iteration of the program management cycle allows regular evaluation of project outcomes, in regards of organizational benefits, and offers the flexibility to readjust expectations, as required by emergent inputs, or to readjust the expected outcomes to fit the circumstances.

The final element of the strategic decision management framework concerns evaluation paradigms. According to Guba and Lincoln (1989), evaluations can be "summative" (to assess) or "formative" (to improve). Quinn (1996) argues that, in complex situations, each stakeholder will construct different evaluation criteria and evaluation measures will therefore need to be negotiated. Guba and Lincoln (1989) and Stake (1975) talk about "responsive" evaluation, where parameters and boundaries are determined through an interactive negotiated process.

Most proponents of a combined learning/performance program management paradigm (Wijnen and Kor, 2000; Görög and Smith, 1999; Murray-Webster and Thiry, 2000) argue for a continuous reevaluation or "reformulation" of the program, in regards of the realization (or nonrealization) of benefits. On the other hand, proponents of the performance paradigm (Bartlett, 1998; CCTA, 1999; Reiss, 1996) argue for summative control, based on performance parameters such as time, cost, resources, risks, and so on.

In the context of complex strategic situations, a valid decision model must offer *both* summative evaluation at the project level and formative evaluation at the value-strategy level, allowing for negotiated evaluation as program objectives are redefined.

A Program Management Life Cycle

There are two characteristics that will make program management the most suitable methodology to ensure successful implementation of strategies:

- The fact that it is a cyclic process, which enables regular assessment of benefits, evaluation of emergent opportunities, and pacing of the process
- The emphasis on the "interdependencies" of projects, which ensures strategic alignment and delivery of strategic benefits

To support strategic decisions, reflect a management discourse, and acknowledge the two preceding characteristics, the program life cycle must be iterative, rather than linear; include periods of stability, where benefits can impact the organization and therefore be measured; and have a learning and systems perspective. The program management life cycle in Figure 6.5 is iterative in nature and reflects the extended and evolving nature of strategic decisions.

It is composed of five phases (Thiry 2002c):

FIGURE 6.5. THE PROGRAM MANAGEMENT LIFE CYCLE.

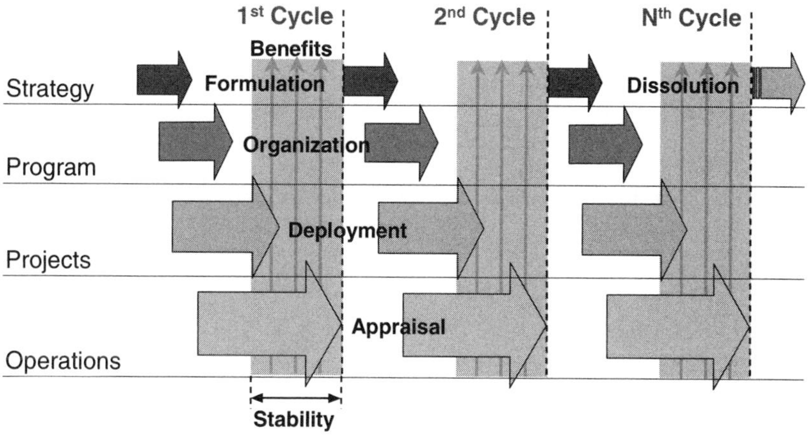

Source: © Michel Thiry (2002).

- *Formulation* is the stage where purpose is defined and stakeholders, along with their needs and expectations, are identified. It is also where program benefits are determined and critical success factors (CSFs) and key performance indicators (KPIs) defined. Contrary to project initiation, it is a complex process, where ambiguity is high. It is the initial learning cycle of programs where sensemaking, ideation, and evaluation of alternatives take place and that ends with the decision to undertake the program. It is iterated regularly during the program and confirms or redefines the direction of the program and the actions required to support it. The main objective of this phase is to identify opportunities and select the best course of actions.
- *Organization* is the process of selecting and prioritizing projects and other actions required to deliver benefits and set up the program team and structures. It includes the installation of operational procedures and structures that will enable project interdependencies and interrelationships to be managed, as well as pacing the program and ensuring ongoing benefits delivery.
- *Deployment* involves the actual initiation of projects and other actions; management of interdependencies and resources; project sponsor type control, including scope verification and closeout; and benefits delivery. It is made up of review, pacing, and approval of project outputs and change/configuration control, including realignment and reprioritizing of resources and projects.
- *Appraisal* essentially concerns the program-level assessment of benefits. It is a process that requires constant reevaluation of the program's circumstances and CSFs and typically

corresponds to a period of stability, enabling benefits to be measured in regards of their impact on the organization.

• *Dissolution* happens when the rationale for the program no longer exists. Uncompleted work, projects, and resources are reallocated to other programs, which are reformulated as needed; a post-program feedback is carried out; and knowledge is recycled. It is a phasing-down process, much more extensive than project closing.

The following five sections detail each of the program life cycle phases.

The Formulation Phase

The formulation phase aims to identify internal or external pressures to change and determine the best way to address them to add value for the stakeholders. For programs to successfully support strategic decisions, it is essential to take a value approach to formulation.

Stakeholders Management

To reduce ambiguity, the formulation phase must first address stakeholder issues. It requires identifying their needs and expectations, estimating achievability of these, gaining agreement on objectives, and in the appraisal phase, assessing results against these and reiterating all the steps on a regular basis. Figure 6.6 illustrates this process.

Benefits Management and Value

Benefits management (Ward and Murray, 1997; CCTA, 1999) is the process of identification, prioritization, quantification, and delivery of expected benefits throughout the program. Benefits are the translation of stakeholders' needs and/or expectations into measurable outputs; the sum of these outputs constitutes the "value" of the program to the organization. *Value management*, and in particular the function breakdown structure (FBS), can be used to manage benefits very successfully (see the chapter on value management in this book). The FBS is to the program what the WBS is to the projects.

Benefits identification is a key process of formulation; both tangible and intangible benefits should be identified, and although it may be more difficult for some, measures of success should be determined for all so that the success of the program can be appraised objectively. Benefits must refer to an organizational-level result. It is a mistake to consider project deliverables as measures of benefits, as it is the impact of the deliverables on the organization, not the deliverable itself, that constitutes the benefit. For example, the delivery

FIGURE 6.6. STAKEHOLDERS MANAGEMENT.

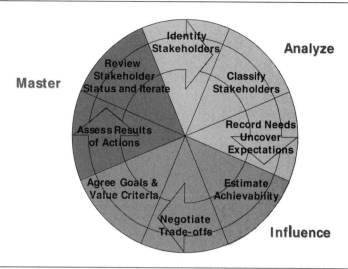

Source: © Michel Thiry (2002).

of training to a number of participants is not a benefit, although it is a deliverable. The benefit occurs when participants apply the knowledge they have gained to their work, and it is the improvement of performance or of work results that is the measure of the program's benefit.

Benefits and their measures of success are best expressed as CSFs and KPIs. *Sensemaking*, as described in the value management chapter of this book, is an effective method to identify, prioritize, and quantify benefits. Once CSFs and KPIs have been agreed and prioritized, the ideation process is used to actively and creatively seek as many as possible alternative courses of actions.

The elaboration process consists of evaluating alternatives qualitatively and quantitatively, combining them and/or further developing them to form options. These options are potential projects or actions that will form the program. When elaborating an option, one must identify both direct and indirect values, and both hard (tangible) and soft (intangible) benefits. *Direct values* are financial or nonfinancial impacts directly related to the choice of an option; they are usually easy to measure. *Indirect values* are consequences of the option, valued by stakeholders and, especially, decision-makers; they are usually more difficult to measure because generally they are not clearly expressed. *Hard benefits* are economic, technical, and operational; *soft benefits* are linked to power, politics, and communications and are more difficult to identify and measure.

Selection/Prioritization of Actions

All the options that are clearly acceptable are ranked in order of their contribution to expected benefits (CSFs) of the program. In addition, actions must be evaluated in terms of their achievability, which is based on factors like financial resource availability (funding, capital cost, cash flow, life cycle costs, etc.), parameters and constraints (size, budget, time-scale, type of work), human resources (expertise, spread, external vs. internal), people factors (availability, competence, customer perception), and complexity (innovativeness, interdependencies, stakeholders). Weighted matrices are typically used to rank actions against CSFs (see the value management chapter for detailed use) and achievability factors. Selection/prioritization matrices are then employed to classify the portfolio of potential actions (see Figure 6.7).

If economic factors are not part of the CSFs or of the achievability factors, options are then assessed in terms of their economic contribution to the business. Typical methods include economic value added (EVA), net present value (NPV), internal rate of return (IRR), and return on investment (ROI).

Choice consists of deciding the best course of action for the program; it is the last step of this phase. The decision is based on the results of options prioritization combined with

FIGURE 6.7. EXAMPLE OF SELECTION/PRIORITIZATION MATRIX.

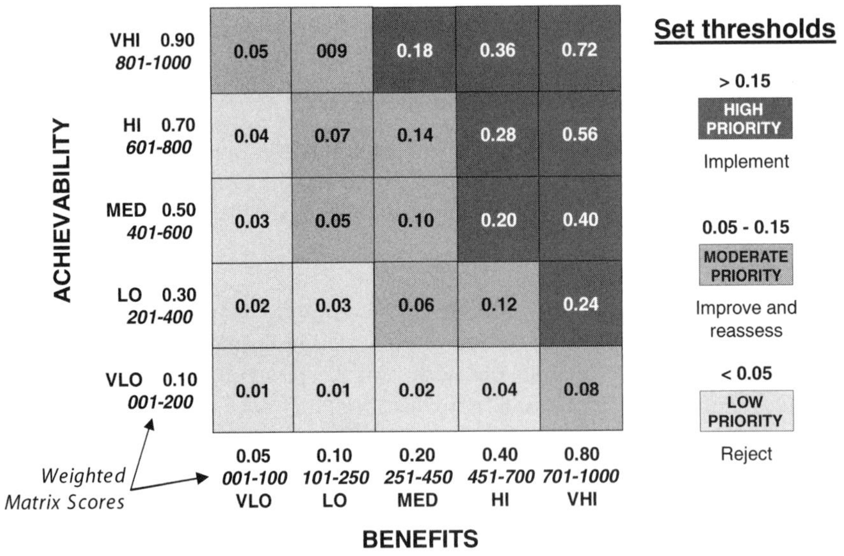

Source: © Michel Thiry (2002).

pacing and synergy factors. Priorities are set between different programs within the organization, and "business cases" are compiled. The choice process also seeks to secure approval and support for the program and allocation of the funding and authority to undertake it.

Especially in programs, it must be acknowledged that decisions are based on available information and expressed value *at the time of the decision*. As these may evolve during the course of the whole program, assumptions and choice factors must be clearly documented and reassessed regularly; this is the main purpose of the appraisal stage.

The Organization Phase

Once funds are allocated to a program, or to the first "cycle" of a program, and the authority of the program manager is secured, the organization phase can begin. It essentially consists of the strategic planning of the program and the definition of actions that constitute its current phase.

Strategic Plan

The objective of strategic planning is to develop a resource management and communications plan, pace the program actions, and develop a change management system. The hierarchy of the FBS shows how each action contributes to a CSF (benefit) and how these support the strategy, thus allowing the program manager to clearly define responsibilities for each CSF and identifying interactions and interdependencies.

The FBS also allows identification of the communication channels for the program; how information is filtered and sorted to suit different organizational levels and how access to systems is allowed or restricted. Using the stakeholder analysis, the program manager develops a marketing plan that clearly identifies expected benefits and goals in measurable terms and, as the program progresses, reports on them. This is a key element of maintaining support throughout the duration of the program.

Resource Planning. The efficient management of resources across the organization is a key component of program management, but this process must also foster the effective use of resources in each project. Resource planning essentially consists of matching required resources with available resources (O'Neill, 1999); it is a value-based concept. "Required resources"—or demand—consists of the prioritized workload necessary to implement the program (see preceding Formulation Phase Section). "Available resources"—or supply— consists of the assessment of the capacity, capability, and availability of existing resources. The program manager's role is to make sure that supply is always equal to or greater than demand, or that demand is adjusted to match supply. Regular reprioritization and reevaluation provides enhanced flexibility and proactive, rather that reactive, resource allocation so that the best resources are used on the most significant assignments. The concepts developed in the book *The Critical Chain* (Goldratt, 1997)—having a holistic view, sequential tasking rather than multitasking, focus on a "drum" resource, and the use of buffers—can also be effectively applied to program resource planning.

Pacing. The distribution and pace of projects, duration of cycles, and the interval between periods of stability are based on a minimum acceptable level of performance, which cannot be crossed (see Figure 6.8). Periods of stability, or lesser pace, during which benefits are allowed to impact the organization, are spread significantly throughout the program to pace it. Their distribution depends on benefits sought and the significance of those benefits for the business. Financial issues, such as cash flow and/or funding, and human resource issues, like the organization's culture (e.g., risk seeking or risk averse) and the resistance to or acceptance of change, determine this minimum level. Pacing, which includes prioritization of actions, must focus on early benefits, positive cash flow, and maintenance of the motivation of stakeholders. The program appraisal process is also built around these program "cycles" (see Figure 6.5).

FIGURE 6.8. THE PACING OF PROGRAMS.

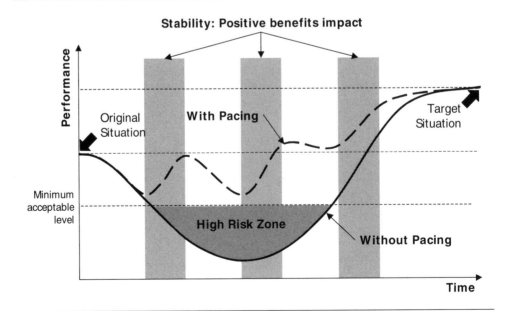

Gateways and Change. Major project changes and benefits appraisal, both of which involve program managers in their his or her project sponsor role, are usually planned around the gateways of projects; those must be significantly spread to correspond both to major project deliverables and to expected benefits delivery for the program. They are not related to the program cycles but rather to project milestones.

The last element of the program's strategic plan is the change management system. In programs, the change process must be value-based. This signifies that, because the program's specific objectives and actions are very likely to be modified as it progresses, emergent opportunities and threats are to be evaluated on a regular basis, as well as the capacity and capability of the program organization to respond to them. This system has to be clearly established from the beginning.

Definition of Actions

The definition of actions is the extension of the formulation phase's preselection process, leading to the initiation of actions. It consists of identifying and prioritizing projects and operational actions for the current phase and developing project initiation documents, if possible in collaboration with the project managers. During this phase and deployment, the program manager acts as the sponsor of projects.

Interdependencies. The selection of actions will take into account the interdependencies between projects, in terms of effects on each other (usually input-output dependencies but also synergy) and in terms of their combined contribution to the benefits sought by the organization. The decision to implement must not be made solely on the individual value of each project or action but also on their capability to contribute to the program as a whole and to align with the strategy. In practical terms, a milestone chart with interdependency activities-on-arrow or a natural network (Marion and Remine, 1997) will effectively represent these interactions.

Prioritization. The prioritization of actions in the overall program must ensure that synergies between projects are taken into account, that there are no overlapping or conflicting actions, and that benefit delivery is optimized. Actions are first prioritized against the critical success factors of the program, but prioritization also takes into account achievability and elements discussed in pacing. Aside from cash flow and funding, financial factors can include payback period, cost-benefit and risk elements. The objective is to make early gains to "fund" the rest of the program and motivate stakeholders, especially investors, to continue supporting the program. In terms of human resources, the program team must take into account not only culture and responsiveness to change but also the perception of benefits by the users, the assessment of expectations of those affected by the change, and the measures put in place to support smooth transition from the original situation toward the target situation. Finally, the prioritization plan will be flexible enough to allow regular reprioritization as the program progresses and priorities or objectives change; if the initial plan has been well documented, this poses no problem.

Constraints and Assumptions. The last point is the identification of constraints and assumptions and their documentation. It is crucial that the constraints and other factors on which the assumptions have been based are well identified, because they will evolve and change as the program progresses and will need to be reassessed regularly. This is an integral part of the risk management process of the program.

The Deployment Phase

This phase includes the actual initiation of actions as well as the management of interdependencies between projects. Once actions are initiated, the program manager will have the role of project sponsor, although in a number of organizations, this role is shared with customers, resource managers, or senior managers. For effectiveness and efficiency purposes, it is advisable that the sponsor's authority is not shared and others play an advisory role. As such, the program managers need to manage project reviews (called *gate keeping*), considering project deliverables against CSFs and KPIs. Other responsibilities include continuous assessment and management of stakeholders' needs and expectations, resource prioritization, and change management, using a formative approach.

Planning and Execution of Actions

During this phase of the program, the main task of the program manager is to allocate or reallocate funds and other resources according to priorities. In developing and approving project management plans, the program manager authorizes or confirms the allocation of resources to undertake the detailed planning of projects.

The program manager also manages the communications to and from project managers and especially acts as a buffer to contain senior management's direct influence of the projects. It is not reasonable to expect project managers and project teams to, at the same time, focus on delivering specific project objectives and develop effective responses to emergent change; the latter is clearly the role of the program team.

The identification of emergent (unplanned) inputs on the program, which could trigger the need for changes, also needs to be monitored and managed in an orderly way. This is where the use of risk and value management techniques becomes essential.

Value Management. During the deployment phase, VM is the process through which project gateways will be handled. Project gateways correspond to go/no-go decisions that are typically situated at significant milestones and concern major project deliverables or phases. At each of these stages, the program manager will apply VM methodology to assess results and decide if changes are required to the project management process.

VM is the preferred way to handle stakeholders' needs and expectations and change management as it is a formative approach based on the definition of CSFs and KPIs.

Risk Management. At the program level, risks generally can be categorized into three levels:

- *Program risks.* Risks that directly affect the program and are handled directly at the program level.
- *Aggregated risks.* Project risks that are common to more than one project and therefore considered from a program-level point of view: project-level response with program support or program-level response with project-level follow-up.
- *Project risks.* Risks that affect only one project.

Risk impact is assessed against all the weighted CSFs to identify those risks that must be dealt with at the program level. The risk response management includes coordination of responses throughout the program in order to optimize results.

Interdependencies and Resource Management

The program manager role requires focus on project interdependencies, rather than on project activities. Even in terms of the project product, the program-level intervention consists of assessing benefits of major project deliverables and output-input relationships between projects—not solving the technical or operational-level problems.

Because program managers are required to manage interdependencies, it is advisable to revert to a variation of the ADM (arrow diagramming method) to represent the program-level interactions between projects. Combined with milestone scheduling and critical chain concepts, it can offer a high level and effective view of a number of projects to program managers and enable them to monitor only major tasks and deliverables, while keeping an eye on both resources and interdependencies (see Figure 6.9).

Resource planning must take into account resources offered (supply) versus resources required (demand); demand should always be less than or equal to supply. This is one of the challenges of good program management, as it is often not the case. Demand consists of the prioritized requirements; supply is an evaluation of the prioritized workload and includes information on resource capacity, availability, capabilities, competence, and signif-

FIGURE 6.9. PROGRAM-LEVEL NETWORKS.

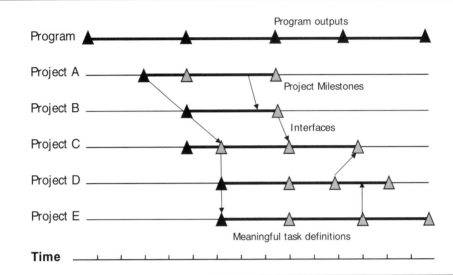

icance. The objective is to create a flexible system that enables quick, objective reaction to emergent inputs or changing priorities.

Project Review and Control

At the program level, project review/control goes beyond baseline control to include continuous reevaluation and reprioritization of projects in regards of the achievement of organizational benefits. Reviews generally correspond to project gateways, or major milestones. Control of projects concentrates on the impact of deliverables on the overall program benefits and strategic alignment, rather than a simple assessment of quality, cost, and schedule—or even earned value. All these measures become valid only as a means to make the necessary adjustments to achieve overall objectives. The program manager controls project progress with a "system's" view of all the actions included in the program, taking into account resources and major project deliverables. The key is to control outputs and milestones, not activities.

Because program management should be formative rather than summative, one of the roles of the program manager is to assess the need for project plan review or readjustment, propose or implement changes in projects, and assess their impact on the critical success factors. Project performance is analyzed with a view to the reallocation or reprioritization of resources and contingencies amongst projects or scope changes required to respond to results. The output of this process is a decision to continue, realign, or stop individual projects. Because of this, a sound aggregated information management system needs to be in place, both at project level and at program level. Program-level information management addresses the information between project- and program-level stakeholders, especially the reporting system, as well as the information circulating among projects that needs to be managed at the program level.

The Appraisal Phase

The periods of stability that mark the end of each cycle are an ideal opportunity to appraise the program. This is the time when the organization must evaluate the need to carry on with the program, review its purpose, or stop it. In this phase, results are compared with expected business benefits; emergent threats and opportunities are managed and overall program performance assessed. One of the key objectives is to evaluate the opportunity to continue and/or repace or reprioritize work and actions if required.

In essence, this learning/formative process outlines the program team's success in achieving expected benefits and produces useful feedback for the next phases. The first step is to ask a number of questions:

- Have the expected benefits for this cycle been achieved?
- Have the program or business circumstances changed?
- Have stakeholders' needs or expectations changed?
- Does the rationale for the program still exist?
- If not, should the program be reviewed; should it be stopped?

The appraisal process requires the program team to loop back to the formulation phase in order to reassess the validity of the original needs in regards of external or internal developments since the program was started; this includes positive or negative impact on the business of outcomes from previous cycles and knowledge management. Any change in the CSFs must be identified and examined to understand how it affects the expected benefits.

Following that exercise, the actual benefits of the cycle are evaluated against the expected benefits and a gap analysis is performed. If gaps are identified, the team defines how the program plan is to be modified to take them into account. Using value management, the team examines alternatives and evaluates options. Finally, a decision is made on how to modify the plan for the next cycle. If everything has gone as planned and nothing has changed in the expected benefits, the decision is made to carry on according to the original plan.

Benefits Assessment

When assessing the delivery of benefits from a program management point of view, the program team must have a broad perspective and examine two different levels:

1. At the organizational level, the team's main feedback is the effective delivery of expected benefits and satisfaction of CSFs, specifically:
   - Management of changing corporate or client objectives
   - Prioritization of shared and/or limited resources
   - Interface between functional and project managers
   - Clear definition of roles and responsibilities and mutual support to achieve corporate goals
   - Effectiveness of project review and approval process
   - Project managers' focus on key business issues
2. At the project level, the team reviews the relevance of projects that are spanning over a number of cycles and the aggregated benefits generated by all the projects that are part of that cycle. In particular:
   - Assess overall performance of projects against business benefits, including emergent factors.
     - Quality and timeliness of deliverables
     - Resource usage and budget
     - Use of contingencies
     - Interfaces and interdependencies
   - Identify new threats and opportunities and implement changes, if required.
   - Replan work and relative priorities, at the business level, for the next projects or phases of projects.
   - Loop back to project definitions and readjust, if required (learning loop).
   - Ensure information is recycled into a feedback loop, for the next phases or future programs.

Program Validation

The elements on which the justification for continuation will be established are as follows: progress toward achievement of expected benefits; response to emergent change in business environment, including changing needs and bottom-up initiatives; overall performance measured against CSFs and KPIs, including identification of threats and opportunities; and efficient and effective management of resources in general and of line and project manager's complementary roles in particular.

Following appraisal, the team reviews work and relative priorities for the next cycle, specifically, resource allocation, compare pace of projects, and the need to replace project managers. Table 6.1 outlines major decisions in regards of evaluation outcomes.

TABLE 6.1. MAJOR PACING DECISIONS.

Evaluation	Decision
Reassess significance of benefits for business (CSFs). —Internal pressures change focus. —External pressures require adjustment.	Reprioritize as needed.
Test integration of change on ongoing basis. —Changes integrated slower/faster. —Resistance to change identified.	Adjust pace as necessary; hold information workshops, training sessions, etc.
Measure delivery of benefits. —Benefits not delivered as planned. —Rate of benefits impact too low.	Reassess distribution of islands of stability as required.
Monitor achievement of minimum level of performance. —General performance drops. —Specific areas are in difficulty.	Review priorities and reallocate funds to urgent matters accordingly.

Programs take place over a period of many months (IT-supported change programs, business process reengineering, etc.) or many years (drug development, transportation infrastructure refurbishment, new product development, etc.). Some programs are even ongoing (performance appraisal schemes, account management, continuous improvement, etc.). The appraisal process is the first step toward dissolution; even in ongoing programs it must be carried out on a regular basis, and, every time, the program team must ask itself: Should the program be stopped or is it worth continuing?

The Dissolution Phase

If the team realizes that the rationale for the program no longer exists, it implements the dissolution process. This decision can be based on a performance paradigm (initial expected benefits have been achieved; cost of the program is greater than the benefits it is bringing to the organization; cost-benefit ratio of program is greater than that of independently run projects) and/or on a learning paradigm (the environment or context have changed and the benefits that the program was seeking to achieve are no longer required, or the implementation of the first cycle or cycles has demonstrated that the program's ultimate purpose cannot be achieved).

As for projects, the closing, or dissolution, of the program is not an easy task; when a decision to stop a program is made, a number of people and funds are reallocated to other ventures, and therefore there is likely to be resistance from the team to "let go." For this reason, some organizations even choose to involve a team external to the program in the closing phase to make it more efficient.

There will always be some uncompleted work that needs to be either completed within a reasonable period of time or transferred to other programs. The team needs to identify and agree on uncompleted work and clarify what can be completed within a reasonable period; it must agree on and secure resources to carry out a post-program feedback, transfer residuals to other programs, allocate outstanding work, reformulate programs, and reallocate resources as required. The program team estimates the resources required to deliver outstanding benefits and the value to do so (benefits/cost analysis). All the documentation must be updated and filed and a post-program review conducted. The data is then fed back into the organization through a learning/innovation loop.

Once all this has been accomplished, the program dissolution team is disbanded and reassigned. This is also the time when overall knowledge, gained from the process, must be collated, although knowledge management takes place at each cycle, especially appraisal.

The Program-Based Organization: Framework and Support

Program Culture

A number of factors currently influence the management of organizations that run projects: Change is accelerating and becoming more complex, influences affecting projects are often of high level, project objectives conflict with one another, projects compete for the same resources, project management is too product-centric, the link between projects and strategies is not apparent, and there is often no coordination of deliverables to realize benefits. The results are ineffectiveness of overall solutions, difficulty to achieve corporate benefits, lack of assessment of the organization's capability, inefficient use of resources, ill response to unplanned (emergent) change, and overruns caused by a lack of coordination. This requires the development of a different view of organizations that includes a strategic/systemic view and the capability to quickly adjust and/or respond to emergent change and to change the relative priority of projects or realign project objectives.

Program and portfolio management help address these challenges, but that often means changing the organizational culture. To succeed in the implementation of a program culture, an organization will require the support of senior management and a clear distinction between project and program paradigms to establish its foundation; structures supporting knowledge management as well as a systemic view of resource management (contingencies, personnel, funding, priorities); and, finally, an attitude aligned on vision and strategy, aimed at stakeholders benefits, and an openness to change.

Merritt and Helmreich (1996) have defined a few keys to successful culture change:

- *Role modeling.* Significant models, senior management
- *Mentoring.* To explain implicit cultural norms
- *Language/discourse.* Manuals, guidelines, logos, mission statements
- *Attractive membership.* Success stories, belonging to a club, fostering pride, and other incentives
- *Proactive approach.* Show concern and willingness to support culture
- *Timing.* Use of emergent inputs and successes to reinforce culture
- *Training and development.* Build culture into training and development; use training and development to reinforce culture

These can be part of the introduction a program culture.

Program-Based Organization

For large programs and organizations that run a number of programs, the organizational structure shown in Figure 6.10 is probably appropriate. For smaller organizations, some roles can be combined.

In this organizational context, the roles of each stakeholder are as follows:

- Top Management and Executive Leadership establish strategic direction for organization.
- The Corporate Program Office is responsible for the whole enterprise portfolio and ensures the strategy is met.

FIGURE 6.10. PROGRAM-BASED ORGANIZATION TYPICAL STRUCTURE.

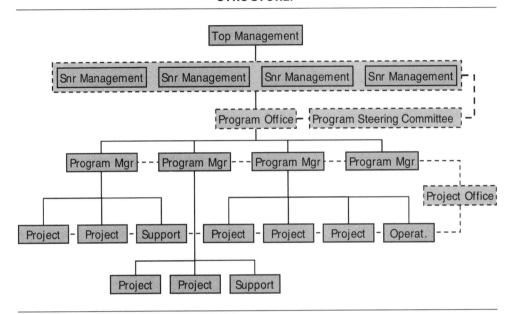

- Individual Senior Managers are usually the program sponsors. Collectively, Senior Management is accountable for the organization's portfolio.
- The Program Steering Committee is the executive group responsible for individual program oversight, guidance, and barrier removal. It links with key senior stakeholders, typically strategy, finance, commercial, and production. It can be combined with the program management office.
- The Program Management Office—or Program-Level PMO—supports the program manager in large, complex programs, prioritizes program resource allocation, and takes a strategic perspective with a focus on pacing and program appraisal. (See the text that follows for the detailed role.)
- Program Managers have accountability for the program, act as project sponsors, and ensure that benefits are delivered. More specifically, their specific responsibilities are listed in the text that follows.
- The Project Management Office provides assistance to project managers on large projects, provides tracking and oversight of projects, coordinates information and reports for the program level, and sets organizational level project management processes and procedures.

- The Project Manager is the single point of accountability for projects, managing project resources within set parameters, as well as day-to-day project activities, and delivering agreed outcomes at milestones.

Concerning the program management office, a series of recent papers (EDS Web site, 2002, Richards, 2001; Kendall, 2001; Moore, 2000; and the chapter by Powell and Young in this book) have redefined its role to give it a more active role. It is now generally acknowledged that this role can cover two major areas: support and delivery. Key areas of support cover the improvement of program and project management efficiency and effectiveness: becoming a corporate program knowledge management center, offering consultancy and advisory services, and supporting program and project managers in their roles and responsibilities. Key areas of delivery include ensuring that corporate strategy is owned and delivered and that it is on track, ensuring delivery of specific measurable outcomes and identifying gaps in portfolio of projects, and analyzing past situations and future trends for knowledge management.

Specific responsibilities of the program manager include, but are not limited to:

- Prioritizing projects and actions in regards of CSFs
- Initiating projects and other actions
- Allocating or reallocating resources within program
- Managing contingencies and coordination of resources
- Balancing project scope and quality versus cost and time (a value ratio)
- Ensuring that benefits are delivered
- Managing projects' interfaces and interdependencies
- Assessing deliverables against expected benefits at gateways

Summary and Conclusions

"Unaligned organization is a waste of energy, whereas commonality of direction develops resonance and synergy."

TSUCHIYA, 1997

Project and program management depend on different paradigms. Whereas project management is subjected to a performance paradigm and has proven effective in delivering short-term tactical-level deliverables, it has not proven its ability to deliver strategic change or improvement programs. Experience with strategic programs or soft organizational change programs has demonstrated that programs need to take into account a learning paradigm that comes from strategic management and value.

A robust program management approach will increase the efficiency of organizational processes (financial, resources, knowledge), support deliberate change (strategies, reengineering), capture bottom-up inputs (innovation, continuous improvement), and enable the proto-typing of emergent strategies by shortening feedback from experience and limiting risks.

Some key issues for program management success are as follows:

- If programs are using a project (performance only) management mind-set and methodology, benefits are lost.
- Programs should exist only if they generate benefits over and above those that projects generate on their own.
- Organizations must understand why program management ought to be implemented and how it can be sustained.

Program management will fall short as a strategic decision management process if the organization implements programs without taking the organizational culture into account, if it does not adequately quantify expected benefits or link them to project deliverables, or if there is a loss of focus of stakeholder expectations over time.

References

Artto, K. A., M. Martinsuo, and T. Aalto. 2001. *Project portfolio management.* Project Management Association, Helsinki, Finland. ISBN: 951-22-5594-4.

Association for Project Management. 2000. *The Association for Project Management Body of Knowledge.* High Wycombe, UK: APM.

Association for Project Management. 2003. Programme management. Chap. 11 in *Project Management Pathways.* High Wycombe, UK: APM.

Bartlett, J. 1998. *Managing programmes of business change.* Wokingham, UK: Project Manager Today Publications.

Carver, C. S., and M. F. Scheier. 1990. Origins and functions of positive and negative affect: A control process view. *Psychological Review* 97:19–35.

Central Computer and Telecommunications Agency. (1999) *Guide to programme management.* London: The Stationery Office.

Center for Business Practices. 2002. Top 500 project management benchmarking forums. *PM Network* (November): 6–8.

Cooke-Davies, T. 2002. Establishing the link between project management practices and project success. *Proceedings of the 2nd PMI Research Conference.* Seattle, July. Newtown Square, PA: Project Management Institute.

de Bono, E. 1990. *Lateral Thinking: A Textbook of Creativity.* 3rd ed. Harmondsworth, UK: Penguin.

Defense Systems Management College. 2000. *Program management 2000, Know the way: How knowledge management can improve DoD acquisition.* Fort Belvoir, VA: Defense Systems Management College Press.

EDS 2002. Program Management Office according to EDS. Retrieved June 2003 from: http://www.eds.com/services_offerings/so_project_mgmt.shtml.

European Foundation for Quality Management. 1999. *The EFQM Excellence Model.* Brussels, Belgium. www.efqm.org/.

Frame, J. D. 2002. How PMI is keeping up with rapid change in the profession. *PMI Today.* Newtown Square, PA: Project Management Institute.

Gareis, R. 2000. Program management and project portfolio management: New competences of project-oriented organizations. *Proceedings of the Project Management Institute 31st Annual Seminars and Symposium.* Drexel Hill, PA: PMI Communications.

Goldratt, E. 1997. *The Critical Chain.* Great Barrington, MA: North River Press.

Görög, M., and N. Smith. 1999. *Project management for managers.* Sylva, NC: Project Management Institute.

Guba, E. G., and Y. S. Lincoln. 1989. *Fourth generation evaluation.* Newbury Park, CA: Sage Publications.

Hatch, M. 1997 Strategy and goals. Chap. 4 in *Organization theory: Modern symbolic and postmodern perspectives,* 101–119.Oxford: Oxford University Press.

Hartman, F., and R. A. Ashrafi. 2002. Project management in the information systems and information technologies industries. *Project Management Journal* 33(3):5–15.

Hurst, D. K. 1995. *Crisis and renewal: Meeting the challenge of organizational change.* Boston: Harvard Business School Press.

Johnson, G., and K. Scholes. 1997. *Exploring corporate strategy.* 4th ed. Hemel Hempstead, UK: Prentice Hall Europe.

Kaplan, R. S., and D. P. Norton. 2000. Having trouble with your strategy? Then map it. *Harvard Business Review.* (September–October): 167–176.

Kendall, G. I. 2001. New executive demands of projects and the PMO. *Proceedings of the 2001 PMI: Seminars and Symposium.* Newtown Square, PA: Project Management Institute.

KPMG. 1997. *What went wrong? Unsuccessful information technology projects.* http://audit.kpmg.ca/vl/surveys/it_wrong.htm.

Kubr, M. 1996. *Management consulting: A guide for the profession.* 3rd ed. Geneva: International Labour Office.

Marion, E. D., and E. W. Remine. 1997. Natural networks: A different approach. *Proceedings of the Project Management Institute 28th Annual Seminars and Symposium.* Newtown Square, PA: Project Management Institute.

Martinsuo, M., and P. Dietrich. 2002. Public sector requirements towards project portfolio management. *Proceedings of PMI Research Conference 2002.* Seattle, July. Newtown Square, PA: Project Management Institute.

Merritt, A. C., and R. L. Helmreich. 1996. Creating and sustaining a safety culture. *CRM Advocate* 1:8–12

Mintzberg, H., and J. A. Waters. 1994. Of strategies, deliberate and emergent. Chap. 10 in *New thinking in organizational behavior, ed.* T. Hardimos, 188–208. Oxford, UK: Butterworth-Heinemann. Previously published in *Strategic Management Journal* 6:257–272.

Mintzberg, H., B. Ahlstrand, and J. Lampel. 1998. *Strategy safari.* London: Prentice Hall.

Moore, T. J. 1999). An evolving program management maturity model: Integrating program and project management. *Proceedings of the PMI Seminars and Conference 1999*. Newtown Square, PA: PMI.

Morris, P. W. G. 1997. *The management of projects*. London: Thomas Telford.

Murray-Webster, R., and M. Thiry. 2000. Managing programmes of projects. Chap. 3 in *Gower handbook of project management, 3rd ed*. Ed. R. Turner and S. Simister. 47–64. Aldershot, UK: Gower.

NASA. 1998. NASA program and project management processes and requirements. NPG: 7120.5A.

Neal. R. A. 1995 Project definition: The soft systems approach. *International Journal of Project Management* 13(1):5–9.

Partington, D. 2000. Implementing strategy through programmes of projects. Chap. 2 in *Gower handbook of project management, 3rd ed*. Ed. R. Turner and S. Simister. Aldershot, UK: Gower.

Patrick, F. S. 1999. 'Program management: Turning many projects into few priorities with TOC. *Proceedings of the PMI-99 30th Annual Seminars and Symposiums*. Newtown Square, PA: Project Management Institute.

Pellegrinelli, S., and C. Bowman. 1994. Implementing strategy through projects. *Long Range Planning* 27(4):125–132.

Porter, M. 1985. *Competitive advantage*. New York: Free Press.

Project Management Institute. 1996. *A guide to the Project Management Body of Knowledge*. Newtown Square, PA: Project Management Institute.

Quinn, J. B. 1978. Strategic change: Logical incrementalism. *Sloane Management Review* 1(20):7–21.

Quinn, J. J. 1996. The role of 'good conversation' in strategic control. *Journal of Management Studies*. 33(3):381–394.

Reiss, G. 1996. *Programme management demystified*. London: Spon.

Reiss, G., and P. Rayner. 2001. The programme management maturity model. *Proceedings of the 4th PMI-Europe Conference*, London.

Reiss, G., and P. Rayner. 2002. The programme management maturity model: An update on findings. *Proceedings of the 5th PMI-Europe Conference*. Cannes.

Richards, D. 2001. Implementing a corporate programme office. *Proceedings of the 4th PMI-Europe Conference*. London.

Spradlin, T. 1997. A lexicon of decision making. *Decision Analysis Society*. http://faculty.fuqua.duke.edu/daweb/lexicon.htm.

Stake, R. E. 1975. *Evaluating the arts in education*. Columbus, OH: Merryl.

Standish Group International. 1996 *A Standish group research on failure of IT projects*. Yarmouth, MA: The Standish Group.

Thiry, M. 2001. Sensemaking in value management practice. *International Journal of Project Management*. Oxford, UK: Elsevier Science.

———. 2002a. Combining value and project management into an effective programme management model. In *International Journal of Project Management*. Oxford, UK: Elsevier Science.

———. 2002b. How can the benefits of PM training programs be improved? *Proceedings of the 5th PMI Europe Conference*. Cannes, June. *International Journal of Project Management*. Oxford: Elsevier Science. 22, 13–18 (January 2004).

———. 2002c. FOrDAD: A program management life-cycle process. *Proceedings of the 5th PMI Europe Conference*. Cannes, June.

Thomas, J., C. Delisle, C., K. Jugdev, and P. Buckle. 2000. Selling project management to senior executives: What's the hook? *Proceedings of the Project Management Institute 1st Research Conference*. Newtown Square, PA: Project Management Institute.

Tsuchiya, S. 1997. Simulation/gaming: An effective tool for project management. *Project Management Institute 28th Annual Seminars and Symposium Proceedings.* Drexel Hill, PA: PMI Communications.

Weick. K. E. 1995. *Sensemaking in organizations.* London: Sage Publications.

Winch G., A. Usmani, and A. Edkins. 1998. Towards total project quality: a gap analysis approach. *Construction Management and Economics* 16:193–207.

Wijnen, G., and R. Kor. 2000. *Managing unique assignments.* Aldershot, UK: Gower.

CHAPTER SEVEN

MODELING OF LARGE PROJECTS

Ali Jaafari

This chapter presents an integrated approach to the conceptualization, planning, and implementation of large, complex projects. The perspective is on the whole project life cycle, which includes creation, definition, initiation, planning and documentation, execution, commissioning and start-up, operation, and recycling. (The operation phase is considered only in terms of managerial decisions that need to be taken during the preceding phases.) On most large projects it is not possible to separate the project's end product from that of project delivery and management activities (PMCC, 2001; Brook, 2000; Forsberg and Mooz, 1996); thus, any reference to the project life cycle is taken to imply both product and project life cycles.

Generally speaking, the project life cycle can be divided into three distinct phases: the project strategic (promotion) phase (all activities up to and including project approval and funding), the project implementation phase (comprising initiation, planning, detailed design, documentation, execution, and commissioning activities), and the project operation phase (including operation and eventual recycling). Some authors divide the life cycle into two phases only: development and operation. The former includes all activities prior to the start-up and operation phase; the latter includes the utilization phase, including project recycling. It is worth emphasizing that the project as whole is the focus, not the functions of individual players within the project life cycle.

Examples of large projects are aerospace, defense, mining, infrastructure, large telecommunication systems, large software, power, and transportation schemes—all must be recognized in terms of their complexity and managed accordingly. Thus, one would expect to see a similar approach to the management of this class of projects regardless of their industry, yet this is not necessarily the case. For example, in aerospace and defense projects, typically, the emphasis has been on systems engineering and procurement functions; in the construction industry, the emphasis has been on contract and resource management; software and information systems projects have tended to be approached from a technical perspective.

Objectives

This chapter portrays the complex and uncertain internal and external environments within which large projects are typically developed and implemented. A broad classification of project types (in terms of both the characteristics of these projects and their environmental complexity) is presented and the position of capital projects highlighted. The chapter will show that project strategies must relate to project types and environmental complexity (uncertainty). While an integrative framework is needed to manage the evolution of the project concept, management of risks and uncertainty will have to guide the entire process (Jaafari, 2001). This discussion leads to the presentation of a framework appropriate for modeling large projects. The critical criteria for successful management of these projects are highlighted and their realization through the adoption of appropriate strategies is demonstrated. The chapter presents a brief overview of techniques that aid the quantitative and qualitative management of large projects. It emphasizes the need for a holistic approach as far as possible.

Characteristics of Large Projects

There is no universal definition for large projects. Complexity is a common feature in these projects. Complexity stems from two sources: the project's external environment and the complex make-up of the project itself. Miller and Lessard (2001) state: "Large engineering projects are high-stakes games characterized by substantial irreversible commitments, skewed reward structures in case of success, and high probabilities of failure." The environmental complexity is normally created because of the changing market and regulatory regimes impacting both implementation and operation of these projects. Project complexity can be understood in terms of relevant interlocking subsystems of hardware, software, of project-specific and temporary human and social systems, of related technical and technological systems, of financial and managerial systems, of specialized expertise and information sets, and so on that are typically created and managed to realize the project objectives (Jaafari, 2001; Yeo, 1995; Yeo and Tiong, 2000). The cost to promote these projects up to the implementation point is high, of the order of 5 to 10 percent of the total capital expenditure (Merna et al., 1993; McCarthy, 1991). A recent study by Hobbs and Miller (1998) puts the front-end costs up to 30 percent of total costs. Risks are high and the project delivery method is normally shaped to achieve a reasonable outcome in respect of the promoters' and community objectives (Hobbs and Miller, 1998; Wang and Tiong, 2000). Many infrastructure projects are nowadays delivered under build-own-operate (BOO) arrangements (see the chapters by Turner and by Ive). In the resources and industrial sectors, projects are normally fully owned and operated by the private sector.

The risk profile on large projects is complex. Some risks arise from the clash of social, political, and commercial interests and values of project promoters and those of the wider stakeholders that surface during the project development phase. Others relate to project functionality and fitness for purpose. A third set relate to project delivery dynamics. Often there is a window of opportunity in which a project can be favorably launched, as delays may see either the project concept becoming less relevant or even obsolete, or competitors moving in to fill the market need. Major projects are often dependent on novel technologies and innovative solutions; this in itself is a major source of risk.

Miller and Lessard (2001) classified risks on large engineering projects as market-related, technical risks and institutional/sovereign risks. See also Yeo and Tiong (2000).

Exposure to risks can change with time; new risks can be encountered and seemingly unimportant risks pose new threats. On the positive side, there can be opportunities, too, that may provide conditions for improving the project's base value (Miller and Lessard, 2001). This narrative suggests that risk management must inform all decisions and guide all strategies adopted for the creation and delivery of these projects. Hobbs and Miller (1998) undertook a study of a sample of 60 projects in 4 continents (31 power, 5 petroleum, 20 urban infrastructure, and 4 technology projects). They found that the front-end part of the project life cycle was particularly risky. This phase was often marred by serious setbacks that put projects as a whole at risk. Some key findings by these authors have been summarized in the following list to shed light on the dynamics of capital projects worldwide:

- No distinct phases (e.g., feasibility studies, design, and construction) could be identified on these projects; instead, a series of milestones were found to be common. In addition, a front-end part (referred to as the *strategic phase* in this chapter) could be distinguished from the engineering-procurement-construction (EPC) part (refered to as the *implementation phase*);
- There were wide variations in terms of the length of time taken to develop the projects in the sample to the point of implementation: 55 percent took more than five years. Projects that had shorter front ends were required to fulfill urgent needs.
- The promotion phase was found to be a dynamic play, which, in addition to those involved in more technical aspects of the project, saw participation of communities affected, environmental organizations, pressure groups, financial institutions, politicians, regulators, and government agencies.
- The decisions made during the front-end and the institutional, organizational, and financial framework that were put in place by and large controlled the success or failure of these projects and profoundly influenced the implementation stage.
- Some projects in the sample had a defined technical solution right from the outset, while for others the technical solution either evolved along the path or was deliberately held off until late to accommodate changes until the implementation stage
- The influence of the environmental, social, political, and community aspects on the sample projects were found to be increasing. This is largely because large projects epitomize the current profound restructuring of the institutions of society and government machinery. The principles of social equity, privatization, user pay, sustainable development, legal

legitimacy, and community ownership all exert varying degrees of influence on the creation and execution of large projects.

According to Hobbs and Miller (1988):

> "In recent years, the process has become much more complex. This increase in complexity is due to several factors including: the globalisation of competition, the trend toward deregulation, the changing role of governments under dual influences of free market doctrine and debt loads that prohibit further borrowing, and the actions of the pressure groups locally and internationally. In their search for solutions in this highly complex context, organisations have developed highly complex solutions which often include some form of coalition building. Often the initiators of projects do not have all the political, social, technical, financial and organisational resources and skills that are needed to deal with the multiple risks that the highly complex context presents. Therefore, they search for partners who can bring needed resources or skills, or that can control or support various risks. In the process of searching for a feasible solution, the skills of managing political and social interfaces, of organisational and financial engineering, and of deal-making were often critical.

> Planning in this context is very difficult. A deductive and linear plan to get to a solution is not workable because the solution is not identifiable at the outset, in fact, the problem is not usually well defined at this stage. At the outset, it is not obvious who the important players will be, and there is some trial and error in the search for partners and solutions to the many problems posed by the project. The activities of risk identification, analysis, mitigation and partitioning among players dominate the process. Negotiations are constant throughout this phase. Further, if the process was not complex and unpredictable enough, it is highly likely that during the search for a solution, a new problem will materialise and send the project off track at least once."

Evidence from other sources is just as revealing; Jaafari and Schub (1990) carried out a field study of large and complex projects in Germany and concluded that the development of such projects was substantially impacted by the resolution of risks and uncertainties. These authors showed that risks were not only increased by a poor choice of concept at the outset but also by community demands, changing regulations, political and social forces, and dynamics of the project environment itself. Morris and Hough (1987) have also shown the complexity and dynamics of these influences in their case studies of a number of large and complex projects executed in the United Kingdom.

This brief review is intended to show that project conceptualization and implementation is a complex, dynamic, and evolving process; that it should be managed on the basis of a set of objectives, which themselves would be subject to change, on a fully fluid and flexible basis (Jaafari, 2001; Chaaya and Jaafari, 2001; Jaafari, 2000, Miller and Lessard, 2001; Yeo and Tiong, 2000; Morris and Hough, 1987). Further, that a holistic and integrative framework is needed in which not only planning and proactive management of technical and financial factors receive attention but equally the social, environmental, political, and community aspects are placed at the center of attention (Jaafari, 2001; Jaafari and Manivong, 2000; Miller and Lessard, 2001). The objectives chosen should embrace the project's viability in its broadest sense, over its entire life and should facilitate management of the process using continuous risk and uncertainty resolution within a fluid and flexible management framework. This is very much an open-systems approach to the management of large projects of this nature (Yeo, 1995). Scott (1992) states that "systems are interdependent activities linking shifting coalitions of participants; the systems are embedded in—dependent on continuing exchanges with and constituted by—the environments in which they operate." For a more detailed understanding of complex systems' theory refer to Scott (1992). See also www.brint.com/Systems.htm.

Environmental Complexity and Influence on Strategic Direction of Large Projects

Large projects occupy the complex side of the project population space (Figure 7.1). Numerous forces impact project environments; viz (a) increased demand from owners and

FIGURE 7.1. CLASSIFICATION OF PROJECTS IN TERMS OF PROJECT AND ENVIRONMENTAL COMPLEXITY.

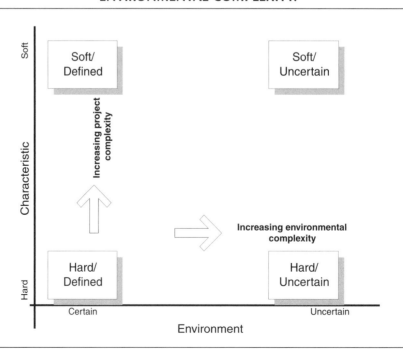

customers for solutions that deliver definitive advantage to them along their business objectives; (b) operation of markets, which nowadays shift a lot in a chaotic manner; (c) rapid rate of change in the underlying technology and scale of operation, which in many instances require novel solutions; (d) the influence of the regulatory bodies, who tend to aim for zero-risk solutions; (e) the information technology revolution enabling global collaboration and streamlined managerial processes; and (f) rising influence of community and pressure groups (Miller and Lessard, 2001; Jaafari, 2001; PMCC, 2001; Dixon, 2000).

As an example of environmental complexity, the following is an excerpt from Byers and Williams (2000). This excerpt illustrates the complex commercial and regulatory environment for electrical utility industries in the United States.

"As the world-wide economy evolves, electric utility companies in the United States and most industrialized nations of the world are under increasing pressure. In the U.S., deregulation of the electric utility industry has led to significant business and management changes. Corporate reorganizations, staff downsizing, outsourcing of services, and reengineering of business processes have had a profound impact on the industry.

The traditionally conservative U.S. electric utility industry, which had previously considered itself almost impervious to outside influences, was feeling the effects of a global economy and was under pressure to become more efficient and cost effective. Clearly, there was a continuing world-wide transformation going on, moving faster than most people had anticipated; now, the world was our market place with new opportunities and new competition.

Nation-wide, a vigorous move toward deregulation had, in just three years, changed the industry's view of its customers, its competition, and itself. Customers formerly bound to the company by geographic monopoly now had to be courted and costs had to be reined in to help meet low competitive pricing."

Other sectors of the economy, too, experience rapid changes in a similar fashion. Commercial and sociopolitical environments will set the scene for projects, as these are often the foundation for reshaping the competitiveness of firms or whole industry. As Struples (2000) writes:

Today's large engineering projects can involve significant organisational and operational complexity covering joint ventures, consortium working, international involvement, conflicting stakeholders, shareholders, special purpose companies, project financing, prime contracting, etc. Often such projects face diverse political pressures with difficult-to-define socio-economic impacts, and can be subject to risk-sharing contractual arrangements with pain/gain sharelines. As the project moves through its life cycle, organisational and operational issues and engineering change together can have catastrophic effects on the project schedules and costs, often with dire consequences. Many recent projects have been reported as experiencing cost overruns in excess of 300% and 150% to 200% is becoming the norm.

The Defence Evaluation and Research Agency (DERA) in the United Kingdom has portrayed a hierarchy of environments impacting projects and systems, shown in Figure 7.2. As seen, the project environment takes place within specific social and organizational environments (DERA, 1997). The social environment defines legal requirements, social norms, fiscal rules, environmental requirements, and business competitiveness. The impacts of the social environment on project formulation and management can be profound over its life.

The enterprise environment comprises both the suppliers and the customers and end users of large projects. This environment has a profound influence on project environment, particularly when projects are sponsored and implemented by a coalition of firms with complementary resources and expertise. There are two main enterprise environments: that of suppliers and those of the sponsors or client/end users. Often the actions and responses of these organizations create a dynamic setting that influences many project decisions and outcomes.

DERA (1997) cites examples of these as

- the business scope of the enterprise; this determines the markets, application areas, and opportunities that systems' ventures pursue
- business policy that defines how, on behalf of the stakeholders in the enterprise, resources will be invested. It influences decisions to bid, invest, or proceed with a product development; it determines the nature and allocation of corporate resources capacity; and so on.
- market strategy that influences product system families, intended product lifetimes and support policy
- investment strategy; this impacts product novelty, introduction or use of technology, supporting infrastructure for system design, the capabilities and training of personnel, the manufacturing locations and capacities, and so on.
- business practices that lead to mandated or recommended process standards, business and technical process improvement actions, common methods, and tools

FIGURE 7.2. THE HIERARCHY OF ENVIRONMENTS.

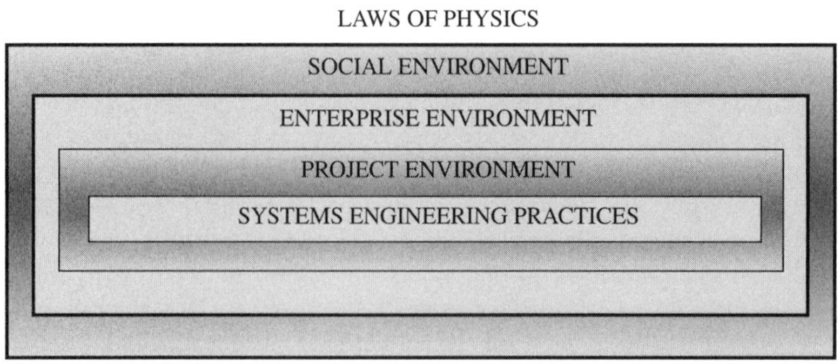

Source: DERA (1997).

FIGURE 7.3. INFLUENCES ON PROJECT DECISIONS.

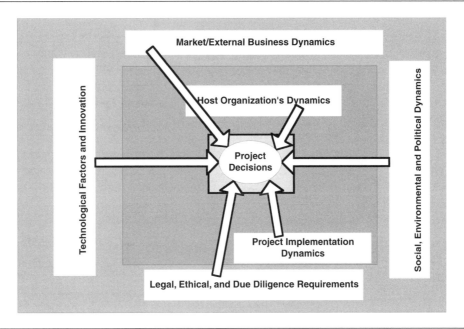

The project environment is in reality a subset of the enterprise environment, though in the case of large projects influenced by different enterprise cultures, the project environment may be more complex. Figure 7.3 indicates the influences on project decisions in a complex setting.

The project environment has to respond to a number of challenges, e.g. the aspirations of the sponsor organisations, the needs of wider stakeholders, the legal and due diligence requirements, the social and environmental requirements and so on (Manivong et al., 2002).

DERA (1997) states that

In the project environment:

- teams are built to provide capacity and breadth of understanding and experience
- plans are devised to guide the technical endeavours
- achievement is monitored to ensure that resources are effectively applied
- decisions are made, selecting alternatives to most successfully achieve objectives
- uncertainty is contained, limiting the commercial exposure of the host enterprise.

Project Life Cycle

Figure 7.4 shows a typical life cycle of large projects. As seen, one or a number of business and/or social needs or changes must be satisfied. A project idea is then born to respond to

FIGURE 7.4. PROJECT LIFE CYCLE PHASES, LARGE PROJECTS.

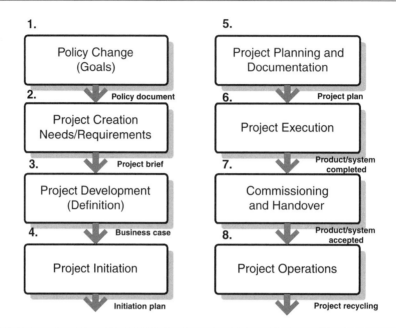

these needs. The responsibility for the project (or more accurately the needs to be fulfilled) is assigned to an operational unit (client) who typically sets up a project directorate to handle the project formally and properly. The next phase is the project development phase; this is the basis for detailed investigation of alternative ways to respond to the project goals or satisfy the stated needs. The outcome of this phase is very critical in the sense that it identifies the optimum way to respond to the relevant business needs and requirements and to formulate a clear business case for the selected approach (which may or may not lead to the scope originally foreshadowed). The outcome is normally captured in a project definition report and is used for approval and funding of the project.

The implementation phase is really about finding an optimum way to deliver the business case (Adams and Brown, 2002). It starts with the initiation phase that formulates a set of strategies that will guide the planning and execution phases. If all decisions taken during the development phase are optimal and provided that there are no major changes in the business or operational environment, one would expect the project to progress to a successful conclusion, leading to the operational phase.

The project life cycle may be portrayed using different terms—for example, that of the Asian Development Bank (see Figure 7.5). This representation is based on project identification, preparation, appraisal, funding (loan negotiation and board approval), implementation, and evaluation.

Another way to look at a project is from the underlying system point of view, where the following can be noted (DERA 1997):

FIGURE 7.5. THE ADB PROJECT LIFE CYCLE REPRESENTATION (www.adb.org/projects).

The Project Cycle

Identification

Preparation

Evaluation

Appraisal

Implementation

Loan Negotiation and Board Approval

- Conception
- Creation
- Utilization
- Disposal
- Whole of life approach

Decisions made during the front-ends of projects have a profound impact upon the success or failure of the project during both the implementation and operation phases, not only in terms of scope, cost, and time but more importantly the underlying operational capability and business viability. So it is important to consider all implementation and operational aspects, and combine upstream and downstream information before formulating the project concept. It is also important that a consistent and integrated framework/model of the project is set up and all decisions evaluated against a set of criteria representing the whole project life cycle. This needs a modeling approach that enables the project team to develop an optimal project solution initially, coupled with a capability for real-time adjustment of the same throughout project life cycle.

Management of Risks and Uncertainties

The challenge of planning, and successfully delivering, large projects principally lies in the effective recognition and management of project risks and uncertainty, given relevant environmental complexities.

Many business and strategic considerations motivate project promoters, including securing a presence in a particular market, entering global competition, and maintaining technological supremacy. However, more often than not, the overriding objective is to gain financial rewards for a relatively small financial outlay. This requires a core competency in risk and uncertainty management. The promoters' first challenge is to obtain permits to construct and operate their schemes. Promoters are not always investors, and the investors' interests may be different in the sense that many institutional and individual investors are not active participants in the management of the process but invest in, or lend funds to, the project in the expectation of future returns. Put differently, the promoters' objective is to create a long-term financially viable and balanced business entity. The eventual facility is a compromise between the promoters' interest and the interests of the community at large. If the objective is to create a viable business entity, then the processes of development and decision making must also be shaped primarily by the same consideration (Jaafari, 2000, 2001).

It is interesting to note that managers on the sample of 60 large engineering projects studied by Miller and Lessard (2001) ranked market-related risks (and uncertainty) at 41.7 percent, followed by technical risks at 37.8 percent and institutional/sovereign risks at 20.5 percent. The latter includes social acceptability risks.

Public sector projects are somewhat different in the sense that they are primarily dependent on budgetary restrictions and stakeholders' consensus. Changing stakeholders' expectations imposes substantial challenges on the project success in terms of scope changes and shifting priorities over the project life cycle. While these projects are subject to different risks and challenges, the need for creation of value and reduction of risks and liabilities over the project life remains unchanged. The value proposition on these projects is often expressed in terms of minimization of the total life cycle costs for each function fulfilled over the project life. Alternatively, another measure representing the service value generated by the project over its life can be defined as the basis for project value creation and optimization.

If the general project environment is subject to change, the project objectives must also change in line with relevant dynamics. This means that the entire project management philosophy must be opportunistic and driven by risk/rewards prospects throughout the project life. The necessity for adopting such a flexible and fluid framework for projects lies partly in the turbulent environments (particularly shifting markets) and partly in the rapid rate at which technology changes. A third factor is the increasing influence of the host communities and stakeholders, as well as the complex requirements often imposed by legal, environmental, social, safety, and fiscal regulations (Hobbs and Miller, 1998). Thus, it is not possible to close a project's options too quickly by freezing everything in the form of fixed design/specifications and/or lump-sum fixed-scope/-price contracts (Miller and Lessard, 2001; Laufer, et al., 1996). Decisions have to be analyzed and optimized continuously using the life cycle objective functions (LCOFs) as the basis of evaluation (Jaafari, 2000, 2001). Miller and Lessard (2001) argue that "Sponsors strategize to influence outcomes by using four main risk-management techniques: (1) shape and mitigate; (2) shift and allocate; (3) influence and transform institutions; and (4) diversify through portfolios." Note that these strategies are not mutually exclusive. Of these, the last strategy, namely diversification through the acquisition of a portfolio of similar projects, is not often open to sponsors. The exception is large multinationals who operate globally (typically in mining, resources, and industrial sectors). While risk management processes will enable project sponsors to approach large, complex projects systematically, criteria for evaluation must always be life cycle objectives, as these projects take a relatively long period to eventuate and then a longer period to operate, to retire the investment and return a profit to sponsors.

Life cycle objective functions are those that

- determine the project's financial status and its profitability;
- represent the operability, quality, or performance of the facility or the utility of the product to customers; and
- will influence the owner's short- and long-term liabilities, including occupational health and safety (OH&S) risks during both construction and operation, environmental impact, and third-party liabilities.

The financial LCOFs vary from one project to another. These may include cost/performance ratio (total life cycle cost/unit output), which typically suits production or extraction facilities; cost/worth ratio, which suits public projects; internal rate of return; and profitability index (ratio of the net present value over capital expenditure). There may be other (secondary) objectives, viz early cash flow generation, debt reduction, and so on (Woodward, 1997). On privately sponsored infrastructure projects, the concession award, the environmental impacts statement (EIS), and the finance deals may well contain appropriate target values that can be used as project life cycle objectives. Targets set for operability, quality, or facility performance will directly or indirectly affect profitability, while OH&S and environmental objectives influence profitability and long-term liabilities.

Traditionally, all LCOFs, objectives are assessed at the time of project planning and definition (or feasibility studies and conceptualization). Also, on major projects, an environmental impact statement is normally required for the issuance of a permit by the relevant authorities. Such documents typically contain recommendations for environmental management and safeguards against adverse environmental impacts. Under a whole of life approach, the project status and its decisions are evaluated continuously in terms of the relevant LCOFs and in comparison with the targets set for the LCOFs at the start of the project.

Under this approach, project time and cost are not to be treated as the main objective functions even for the management of the implementation phase, as these do not directly represent the LCOFs. As an example, a modest rise in the capital expenditure on a project may well be justified if it leads to a shorter delivery timescale and an increase in the project's internal rate of return. In general, the status of the life cycle objectives, including compliance with the statutory requirements and exposure to risks and/or future liabilities, will determine project success and should therefore be the basis for ongoing evaluation. Put differently, project decisions must at all times be aimed at improving the base value of the project, its fitness for purpose, and due compliance, while at the same time minimizing the impact of potential risks and uncertainties.

Critical Success Criteria and Framework for Adoption

Successful synthesis, development, and delivery of complex projects typically involve a large number of professional people working within a range of organizations who are party to the project via contracts or other means. The owner or sponsor has, however, the highest stakes in the project and must adopt a set of criteria that will enable him or her to solicit and integrate the diverse inputs from all participants and steer the project to a successful completion in an optimum manner. The following criteria are recommended:

- *Focus on the whole business of which project hardware and software are only parts.* This means that all decisions throughout the project life should be judged in terms of their impacts, whether positive or negative, on the business and strategic objectives that the project is required to deliver. Thus, intermediary objectives such as scope, time, and cost should be used for communication and expediting of the project implementation, not as the ultimate criteria for decision making.
- *Maximize opportunities for value creation.* As the project moves from the concept to development and implementation phases, the opportunity to add value and reduce exposure to risks and liabilities will decrease and the cost to implement change will increase exponentially (see Figure 7.6). It is important that the project team is encouraged and given due assistance to develop breakthrough solutions early in the life of the project in a manner that maximizes the project's chance for success and minimizes its exposure to

FIGURE 7.6. VALUE CREATION AND RISK REDUCTION POTENTIAL VERSUS PROJECT PHASES.

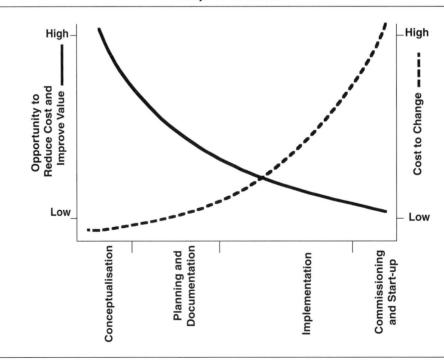

risks and liabilities. Value engineering is most effective at the concept and design phase; though it should be the basic focus of project team in all phases and in a creative manner (Merna, 2002). (See the chapter by Thiry.)

- *Institute a proactive approach to handling risks and uncertainties.* Large, complex projects are subject to multiple risks and uncertainties, particularly in their formation phase (Miller and Lessard, 2001). Most projects suffer significant cost and time overrun, or experience performance setbacks because of the unresolved risks surfacing or new risks arising unexpectedly. Prudent risk and uncertainty management is the key to successful project management. (See the chapters by Chapman and Ward, and by Simister.)
- *Incorporate and manage community and stakeholders' interests.* While this sounds like a self-evident statement, its implementation is quite complex. Many projects touch the life and economic well-being of many people in their host communities and have multiple stakeholders who are not necessarily financial party to the project. The interest and influence of these parties have to be recognised and factored in at the time of project conceptualization and planning. These have to be managed systematically throughout project life. Appropriate resources must be allocated to ensure success on this front. (See the chapter by Winch.)
- *Create synergy among participants.* A successful project needs to capture and effectively utilize the energy and intellect of all its participants in an effective manner. This is not easy, as project participants often come from different organizations, each with its own unique culture, norms, and standards. Synergy must be created in terms of congruence of project objectives and contractual terms that commit participating contractors and consultants to project objectives. Formulation of the actual terms and conditions of contracts is strategically very critical to the project success. In recent times, contractual terms have even included obligations to attend team facilitation workshops in order to create a teamwork spirit.

Modeling and Integrated Life Cycle Planning

Figure 7.7 is a simplified representation of modeling for a large project. As can be seen, the model should ideally

- provide a consistent and efficient framework for development of the project from concept to completion and through to facility operation;
- integrate project information related to all project life cycle phases;
- integrate all project management functions, including both hard and soft functions;
- support scenario analysis and offer an integrated environment to effectively and interactively apply what-if planning;
- have the potential to accommodate modeling and simulation of the operation of the end facility (Jaafari & Doloi, 2002);
- provide graphical simulation of the proposed implementation plan in a manner that helps the optimization of relevant work plans;
- provide a means for interdisciplinary communication and teamwork;

FIGURE 7.7. SIMPLIFIED REPRESENTATION OF PROJECT MODEL.

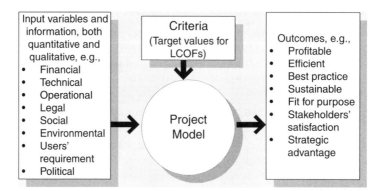

- provide an effective means for conveying the planning results to the field;
- allow early problem detection, including removal of clashes and interferences; and
- integrate the processes of planning, engineering, documentation, procurement, and execution throughout the project life cycle (dynamic process).

Realization of such sophisticated modeling systems is still some way off. The project model has to expand from the concept to the implementation phase in terms of information sets and linking of these within the model so as to enable the holistic evaluation of all project decisions. What follows is a study of large projects in terms of the different phases—that is, the strategic phase and the implementation phase (see Figure 7.8).

Project Strategic Phase

Phases 1 to 3 inclusive in Figures 7.4 and 7.8 are part of the strategic planning phase (project promotion phase). The strategic planning happens at two levels: organizational and project.

Project Creation. Figure 7.9 shows the organizational planning framework. Very often policy changes (particularly at government and large corporations) will give rise to the project concept (Beder, 1991; Kelley, 1982). Discussion regarding grounds for such policy changes and the processes that are followed to introduce such changes are outside the scope of this chapter. These are typically the domain of strategic planning for the whole organization, government department/agency, and/or relevant communities.

Generally speaking, the need must be expressed in terms of business or strategic needs of the organization, not acquisition of new assets or increased capacity (Artto et al., 2001). The criteria for fulfilment of this need should also be spelled out in terms that can easily be cast into target values for LCOFs. This approach fulfills two specific purposes: It provides

FIGURE 7.8. PROJECT LIFE CYCLE.

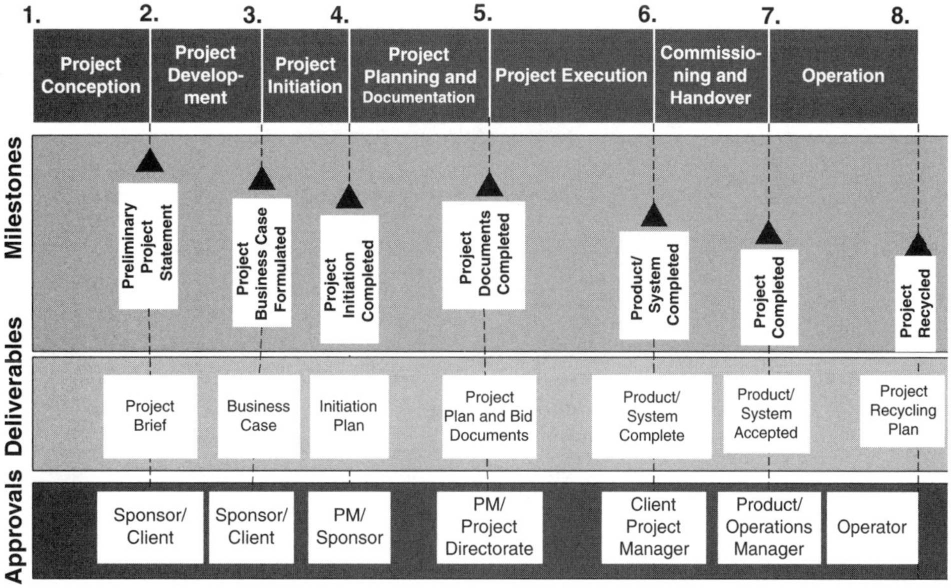

	1.	2.	3.	4.	5.	6.	7.	8.
	Project Conception	Project Development	Project Initiation	Project Planning and Documentation	Project Execution	Commissioning and Handover	Operation	

Milestones	Preliminary Project Statement	Project Business Case Formulated	Project Initiation Completed	Project Documents Completed	Product/ System Completed	Project Completed	Project Recycled	
Deliverables	Project Brief	Business Case	Initiation Plan	Project Plan and Bid Documents	Product/ System Complete	Product/ System Accepted	Project Recycling Plan	
Approvals	Sponsor/ Client	Sponsor/ Client	PM/ Sponsor	PM/ Project Directorate	Client Project Manager	Product/ Operations Manager	Operator	

a basis for search and/or development of alternative business solutions that may or may not involve a new project, and it enables risk and uncertainty evaluation be carried out from the outset in terms of LCOFs. (See the chapters by Artto and Dietrich, and by Jamieson and Morris.)

Some of the activities in this phase are (1) determination of organizational and decision making processes and structures at owner level, (2) conduct of high-level business/strategic consultations and deliberations to ensure that business/strategic needs are correctly determined and documented, (3) setting of initial targets for LCOFs, and (4) notional budget and designated time frame to fulfill the stated needs. Thus, the main purpose is to define both the needs and the required business hurdles (target LCOFs) that must be met in subsequent project development and implementation phases. The study and decisions outcomes are captured in a document typically referred to as a *project brief*.

A project brief may or may not contain several suggested options for responding to the stated needs, but it should really avoid giving a prescriptive solution or preempting the creative process that needs to be followed subsequently to locate and develop an optimum solution to relevant business/strategic needs. Generally speaking, a brief is the basis for conducting a systematic project definition studies and delineation/optimization of the project business case for the preferred solution. Many alternatives (including non-project options) should be considered, developed, and appraised to locate the optimum business solution.

FIGURE 7.9. BROAD REPRESENTATION OF PROJECT CREATION PHASE.

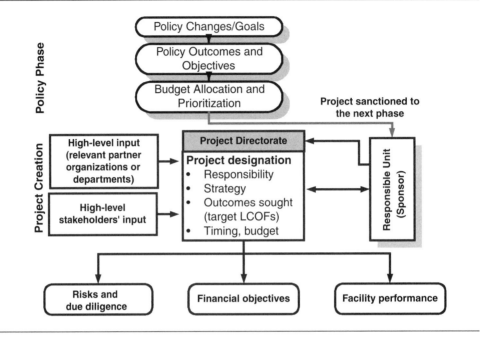

Some owners assume that the project brief is a sufficient basis for proceeding to the project implementation stage. This has to be resisted, as it may lead to suboptimal solutions or increased exposure to hidden risks and other traps (Miller and Lessard, 2001; Jaafari, 2001). The purpose of a project brief is to have enough information on the problem (business needs and requirements) so that a multi-disciplinary team of experts can be engaged to conduct project definition studies; locate, define, and refine an optimum solution; and evaluate its value proposition vis-à-vis target LCOFs. As a minimum, a project brief should contain the following key information:

- Strategic needs and business requirements
- Commercial opportunities
- Objectives (target LCOFs)
- Perceived constraints
- Funding options
- Value proposition in terms of client's business needs
- Nominated project organization and governance
- Anticipated project budget and duration
- Major resources required to realize the objectives and perceived risks
- Proposed plan to conduct project definition studies and establish the project business case

Project Development (Definition). Figure 7.10 shows a broad representation of this phase, whose aim is to locate, define, and refine an optimum solution with a firm business case vis-à-vis targets set for LCOFs. The project business case is the basis for subsequent project implementation (initiation, planning, execution, and commissioning) (Gray et al., 1985; Morris, 1983).

The project development phase is sometimes referred to as the *project definition phase*. This phase results in a project definition report, which is often the basis for project funding and approval. The project development phase is the most creative phase of the project life cycle. The project study team needs to explore all plausible options that can be thought of in order to satisfy target LCOFs and stated business needs.

Typically a project business case must provide the following key information:

- Executive summary
- Needs (of end users and customers) assessment and facility/product definition, priorities
- Target LCOFs to be achieved and means of assessment
- Background
- Stakeholders' requirements
- Options generated and evaluated, and selection of the preferred option
- Product description
- Financials
- Project organization and governance
- Commercial risks
- Delivery strategies

FIGURE 7.10. BROAD REPRESENTATION OF PROJECT DEVELOPMENT PHASE.

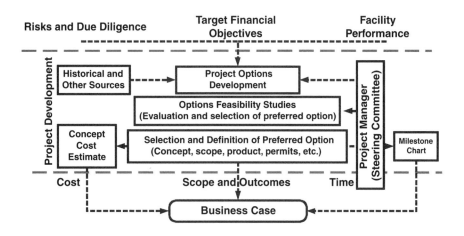

- Project Quality Management model
- Intellectual property and licensing issues
- Resources
- Statutory requirements and due diligence
- Operational issues
- Handover issues
- Time-to-market
- Knowledge management
- Community and stakeholders' interests
- Financing
- Teamwork
- Value creation opportunities

It must be noted that while Figure 7.10 shows a linear process for obtaining an optimum solution as the basis for project scope, in reality the whole process is recursive—that is, evaluation of potential options may necessitate reverting to the original assumptions, target LCOFs, and business needs or strategic goals. In the light of the knowledge gained from the first cycle of evaluation, it may be necessary to clarify or modify the original assumptions, hurdles, and business objectives, including targets set for LCOFs, and then start the process again to see if the solution will work. The test as to whether or not the solution is optimum is the extent to which target LCOFs can be met assuming a successful project completion. However, there may be secondary objectives or constraints that must also be met, and these are typically addressed in the business case statement.

As found by Miller and Lessard (2001), the front-end of large projects takes a considerable amount of time (typically ten years on the sample of 60 large engineering projects they studied). During this period, the project may be subject to several evaluations at different junctures with regard to new political developments, market shifts, stakeholders' shifting priorities, and so on. This period often provides an opportunity to develop creative solutions that may also make the project ultimately feasible and attractive to invest in. An example is the Northwest Shelf Liquefied Petroleum Gas (LPG) project, constructed in the 1980s in Australia. The original design proposed construction of a cooling tower requiring a large steady supply of fresh water that was not available locally. The project was delayed considerably, and in the meantime, the dry cooling technology developed further. The eventual solution was a dry cooling process that made the LPG project attractive to invest in. A major design change of this type will normally trigger a fresh round of project evaluation and replanning.

Project Implementation Phase

The project implementation phase can be thought of as the part of the project life cycle that starts after the project funding and approval and concludes with the successful handover of the end product to the client organization, including the contractual closeout of the project, lessons-learned documentation, and archiving of the project documents. Ideally an integrated project team oversees the entire implementation phase (Chaaya and Jaafari, 2001; Jaafari, 2000; Jolivet and Navarre, 1996).

One of the theses advanced by this chapter is that all project implementation decisions must ultimately satisfy the project's business needs and requirements. Implementation strategies and scenarios must be evaluated continuously using target LCOFs as the criteria. This is so even though, traditionally, scope, time, and cost management have been the focus of the project implementation phase. When the project environment is dynamic, it is necessary to regularly review the criteria against which decisions must be evaluated. In the case of a life cycle project management framework, the appropriate criteria are the targets set for LCOFs; these should be continuously reviewed and revised downward or upward as deemed appropriate to ensure alignment with the market, realism in terms of what is achievable, optimality in terms of balancing needs and requirements, and consistency across the project life cycle.

Why should decision criteria be reviewed and adjusted continuously? The reason is that unlike small to medium-size projects, which take a relatively short period of time to conceptualize and deliver, large, complex projects span many years and are subject to change (Jaafari and Schub, 1990; Morris and Hough, 1987). Thus, a chief function of the project management team (PMT) is to continuously monitor the project's underlying business appeal from an owner's and operator's overall perspective and implement changes to the target LCOFs as well as the project business case in response to shifts in the business, social, political, and regulatory environments.

It is ironic that few large projects are managed in such a dynamic and systematic fashion. There is still a belief that during the implementation phase, project managers should focus on the management of the delivery phase as generally characterized by the management of cost, time, scope, and quality. So it is not uncommon to read reports in the press on cost and time overruns experienced on public projects. Very little discussion centers on whether or not the project's business case has been enhanced because of positive changes introduced, notwithstanding cost and time overrun. The emphasis on value creation extends to contractors who can also search for a better outcome from their perspective purely for self-interest, such as adoption of an accelerated completion strategy that may increase the direct costs somewhat but lead to a significant reduction in the total indirect costs (Jaafari, 1996a, 1996b). Indirect costs are generally a function of project duration (Jaafari, 1996a, 1996b). In addition to self-interest, contractors will need to compete increasingly on the basis of their capabilities to deliver a strategic advantage to their client organizations through value creation opportunities that often come from the reengineering of projects early in their implementation phase. In this way, they can enter into partnership deals with major clients and share any potential gains.

The competency of the PMT is of paramount importance to the project implementation success. While assessment, acquisition, and enhancement of the team's competencies, and the delineation of competency gaps, are outside the scope of this chapter, it is important to note that key team competencies fundamentally determine the fate of the project. Note that the emphasis should be on *team* competencies not just individual competencies. The array of competencies required normally includes technical and commercial acumen, people skills, and project and organizational management, to mention a few. If the PMT lacks the necessary competencies to respond to the project challenges optimally, then it must set about to remedy its deficiencies in an appropriate and timely manner, such as acquisition of new staff, intensive training, hiring of competent consultants, and so on. One way to acquire this is to bring the range of expertise needed on board through an alliance mode of project delivery (Scott, B., 2001; Black et al., 2000; Halman and Braks, 1999).

Life Cycle Project Management

The life cycle project management (LCPM) approach for the entire implementation phase of large, complex projects is recommended. Traditionally, projects are packaged into multiple contracts, and each package is given out to a contractor or consultant to deliver. Each contractor sees his or her role as that of delivering the scope of the contract with minimal concerns about the impacts his or her work may have on the rest of the project. A contractor's main focus typically centers on achieving the target profit margin while capping or eliminating the corresponding risks and liabilities. Integration of works delivered by a multitude of contractors and consultants is a major challenge to the PMT/owner and is often prone to serious errors and omissions, delays, and cost overruns. Under the traditional mode of project delivery, the energy and intellect of the owner or his or her PMT are typically absorbed on the management of contracts and interfaces, and not necessarily spent on the attainment of the best overall project results (Halman and Braks, 1999).

LCPM addresses all of these shortcomings, as it shifts the focus of decision making and optimization from traditional scope, time and cost management in each contractual package to reaching or exceeding targets set for life cycle objective functions for the whole project. It also provides a firm basis for a more efficient management and integration of the entire implementation process. Life cycle project management is based on the following components:

- A culture of collaboration based on strategic partnership and unity of purpose (partnership for achieving or exceeding target LCOFs and sharing the resultant rewards)
- A life cycle philosophy and framework and an integrated single-phase approach for the entire implementation activities, covering initiation, planning and documentation, execution, commissioning, and finalization
- A concurrent teamwork approach, facilitated by a real-time communication system to cut the project delivery timescale
- A fully integrated project organization structure, run by a *project board* constituted from executives of the relevant project participating organisations
- An integrated project management information system that facilitates real-time evaluation of LCOFs as the basis for decision making

Fundamental to the success of the life cycle project management approach is the selection of the right partners (consultants and contractors) who can augment the owner and his or her PMT in terms of the missing competencies. Normally the project board has the ultimate decision-making oversight (strategic role) over the entire project implementation activities, including negotiation with the client body and relevant competent authorities. It appoints its own PMT with delegated authority to run the project on a day-to-day basis.

It must be noted that in recent times there has been a prevalent shift to partnering and alliance mode of project delivery. However, most alliance deals are still based on delivery objectives of time and cost. Also, integrated product development teams are generally missing in these arrangements. A full life cycle project management approach requires true open collaboration and working in terms of a true integrated product development teams.

Implementation of the life cycle project management approach also requires a project management information system (PMIS) that has real-time capability and can facilitate concurrent teamwork that underpins LCPM methodology. Such a system will assist the PMT to stage, run, and effectively integrate the contributions by the relevant consultants and designers organized in specific integrated (product development) teams. The PMT has a central and pivotal role in the evolution of the project, accumulation of parts, and integration of these into a whole viable project outcome. In this respect, the PMIS does not remove this responsibility from the project management team but facilitates this process by real-time analysis of the status of the LCOFs versus submissions of the suggested solutions received from teams. However, the PMIS can automate or expedite a number of tasks—for instance, sharing of input data, estimation of costs and LCOFs values, and production of numerous reports. It can also reduce redundant data entries, replace multiple pieces of software, and economize on human resources.

In summary, the ideal LCPM approach is based on the following key strategies:

- An integrated organization structure of the key partners on the project (owner, project manager, operator, contractors, designers, and major suppliers) whose executives make up the project board, responsible for major and strategic decisions and approvals, and appointing an integrated project management team headed by a competent project manager for the day-to-day management of the project.
- A continuous and integrated approach to the management of the implementation phase of the project as a whole (working back from the project business case).
- Simultaneous inclusion of the relevant information for decision making, including engineering, design, approval, manufacturing, construction, commissioning, operation, recycling, and so on (Jaafari, 1997; Laufer et al., 1996).
- Formation of integrated product development teams, each having a representative from the pertinent parties to the project, including, where appropriate, the appointed architect, engineers, manufacturers, constructor, operator. and facility manager. In some cases, representatives of governments and statutory authorities may also be considered.
- Division of the work into defined parts (product) and allocation of each to a single integrated product development team.
- Proactive management of the project and its parts, specifically, planning, staging, and managing all project activities continuously to maximize attainment of the target LCOFs.
- Integration of the work of the teams into a single project.
- Establishment of direct and real-time intra- and inter-team communication and document integration systems to facilitate the whole process.

As can be seen from Figure 7.11, the project board and the PMT need to drive the whole implementation phase in an integrated and systematic manner. They have the highest influence on the outcome of the implementation phase and, as such, must set up to discharge their responsibility objectively and optimally. If the LCPM approach is applied prudently and provided project participants possess appropriate competencies and do not behave opportunistically during the course of project implementation, it can be expected that optimal results will be achieved through the LCPM approach. The most important gains will be the enhanced project value and its fitness for purpose (in terms of meeting the underlying business objectives), as the entire project organisation will work in harmony to achieve breakthrough solutions that can meet or exceed targets set for LCOFs. This is because many non-essential (duplicate) activities, typical in multiple phase delivery, are eliminated, information and decisions are integrated and optimized against LCOFs, and commercial objectives of the partners are aligned with those of project objectives. This state of tightened

FIGURE 7.11. BROAD REPRESENTATION OF PROJECT IMPLEMENTATION.

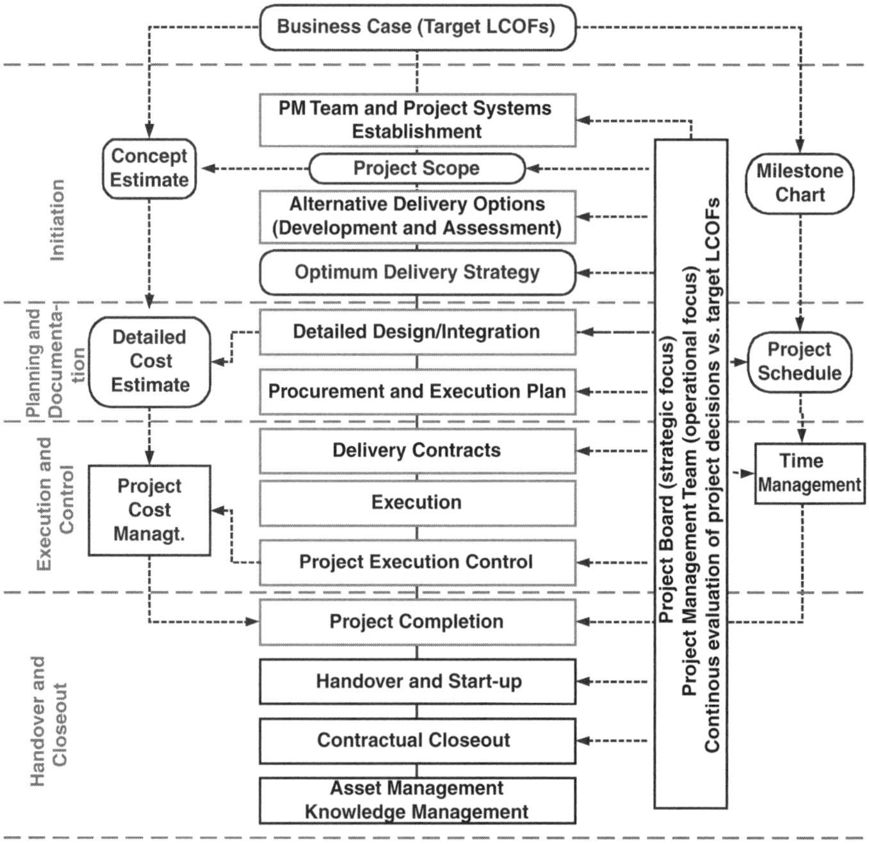

collaboration is often referred to as *cocreation* because of its emphasis on value creation and waste elimination.

Once the owner selects the relevant contractors and consultants, the owner should take them on board as project partners and tie their fortunes on the project to the attainment of target LCOFs through an appropriate gain-share/pain-share scheme. Note that each party comes to the project with a different mind-set and from a different corporate culture; so it is necessary to forge the parties into a unified project organization and develop the project culture as the dominant culture. The parties are then formed into multidiscipline teams with tough targets to achieve in terms of LCOFs (known as *stretched targets*). An example is asking a team to deliver a solution for a power station cooling system that not only meets relevant environmental and permit requirements but is also more efficient in terms of total life cycle cost per unit load.

It should be noted that despite the welcome trend from hard-dollar contracting to relationship-based contracting in the construction industry, there is still some distance to go to set up true alliances based on life cycle objectives and true integrated product development teams, rewarded on the basis of the life performance of the project. Most alliances are based on cost and schedule performance of the project and following a functional project organization structure. For example, on Wandoo B-Platform Oil Project, completed in 1997 in the Western Australia's Northwest Shelf, an alliance scheme based on total project cost was devised and applied, as seen in Figure 7.12. However, in order to encourage solutions that were cost-effective in terms of life performance of the project, this scheme contained a

FIGURE 7.12. THE WANDOO ALLIANCE PAIN-SHARE/GAIN-SHARE SCHEME.

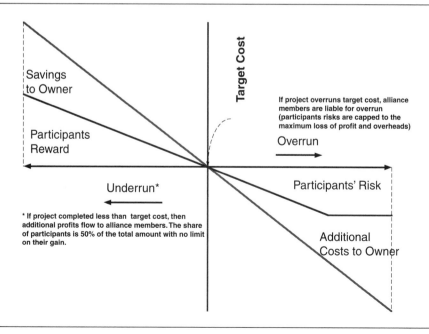

bonus for the alliance parties payable during the operation phase, provided that the financial performance of the project exceeded certain thresholds.

Figure 7.13 shows typical alliance organizational arrangements used on most large projects, which were also applied to the Wandoo project. As seen, this is far from a true integrated product development approach needed to foster creativity and development of break-through solutions in terms of true life performance of the project. For more information on alliance and relationship contracting, see Scott, B., 2001; Walker et al., 2000; Black et al. 2000; and Halman and Braks, 1999.

Integrated product development teams can then go through a creative process to come up with optimal solutions that can be evaluated holistically at project level, using LCOFs as the basis. To achieve breakthrough solutions, one needs to consider all stages—in other words, not only the operation phase but also the implementation phase. Together, these comprise initiation, planning and documentation, execution and control, and commissioning and handover. The recycling of the end facilities at the end of the project's useful life should also be considered.

Project Initiation

Traditionally, projects go through distinct phases; for example, the PMT typically conducts the initiation phase; goes out to procure the services of consultants for project design, planning, and documentation; and then goes out to tender to select and appoint contractors to deliver the project in the manner foreshadowed in the contract documents (Jaafari, 1997).

The LCPM model is somewhat different. Project initiation is conducted centrally by the project board and coordinated by the PMT. It is the project board that makes strategic decisions on the implementation phase as a whole, with the input coming from all partners. For planning, documentation, and execution (even pre-installation commissioning), the preference is to engage dedicated teams made up of the participating organizations (see Figure 7.14). Each will be required to come up with its own breakthrough solution in a creative manner to meet simultaneously the global criteria of LCOFs and the specific criteria set for the part under consideration. The project board together with the PMT have the responsibility to preside over the evolution of the project as a whole, including integration of all solutions forwarded by teams and oversight of project commissioning and finalization activities.

Decisions made by the project board and the PMT at the initiation phase have a profound impact on the subsequent success or failure of the whole project implementation process. The main purpose of project initiation is to determine an optimum strategy that will achieve the intent of the business case of the project. This stage also considers the policy and regulatory issues, risks, and due diligence. Successful project initiation will require the generation of multiple options for the realization of project business case, evaluation of these, and selection of an optimum project implementation strategy. Note that while its chief function is to locate or develop an optimum solution to realize the intent of the business case, project initiation activities may lead to further adjustment of the business case as part

FIGURE 7.13. TYPICAL ALLIANCE ORGANIZATION STRUCTURE.

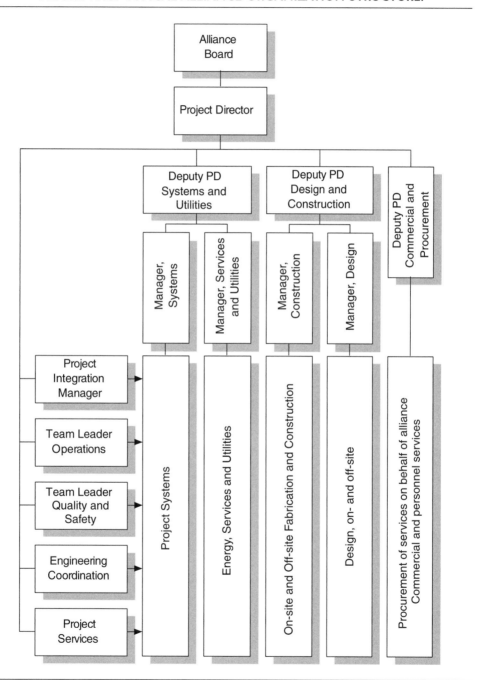

FIGURE 7.14. LCPM PROCESS FOR DESIGN AND DOCUMENTATION PHASE.

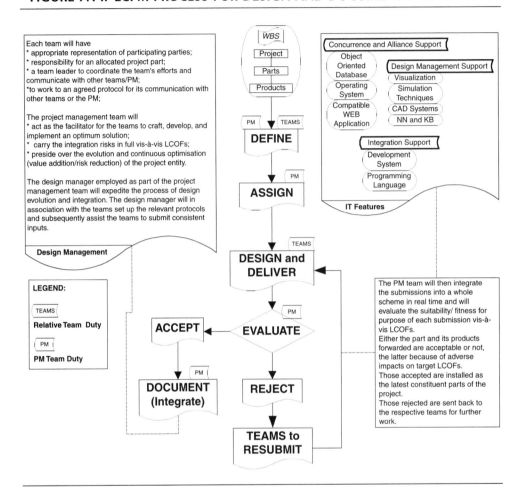

of a dynamic process and because of learning that comes with the first cycle of planning. This will lead to the identification of critical determinants and their influence on the project concept. However, the temptation to totally modify the project business case should be avoided unless changed circumstances demand a total revisit of the business case.

For example, if there has been a significant delay between the conclusion of project definition studies and the start of the implementation phase, it may be necessary to reconfirm the project business case first before proceeding to project initiation. In particular, target values set for LCOFs need careful review to ensure that these are still achievable and the business climate has not experienced significant shifts since the conclusion of project definition studies. Other situations that can give rise to a changed business case are as follows:

- Product changes or changed business priorities
- Changed financial model and/or funding model
- New commercial risks
- Changed legislation and permit conditions
- Constraints because of the intellectual property and technology licensing matters
- Shortage of resources because of other new developments under way or expected
- Stricter due diligence
- Uncovered new handover and operational issues
- New community and stakeholders' demands

Project initiation starts with the establishment of project office, setting up of the relevant project development and management systems, hiring, and formation of the project organization structure, project board, and project management team. The board and the PMT need to thoroughly review, confirm, and internalize the business case requirements and then through a creative process generate options for delivering these. A multicriteria approach may be necessary to evaluate the available options. However, the three main determinants are financial attractiveness (not necessarily least capital expenditure but highest total gain or lowest total life cycle cost over the entire project life cycle; Woodward, 1997), quality and/or fitness for purpose, performance, and likely impacts because of pertinent risks and uncertainties. The outcome is a preferred (optimum) implementation strategy that shows maximum potential and is least risky. As has been emphasised, risk and uncertainty management will drive the whole process.

Project initiation phase typically results in the following outcomes:

1. Adoption of a set of policies and standards that will meet the project needs and requirements cost-effectively (e.g., standards for design, quality, risk, and procurement management)
2. Specification of key performance indicators (from the target LCOFs), framework, and criteria for evaluation of project implementation decisions
3. Details of the preferred project delivery method, including division into parts, criteria or targets for delivery and acceptance, integration strategy, and associated managerial structures
4. Assignment of each part to a team with an appropriate set of protocols for teamwork
5. Design of a system for life cycle integration of all decisions, information, and functions on the project, including systems for project quality management, communication management, and procurement management
6. Methods/strategies for the evaluation and mitigation of risks and uncertainties at both global and local levels
7. Articulation of these in a project initiation report that can be a main source document for subsequent planning, documentation, and management of the project

Project Planning and Documentation

As stated, project planning and documentation is generally conducted by teams and facilitated by the PMT/project board in response to the stretched targets set during the project initiation stage in a concurrent manner. Figure 7.14 shows the process used under the LCPM approach, viz:

1. All teams will convene separately; each team will generate and develop its own alternative solutions, carry out some preliminary investigation, and come up with a preferred solution.
2. These preferred solutions are assembled by the relevant design discipline and analyzed to ensure the system's integrity (at discipline level) and conformity with the relevant statutory requirements.
3. The results from discipline investigation/detailing are fed back to teams to integrate the same into their solutions in the form of products and submit their products to the PMT, including the relevant information on life cycle aspects.
4. The project manager evaluates the LCOFs, installing those products that meet all the LCOFs and returning the remainder to the relevant team for further consideration.
5. The preceding steps are repeated in a dynamic process until a baseline design is evolved. The baseline design is the basis for the documentation of the whole project and procurement of the products/parts and monitoring of the execution process.
6. During the life of a project, the preceding process is maintained and managed continuously, as the responsibility for the eventual realization of the LCOFs will remain collectively with the project board, respective teams, and the PMT right up to the time of facility operation and beyond (when operation and maintenance are also part of the scope).

Project Execution and Control

The LCPM model applies the principles of integrated teamwork to the execution phase and on to commissioning and handover. The following general strategies expedite the application of LCPM methodology to the execution and control phase:

1. Apply the decision-making and risk management methods that the project board/PMT has decided upon for the execution phase of the project; this means that the intention of the execution phase is to apply the plans and strategies that the partners have collectively developed in the preceding phases in a united manner with the reporting and responsibility allocation unchanged throughout the implementation phase.
2. Develop a proactive management philosophy. The execution and control framework must be designed to measure the effectiveness of plans and decisions throughout the project life cycle by evaluating their impacts on the LCOFs. As an illustration, the potential benefit associated with shortening of the execution timescale must be evaluated through the impact it has on the target LCOFs.
3. Use a continuous objective-focused approach for major parts and the project as a whole. Such an approach embodies maximum flexibility in terms of innovation in the underlying concept, timing, resources, and other factors.

4. Focus should be on problem anticipation and resolution. Tap opportunities and monitor the status of risks continuously. In the event of a risk materializing, put in place recovery plans and minimize the impacts on the project outcome (LCOFs).

5. Employ an integrated information management system to assess progress and to provide feedback on the performance of the execution phase for each part and/or the project as a whole using appropriate key indicators but generally using the target LCOFs for project performance and project control functions, such as time and cost and/or earned value for project progress monitoring and reporting.

6. Use a dynamic scheduling and real-time reporting system so as to facilitate the management of anticipated changes. Maintaining maximum flexibility and attempting to add value to the project base value as a whole throughout its life cycle may mean many significant changes during the execution phase. However, most changes in this phase are due to imperfect design information, changed product specification, and so on.

7. Source and apply knowledge relevant to the optimal execution of the parts and the project as a whole.

Project Handover and Closeout

In LCPM the handover and closeout phase is generally planned both strategically and in detail during the project initiation and planning phases. The activities associated with project handover and contractual closeout should be planned as an essential part of an integrated approach to the entire project implementation phase. This means determining early in the initiation phase what strategies need to be put in place for project handover and contractual closeout. In the planning stage, relevant activities to apply the preceding strategies are planned, and these are applied in parallel to the execution phase. The major emphasis in this phase is the project end result or facility performance verification, OH&S and environmental compliance, and due diligence.

Typical activities in this phase include operational strategies, operator training, handover, start-up, commissioning and testing, defects identification and rectification, performance demonstration and validation, operational manuals, parts catalogue, as-built drawings, maintenance of project documents and records including materials and manufacturing records, contractual closeout (i.e., between the project alliance board and the client and between the alliance and the parties forming the same), financial settlement, asset management, and knowledge feedback.

Under the LCPM approach, project handover and closeout can be conceived as comprising two distinct phases: hand over of parts and hand over of the entire project. As noted, each major part will have been entrusted to a team who will deliver the same to the project board and the PMT and through them to the client. Where relevant, each part can be precommissioned and tested, and after meeting all the required performance hurdles, it can be certified as meeting the relevant performance criteria. The project board and the PMT will need to develop a clearly articulated and systematic acceptance scheme that spells out how parts are to be tested and accepted. There is no doubt that eventual facility performance criteria must underline the approach. Pretesting and acceptance of parts saves valuable time and minimizes the incidence of errors and last-minute discovery of major defects; it is also easier to implement. Once relevant parts are delivered, the entire facility can be commissioned and tested for performance verification and certification. Normally, a team of commissioning experts will take over this task, which may take anything from one to three months.

Configuration and asset management practices need to be addressed from the outset and pursued to completion in this phase. The project solution will have embodied significant technical and managerial innovations; these must be captured systematically for future reference. Knowledge thus created should be recognized as a valuable asset in its own right and managed accordingly.

Summary

Creation, structuring, optimization, and implementation of large, complex projects require a systematic approach within a whole of life perspective. Large projects touch the lives of communities, require considerable investment and execution resources, are subject to regulatory and political pressures, and generally involve multiple stakeholders, with different needs and aspirations.

The author has focused on developing an integrated approach to whole of life planning, evaluation, and implementation of these projects. The result is presented as life cycle project management (LCPM) philosophy and framework in contrast with many contemporary project management approaches.

References

Adams, J. D., and A. W. Brown. 2002. Does project management add value to public sector construction projects: A critical perspective. *Proceedings of the International Conference on Project Management—ProMAC 2002.* 117–124. Singapore, July 31–August 2.

Artto, K. A., J. M. Lehtonen, and J. Saranen, 2001. Managing projects front-end: incorporating a strategic early view to project management with simulation. *International Journal of Project Management* 19:255–264.

Beder, S. 1999. Controversy and closure: Sydney's beaches in crisis. *Social Studies of Science* 21 (May 1991): 223–256.

Black, C., A. Akintoye, and E. Fitzgerald. 2000. An analysis of success factors and benefits in construction. *International Journal of Project Management* 18:423–434.

Brook, P. 2000. Project management and systems engineering: An evolving partnership. *15th World Congress on Project Management Organised by the Association for Project Management and the International Project Management Association.* London, May 22–25.

Byers, M. P. and F. L. Williams. 2000. Transforming electric utility project management for the new millennium business environment. *15th World Congress on Project Management organised by the Association for Project Management and the International Project Management Association.* London, May 22–25.

Chaaya, M., and A. Jaafari. 2001. Cognisance of visual design management in life cycle project management. *Journal of Management in Engineering* 127(1):66–75.

Defence Evaluation and Research Agency. 1997. *DERA systems engineering practices model.* Farnborough, UK: Defence Evaluation and Research Agency, GU14 6TD.

Dixon, M., ed. 2000. *APM Body of Knowledge.* 4th ed. High Wycombe, UK: Association for Project Management (www.apm.org.uk).

Gray, K. G., A. Jaafari, and R. J. Wheen, eds. 1985. *Macroprojects: Strategy, planning, implementation.* p. 146. Sydney: The Warren Centre for Advanced Engineering, The University of Sydney.

Halman, J. I. M., and B. F. M. Braks, 1999. Project alliancing in the offshore industry. *International Journal of Project Management* 17 (2, April): 71–76.

Hobbs, B., and R. Miller. 1998. The international research programme on the management of engineering and construction projects. Vol. 1. 302–310. *14th World Congress on Project Management,* Ljubljana, Slovenia, June 10–13.

Jaafari, A., and H. K. Doloi. 2002. A simulation model for life cycle project management. *Journal of Computer-Aided Civil and Infrastructure Engineering* 17:162–174.

———. 2001. Management of risks, uncertainties and opportunities on projects: Time for a fundamental shift. *International Journal of Project Management* 19:89–101.

———. 2000. Life cycle project management: A new paradigm for development and implementation of capital projects. *Project Management Journal* 31(1):44–53.

———. 1998. Perspectives on risks specific to large projects. Keynote paper p. 20 presented to National Infrastructure Strategy 98. August 20–21, 1998, The Institution of Engineers, Australia.

———. 1997. Concurrent construction and life cycle project management. *Journal of Construction Engineering and Management* 123 (4, December): 427–436.

———. 1996a. Twinning time and cost in incentive-based contracts. *Journal of Management in Engineering* 12 (4, July/August): 62–72.

———. 1996b. Time and priority allocation scheduling technique for projects. *International Journal of Project Management* 14 (5, October): 289–299.

Jaafari, A., and K. K. Manivong. 2000. Synthesis of a model for life cycle project management. *Journal of Computer-Aided Civil and Infrastructure Engineering* 15(6):26–38.

Jaafari, A., and K. K. Manivong. 1998. Towards smart project management information systems. *International Journal of Project Management* 16:249–265.

Jaafari, A., and A. Schub. 1990. Surviving failures: The lessons from a field study. *Journal of Construction Engineering and Management* 116 (1, March): 68–86.

Jolivent, F., and C. Navarre. 1996. Large-scale projects, self organising and meta-rules: Towards new forms of management. *International Journal of Project Management* 14(5):265–271.

Kelley. 1982.

Laufer, A., G. R. Denker, and A. J. Shenhar. 1996. Simultaneous management: The key to excellence in capital projects. *International Journal of Project Management* 14 (4, August): 189–199.

Manivong, K. K., A. Jaafari, and D. Gunaratnam. 2002. Games people play with proactive management of soft issues on capital projects. *Proceedings of the International Conference on Project Management, ProMAC,* 141–146. Singapore, July 31–August 2.

McCarthy, S. C. 1991. BOT and OMT contracts for infrastructure projects in developing countries. PhD diss., University of Birmingham.

Merna, T. 2002. Value management. Chap. 2 in *Engineering Project Management, ed.* N. J. Smith. 2nd ed. 16–28. Oxford, UK: Blackwell Science.

Merna, A., and N. J. Smith. 1993. Guide to the preparation and evaluation of build-own-operate-transfer (BOOT) project tenders. Manchester, UK: University of Manchester Institute of Science and Technology.

Morris, P. W. G. 1983. Managing project interfaces: Key points for project success. In *Project management handbook*. D. I. Cleland and W. R. King. New York: Van Nostrand Reinhold.

Morris, P. W. G, and G. H. Hough. 1987. *The anatomy of major projects*. New York: Wiley.

PMCC. 2001. *A guidebook of project and program management for enterprise innovation*. www.enaa.or.jp/PMCC/.

Scott, B. 2001. *Partnering in Europe*. London: Thomas Telford/European Construction Institute.

Scott, W. R. 1992. *Organizations: Rational, natural, and open systems*. 3rd ed. Upper Saddle River, NJ: Prentice Hall.

Stupples, D. W. 2000. Using system dynamics modelling to understand and address the systemic issues on complex engineering projects. *15th World Congress on Project Management Organised by the Association for Project Management and the International Project Management Association*, London, May 22–25.

Walker, D. H. T., and K. D. Hampson, eds. 2002. *Procurement strategies: A relationship based approach*. Oxford, UK: Blackwell Science.

Walker, D. H. T., K. D. Hampson, and R. Peters, R. 2000. *Relationship-based procurement strategies for 21st century*. Canberra, Australia: AusInfo.

Wang, S. Q., and L. K. Tiong. 2000. Case study of government initiatives for PRC's BOT power plant project. *International Journal of Project Management* 18 (1, February): 69–78.

Woodward, D. G. 1997. Life cycle costing:- Theory, information acquisition and application. *International Journal of Project Management* 15 (6, December): 335–344.

Yeo, K. T. 1995. Planning and learning in major infrastructure development: systems perspectives. *International Journal of Project Management*. 13(5):287–293.

Yeo, K. T., and R. L. K. Tiong. 2000. Positive management of differences for risk reduction in BOT projects. *International Journal of Project Management* 18:257–265.

CHAPTER EIGHT

HOW PROJECTS DIFFER, AND WHAT TO DO ABOUT IT

Aaron J. Shenhar, Dov Dvir

Much of the project management literature and training treats projects as universal; assuming one set of techniques and tools applies to all situations. In reality, however, projects differ in many ways, and "one size does not fit all." Many writers have suggested in recent years that the time has come to develop a standard framework in project management that would help managers, researchers, and teachers distinguish among projects, identify appropriate management styles for different types of projects, and direct the further development of project-specific techniques and tools. This chapter summarizes ten years of research on the differences among projects and project management styles. It suggests three frameworks for distinguishing among projects, and in each framework several dimensions for project classification. Each framework can be used by different managerial levels and for different purposes. And within each framework we provide guidelines for management on how to treat different projects in different ways.

Introduction

Ask any project manager about his or her project, and the project manager will tell you why it is unique, and how the project team must adapt their style to their specific challenge and problems. Indeed, no two projects are alike. As any experienced manager will tell you, you must adapt yourself to the situation, the circumstances, the environment, and the people—and you should not try adapting the environment to you.

As a formal managerial discipline, project management is still relatively young. It started only in the middle of the twentieth century, when the first PERT charts marked the beginning of a new discipline (Morris, 1997). Over the years, however, very little attention has been given to the differences between projects, and there is still no universal framework for distinguishing among them. According to most books, all projects are the same, and most project management training still uses an approach of "a project is a project is a project."

We have seen this problem in our ongoing study of project management around the world. Over the years we have studied more than 600 projects and conducted hundreds of interviews. We have come to realize that project success depends greatly on the proper project management style and on adapting the right style to the right project. We have seen projects fail because managers assumed that their current project would be the same as their previous one. And we realized there is a need for a framework that will help managers look at a project by first assessing the project type and then selecting an appropriate management style. After a decade of research, we propose here such a framework and show how to make it work for various project settings and environments.

As you will see later, different managerial levels will use different frameworks to distinguish among projects and for different needs. Therefore, we will present not one, but three frameworks. However, there is an overarching concept to all. It is described in the next theoretical section, which will be followed by the outline of the frameworks and what each framework means for project management.

The Theory of Contingency

Classical contingency theory asserts that different external conditions might require different organizational characteristics, and that the effectiveness of any organization is contingent upon the amount of congruence or goodness of fit between structural and environmental variables (Lawrence and Lorsch, 1967; Drazin and van de Ven, 1985; Pennings, 1992). Burns and Stalker (1961) were the first to introduce the concept of contingency to organizational theory. They presented what is now accepted as the traditional distinction between incremental and radical innovation, and between organic and mechanistic organizations. Organic organizations would better cope with uncertain and complex environments, while mechanistic organizations predominate in simple, stable, and more certain environments.

The project management literature has often ignored the importance of project contingencies, assuming that all projects share a universal set of managerial characteristics (Pinto and Covin, 1989; Shenhar, 1993; Yap and Souder, 1994). Yet, projects can be seen as "temporary organizations *within* organizations" and may exhibit variations in structure when compared to their mother organizations. Indeed, various authors have often expressed disappointment from the universal, one-size-fits-all idea and recommended a more contingent approach to the study of projects (Ahituv and Neumann, 1984; Pearson, 1990; Wheelwright and Clark, 1992; Yap and Souder, 1994; Balachandra and Friar, 1997; Brown and Eisenhardt, 1997; Eisenhardt and Tabrizi, 1995; Song, Souder, and Dyer, 1997; Souder and Song, 1997).

Although contingency studies have had only a limited impact on the literature of project management, some exceptions exist. Most have focused on the impact of uncertainty and change on the way organizations are conducting their project operations. For example,

Wheelwright and Clark (1992) have mapped in-house product development projects according to the degree of change in product portfolio. Some have adapted the radical versus incremental distinction (e.g., Yap and Souder, 1994; Eisenhardt and Tabrizi, 1995; Brown and Eisenhardt, 1997; Song et al., 1997; Souder and Song, 1997), while others suggested more refined frameworks (e.g., Steele, 1975; Ahituv and Neumann, 1984; Cash et al., 1988; Pearson, 1990).

The second dimension of focus is complexity. To deal with complexity, the hierarchical nature of systems and their subsystems has long been the cornerstone of general systems theory (Boulding, 1956; Van Gigch, 1978; Rechtin, 1991), and hierarchies in products are almost always addressed in practitioners' books and monographs, which deal with engineering design problems (e.g., Pahl and Beitz, 1984; Lewis and Samuel, 1989; Rechtin, 1991).

Finally, while classical theory was focused on sustaining organizations, projects as temporary organizations must be studied in the context of time and the constraints they put on project management. Given the high velocity with which decisions are made and the shortened life cycles of products and markets, time and urgency become central factors in any modern look at the organization (Eisenhardt, 1989; Eisenhardt and Tabrizi, 1995; Brown and Eisenhardt, 1997). Based on these observations, our conceptual model is discussed in the next section.

The Basic Frameworks: The UCP Model

Regardless of the industry or technology involved, our research identified three dimensions to distinguish among projects: uncertainty, complexity, and pace (Shenhar and Bonen, 1997; Dvir et al., 1998; Shenhar, 2001a). Together, we call them the UCP Model, and they form a context-free framework for selecting the proper management style (see Figure 8.1). As you will see later, almost all frameworks are based on at least one of these three dimensions.

FIGURE 8.1. THE UCP MODEL.

Uncertainty —
At the moment of
project initiation

Complexity —
Size,
number of elements,
variety,
interconnectedness

Pace —
Available time frame

- *Uncertainty.* Different projects present, at the outset, different levels of uncertainty, and project execution can be seen as a process that is aimed at uncertainty reduction. Uncertainty determines, among other things, the length and timing of front-end activities, how well and how fast one can define and finalize product requirements and design, the degree of detail and extent of planning accuracy, and the level of contingency resources (time buffer and budget reserve). Uncertainties could be external or internal, depending on the environment and on the specific task.
- *Complexity.* Project complexity depends on product scope, number and variety of elements, and the interconnection among them. But it also depends on the complexity of the organization and the connections among its parties. Complexity will determine the organization and the process, as well as the formality with which the project will be managed.
- *Pace.* The third dimension for distinction among projects involves the urgency and criticality of time goals. The same goal with different time constraints may require different project structures and different management attention.

To select the appropriate management style, managers could follow a three-step process: First assess the environment, the product, and the task; second, classify a project by the levels of uncertainty, complexity, and pace; and third, select the right style to fit the specific project type (see Figure 8.2).

Here are some major factors that would impact each domain:

- *Environment.* The external environment includes the market, the industry, the customers, and the competitors. It may also involve the economical environment, as well as the political and the geographical environment where the project or task is being performed. But the environment is also the internal environment of the organization—comprising the organizational culture, the people, the procedures, the way projects are typically being managed, and available resources.
- *Product.* What is the exact product that this project is related to? What does the product do? How does it do it? What are the product operational requirements, and what are its specifications?

FIGURE 8.2. SELECTING PROJECT MANAGEMENT STYLE.

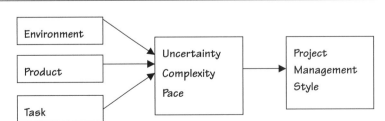

- *Task.* This refers to the exact work that needs to be done and the deliverables of the project. How difficult is this task? How well known is it? Have similar tasks been done before? How complex is it and how much time is available?

Mapping the Frameworks for Project Distinction

Since different managerial levels have different needs, we will use different frameworks for distinction among projects, and each framework will have several dimensions for classifying projects (see Table 8.1). The goal is to identify specific managerial activities, decisions, and styles that are best appropriate for each level and each project type (Shenhar, 2001b; Shenhar et al., 2002).

Although these frameworks proved the most useful, in no way are they unique. Some organizations may find it useful to develop their own frameworks that will fit their specific needs.

Strategic Portfolio Classification

This framework is based on the need to select projects in accordance with their strategic impact and to form a policy for project selection, since many projects compete for the same resources. To make the selection as rational as possible, we suggest (1) dividing projects into groups, based on their strategic impact, (2) allocating resources to each group based on the company or business strategic direction, and (3) selecting the individual projects in each group according to a set of criteria that was created in advance and carefully discussed.

We identified two dimensions to divide projects: the strategic goal dimension, which includes operational and strategic projects, and the customer dimension, which involves

TABLE 8.1. SUMMARY OF FRAMEWORKS TO DISTINGUISH AMONG PROJECT TYPES.

Framework	Major Users	Dimensions	Use
Strategic Portfolio Classification	Top management	• Strategic goal • User	Portfolio management
NCTP	Project managers	• Novelty • Complexity • Technology • Pace	Selecting project management style—leader, team, structure, processes
Work Package	Project teams Subcontractors	• Product type • Work Type	Assessing risk and time to completion of work packages

TABLE 8.2. STRATEGIC PORTFOLIO CLASSIFICATION.

	Operational	Strategic
External	• Product improvement	• New product development
Internal	• Maintenance • Improvement • Problem solving	• Utility and infrastructure • Research

external and internal customers (Shenhar et al., 2002). This results in four major groups of projects (see Table 8.2):

- *Operational projects* deal with existing businesses. They involve improvements in products, line extension, and cash cow projects, to gain more revenue from existing businesses.
- *Strategic projects* relate to new business. These are prime efforts that are made to create or sustain strategic positions in markets and businesses. Typically, strategic projects are initiated with a long-term perspective in mind. Examples are major automobile models introduction, new aircrafts such as Boeing 777, and so on.
- *External projects* are made for external customers, contracted or noncontracted.
- *Internal customers* are done within the organization, for internal departments or units.

FIGURE 8.3. THE NCTP FRAMEWORK.

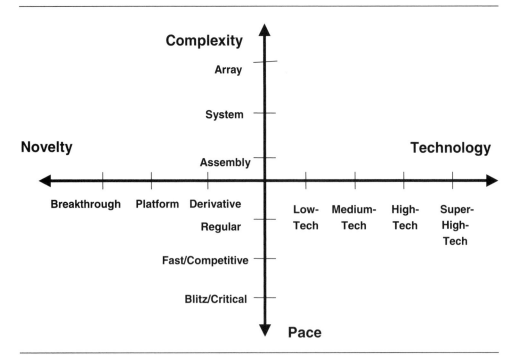

The policy for resource allocation among the groups is based on the expected impact from each group on the business, current strategy, infrastructure needs, and so on. The criteria for individual project selection are based among other things on risk and opportunity, difficulty, and available resources and skills.

The NCTP Model: Novelty, Complexity, Technology, Pace

The NCTP Model is the central framework that has evolved from our studies for distinction among projects (Shenhar and Dvir, 1996; Shenhar, 1998; Shenhar, 2001a; Shenhar, 2001b). It can guide project managers in selecting their project management style during project initiation, recruiting team members, determining structure and processes, and choosing the right tools. This framework involves four dimensions: novelty, complexity, technology, and pace. Each dimension includes at least three different project types, as shown in Figure 8.3.

Product Novelty

The first dimension is product novelty. It is defined by how new the product is to its potential users. It represents the extent to which customers are familiar with this kind of product, the way they will use it, and its benefits. Product novelty will have an impact on project management, especially on product definition and market-related activities. It affects the accuracy and confidence with which one can define the end product—the "what" needs to be done. Different levels of product novelty will determine how accurate market research can be, how well the product can be defined upfront, and what kind of different marketing techniques will be required. When Sony considered introducing its first Walkman, the product presented a new concept to the market, and market research proved inconclusive. The decision to launch this most successful product was based on managerial intuition.

Wheelwright and Clark suggested three major new product categories to manage the company's product portfolio and to create an aggregate project plan: derivatives, platforms, and breakthroughs (Wheelwright and Clark, 1992). These categories will determine different management behavior regarding the marketing of new products and their impact on project management:

Derivative products are extensions and improvements of existing products. Projects that produce derivative products include cost reduction, product improvement, product modifications, and additions to existing lines. Previous products are well established in the marketplace, and market data is readily available. Predictions about product cost, as well as other product requirements, can be fairly accurate, and product requirements should be fixed as early as possible. Finally, marketing of derivatives is focused on product advantage in comparison to previous models, trying to serve existing customers, as well as potential new customers, with added product features and varieties.

Platform products are new generations in existing product families. Projects for platform products typically create new families of products to form the basis for numerous derivatives. Such products replace previous products in a well-established market sector. New combat aircraft designs or new automobile models of are typical platform examples. For these projects, companies should perform extensive market research, study data of previous generations, and make careful planning of product price. The final setting of product requirements will therefore take much longer, and it should be planned well into the project execution period. However, it should not be delayed too long to ensure timely product introduction and reasonable profitability. Marketing efforts for platforms should be focused on creating the product's image, emphasize product advantages, and differentiate it from its competitors.

Breakthrough products are new-to-the-world products. Breakthrough projects introduce a new concept or a new idea, or a new use of a product, that customers have never seen before. The first Post-It notes or the first Walkman are just two examples; so is the first personal computer. Breakthrough products may use new or mature technologies, but in each case the market does not know anything about the new product, nor do customers know how to use it. Market studies are therefore ineffective and product definition must be based on guessing, intuition, and market trial and error. Requirements must remain flexible until first market introduction is possible and until customer feedback is available. Fast prototyping is necessary and is much more critical than extensive market research. Changes in initial product specifications are inevitable, and while clear product definition is perceived to be critical to any project, management of breakthrough projects must realize that product definition will most likely change in time after initial market trials. Marketing of break-through products is completely different from marketing of previous types. It is focused on getting the attention of the "early adopter" customers. Its goal is to educate customers about the potential of the new product and often articulate hidden customer needs.

Table 8.3 summarizes the definitions of the three levels of product novelty and the major impact they will have on product definition and market-related activities.

Technological Uncertainty

The major source of task uncertainty is technological uncertainty. (Other types may involve team experience or tight budget constraints). Higher technological uncertainty at the time of project initiation requires longer development phases, more design cycles, more testing, and later design freeze. In general, we associate such uncertainty with the degree of using new (to the company) versus mature, or well-known, technology within the product or process produced. We found four distinct levels of technological uncertainty associated with different project categories: low-tech projects, medium-tech projects, high-tech projects, and super-high-tech projects (Shenhar, 1991; Shenhar and Dvir, 1996; Dvir et al., 1998; Shenhar, 1998; Shenhar, 2001a).

Type A: Low-Tech Projects. These projects rely on existing and well-established technologies. The most typical examples of such projects are certain types of construction or "build to print" efforts, namely, rebuilding an existing product. Projects in this category require no development work; their architecture, design, and resource planning are all completed prior to the project's implementation phase. In such projects, the product is entirely shaped, and the design frozen, prior to the project's formal inception. Management style in such projects must be firm, since profit margins are typically slim. Project managers should therefore stick to the initial plan, with a *no-nonsense, no-changes, get-the-job-done* attitude.

TABLE 8.3. PRODUCT NOVELTY LEVELS AND THEIR MAJOR IMPACT ON PROJECT MANAGEMENT.

Product Novelty	Derivative	Platform	Breakthrough
Definition	An extension or improvement of an existing product	A new generation in an existing product family.	A new-to-the-world product.
Data on Market	Accurate market data exists	Need extensive market research. Careful analysis of previous generations, cocmpetitors, and markets	Nonreliable market data. Market needs not clear. No experience with similar products
Product Definition	Clear understanding of required cost, functionality, features, etc. Early freeze of product requirements	Invest extensively in product definition. Involve potential customers in process. Freeze requirements later, usually at mid-project.	Product definition based on intuition and trial and error. Fast prototyping is necessary to obtain market feedback. Very late freeze of requirements
Marketing	Emphasize product advantage in comparison to previous model. Focus on existing as well as gaining new customers based on added product features and varieties.	Create product image. Emphasize product advantages. Differentiate from competitors.	Creating customer attention. Educating customers about potential of product. Articulate hidden customer needs. Extensive effort to create the standards.

Type B: Medium-Tech Projects. These use mainly existing or base technology, yet incorporate some new technology or a new feature that did not exist in previous products. Examples include improvements and modifications of existing products (derivatives), but also new generations of products in industries where technology is relatively stable, such as appliances, automobile, or heavy equipment. Although most of the technologies are not new, the project will involve some development and testing. Changes to the design, however, should be confined to a limited period, and the design should be frozen quite early, typically after two cycles. While still firm, management style could be more flexible at the initial phase of the project, since some changes and adjustment in design are needed. Later changes should only be allowed if absolutely necessary for success or for safety reasons. The policy and attitude could be "limit changes to minimum and freeze as early as possible."

Type C: High-Tech Projects. These projects represent situations in which most of the technologies employed are new but nevertheless exist when the project is initiated. Such technologies had been developed prior to project inception, and the project represents the first effort to integrate them into one product. Most defense development projects belong to this category, but also new generations of computers and many products in the high-tech industry. Such projects are characterized by long periods of design, development, testing, and redesign, and they require at least three design cycles. Design freeze must be scheduled, therefore, to a much later phase, normally during the second or even the third quarter of the project execution period. Management style must be more flexible in these projects, since numerous designs must be tested and will lead to changes and improvements during much longer periods.

Type D: Super High-Tech Projects. These projects are based on new technologies that do not exist at project initiation. While the mission is clear, the solution is not. This type of project is relatively rare and is usually carried out by only a few, probably large organizations or government agencies. One of the most famous examples of this type was the Apollo moon-landing program (Shenhar, 1992). At project inception, it had a well-defined mission and timetable, but no available technology was at hand, and nobody really knew how to get to the moon (see Case Study 1).

Case Study 1

Managing Uncertainty in Space Programs

Two of the most famous engineering efforts of this century belong to NASA. They are the moon-landing Apollo program in the 1960s and building the Space Shuttle in the 1970s and early 1980s. While these two programs had much in common in terms of uncertainty, complexity, and risk, they were managed in two different styles and with two different attitudes. Apollo was initiated in 1961 by President Kennedy, who set the goal of reaching the moon before the end of the decade. From inception, it was clear to NASA that the Apollo program was highly uncertain and much more risky than anything it has done before. Uncertainties ranged from radiation and meteoroid hazards to lunar surface extreme environment to unknown launch configurations. In the context of this chapter, the Apollo program was perceived and managed as a super-high-tech, Type D project. NASA spent enormous time and resources to develop the nonexistent technologies. It did so while embarking on a low-scale prototype, called the Gemini program. It had, in addition, installed numerous safety mechanisms to make sure everything was tested and guaranteed before freezing the system design and configuration, well into the program's execution period. The successful moon landing of Apollo 11 in July of 1969 symbolized a victory for humankind, not only on unknown space territories but also on new and far-reaching technologies.

It seems, however, that many of the lessons learned during the Apollo era were forgotten during the development of the Space Shuttle. This project was managed in a different way. Pressured by the Administration to limit the budget and making everything possible to win Congress approval, NASA preferred to treat the program as a safe bet, based on existing technologies and previous skills. The configuration and design were frozen as early as three month after project approval, and a final commitment was made to new and yet nontested technologies. In reality, project uncertainties were higher than in the Apollo program. Developing the Space Shuttle turned out to be one of the most difficult and exasperating engineering challenges of the space age, and the task essentially proved too difficult to accomplish using the "success-oriented" attitude employed by NASA. Eventually, the program suffered extreme overruns, amounting to almost three years in schedule delays and 60 percent in unexpected costs, all accumulated even before the tragic Challenger accident. The second loss of a shuttle in 2003 only added to the doubt whether the development project was managed correctly.

The Presidential Commission that investigated the Challenger accident concluded that the failure was the result of faulty joint design; that the decision to launch the shuttle on a cold January morning was wrong, given early warnings concerning the design problem and recommendations. Using the UCP framework, it is clear that NASA's management style toward the project was one that in our notation would be classified a Type C. In retrospect, it seems that the program had to be conducted using a different, Type D philosophy. In that case, the final configuration, the freezing point of the design, and the operational phase declaration would have been scheduled for a much later point, leaving open possibilities for using other, more thoroughly tested, and maybe improved technologies. In addition, a Type D style would have vested the program with a different attitude toward risks, possible development problems, and probability of failure and would have created the need for a much higher level of communication among various parties involved (Shenhar, 1992).

Unlike previous, lower-uncertainty-type projects, a great deal of the effort in this type is devoted to the development of completely new technologies and to testing and selecting among various alternatives. As our studies found, projects in this category use similar techniques to resolve the issue of unknown technologies. They institute an intermediate program to build a *small-scaled prototype* on which new technologies are built and tested. Type D projects thus require extensive development periods, very late design freeze, and intensive management of ideas and change until the final configuration is selected.

Management style must be extremely flexible, yet cautious. The attitude in these risky projects must be "look for trouble; it's there." And since things change so fast, extensive communication is essential among project teams and team members. The formal system of reporting could simply not accommodate the degree of changes and new information that is created.

A summary of characteristics and managerial styles of different levels of technological uncertainty is described in Table 8.4.

Project Complexity (System Scope)

In our study of project complexities, we observed three typical management styles. They are distinguished by the way the project is organized and its subelements are coordinated,

TABLE 8.4. PROJECT CHARACTERISTICS AND TECHNOLOGICAL UNCERTAINTY LEVELS.

Variable	Low-tech A	Medium-tech B	High-tech C	Super High-tech D
Technology	No new technology.	Some new technology.	New, but existing technologies.	Key technologies do not exist at project's initation.
Typical industries	Construction, production, utilities, public works.	Mechanical, electrical, chemical, some electronics.	High-tech and technology-based industries; computers, aerospace, electronics.	Advanced high-tech and leading industries; electronics, aerospace, computers, biotechnology.
Type of products	Buildings, bridges, telephone installation, build-to-print.	Non-revolutionary models, derivatives or improvement.	New, first of its kind family of products, new military systems (within state of the art).	New, nonproven concept beyond existing state of the art.
Development and testing	No development, no testing.	Limited development, some testing.	Considerable development and testing. Prototypes usually used during development.	Develop of key technologies needed. Small-scale prototype is used to test concepts and new technologies.
Design cycles and design freeze	Only one cycle. Design freeze before start of project execution.	One of two cycles. Early design freeze, in first quarter.	At least two to three cycles. Design freeze usually during second quarter.	Three to five cycles. Late design freeze, usually during third or even forth quarter.
Communication and interaction	Mostly formal communication during scheduled meetings.	More frequent communication, some informal interaction.	Frequent communication through multiple channels; informal interaction.	Many communication channels; informal interaction encouraged by management.
Project manager and project team	Administrative skills. Mostly semiskilled workers, few academicians.	Some technical skills. Considerable proportion of academicians.	Manager with good technical skills. Many professionals and academicians on project team.	Project manager with exceptioinal technical skills. Highly skilled professionals and many academicians.
Management style and attitude	Firm style. Sticking to the initial plan.	Less firm style. Readiness to accept some changes.	More flexible style. Many changes are expected.	Highly flexible style. Living with continuous change, "looking for trouble."

and by the extent of formal versus informal interaction and documentation. As we found, a simple way to define and distinguish among different levels of complexity is to use a hierarchical framework of systems (Shenhar and Dvir, 1996; Dvir et al., 1998; Shenhar, 1998; Shenhar, 2001a). We call it *system scope*, and in most cases a lower scope level may be seen as a subsystem of the next-higher level. Project management practices, however, can be distinguished by three typical levels as follows.

Level 1: Assembly Projects. Assembly projects involve creating a collection of elements, components, and modules combined into a single unit or entity that is performing a single function. Assembly projects are relatively simple; they may produce a stand-alone product such as a CD player or a coffee machine, or create a subsystem of a larger system such as a computer hard drive or a radar receiver. They may also involve restructuring a functional organization or building a stand-alone service (see the discussion in Case Study 2 on reengineering projects).

Case Study 2

The Case of Reengineering

Reengineering projects to restructure business and service process have become a common part of organizational life. Yet not all engineering projects are the same, and their execution should be adapted to the specific problem. Since typical reengineering projects build a new generation of business process, they can be classified as platform from the point of view of market uncertainty, their internal uncertainly is Type A (or at most B), and their pace is regular (time is generally not a critical factor, and it is not associated with a window of opportunity or meeting market demand). the distinctive measure in these projects, however, is complexity, and it should be determined by the level of the reengineered organization: When you are dealing with one department or a single process within a local unit, such as a bank's branch office, the effort is an *assembly* project. People know each other well, the process is simple to depict, and not much information technology is needed to construct and run the process. A reengineering effort of an entire business with various integrated functions, such as engineering, manufacturing, sales, and distribution, will be classified as a *system project.* In this case, coordination becomes much more critical, formal tools and extensive documentation must be employed, and information technology and software is unavoidable. The highest level of complexity involves large corporations, normally spread over the country or even the entire world. Reengineering, then, becomes an *array project.* It must be carried out in a very formal and bureaucratic way, with many subprojects devoted to different parts of the company.

Assembly projects are typically carried out by a single organizational function or a small cross-functional team, often within one organization and with a low level of formality or bureaucracy. The number of project activities or subtasks is normally in the range of tens, and as we learned, the planning of budget and schedule in this type of projects is simple, requires only basic tools of project management, and is often done manually. Interaction, communication, and much of the decision making is, to a large extent, informal; documentation and reporting is minimal, as is the need for administrative staff within the project.

Level 2: System Projects. System-type projects involve a complex collection of interactive elements and subsystems, jointly dedicated to a wide range of functions to meet a specific operational need. System projects may produce aircraft, cars, computers, or buildings, but they may also involve reengineering efforts of entire businesses (see Case Study 2). A main contractor or program office typically leads system projects, and the total effort is divided among numerous subcontractors—some in-house and some external. The program office is responsible for the final integrated result and for meeting overall performance, quality, time, and budget goals. Such projects are managed in a more formal way than assemblies, with extensive documentation and contracting between main and subcontractors.

Since most system projects extend beyond organizational borders, they usually need special administrative staff to handle the planning, budget management, contracting, and controlling issues. The number of project activities is in the range of hundreds to a few thousands. While there are many existing project management tools and applications, system projects and their corporations often find it necessary to develop their own tools and documents to meet their specific requirements.

Level 3: Array Projects. Array-type projects deal with large, widely dispersed collections of systems (sometimes called "supersystems") that function together to achieve a common purpose, such as city public transportation systems, national air defense systems, or interstate telecommunication infrastructures. Arrays are clearly large-scale projects, yet in most cases, they do not involve building the supersystem from scratch. Rather, such projects often involve gradual growth, addition, or modification to an existing infrastructure. Notable array projects were the upgrading of the New York Subway system in the early 1990s, the preparation of the city of Atlanta to the Olympic games of 1996, and building the Anglo-French Channel Tunnel.

Array projects require the administration of many separate programs, each one devoted to a different component or system. Projects are typically organized in the form of a central "umbrella organization" that is set up as a separate entity or company and that formally (and legally) coordinates the efforts of numerous subprojects in other organizations. The actual technical work is executed, however, within the suborganizations. The focus is on extensive documentation, contracting, and tight financial controls. The dispersed nature of the end product and the extent of subcontracting make it necessary to manage these programs in a highly formal way and to put a lot of effort into the legal aspects of the various contracts. Ordinary tools of planning and control are even less relevant here, and management must often develop its own system of contract coordination and program control.

A summary of characteristics and managerial styles of different levels of system scope is described in Table 8.5, and an example of different scope levels of reengineering projects is discussed in Case Study 2.

TABLE 8.5. PROJECT CHARACTERISTICS AND SYSTEM SCOPE LEVELS.

	Assembly 1	System 2	Array 3
Definition	A collection of components and modules in one unit, performing a single function.	A complex collection of assemblies that is performing multiple functions.	A widespread collection of systems functioning together to achieve a common mission.
Examples	A system's power supply; a VCR, a single functional service.	A complete building; a radar; an aircraft; a business unit.	A city's highway system; an air fleet, a national communication network; a global corporation.
Customers	Consumers or a subcontractor of a larger project.	Consumers, industry, public, government or military agencies.	Public organizations, government or military agencies.
Form of purchase and delivery	Direct purchase or a simple contract; contract ends after delivery of product.	Complex contract; payments by milestones; Delivery accompanied by logistic support.	Multiple contracts; sequential and evolutionary delivery as various components are completed.
Project organization	Performed within one organization, usually under a single functional group; almost no administrative staff in project organization.	A main contractor, usually organized in a matrix or pure project form many internal and external subcontractors; technical and administrative staff.	An umbrella organization—usually a program office to coordinate subprojects; many staff experts: technical, administrative, finance, legal, etc.
Planning	Simple tools, often manual; rarely more than 100 activities in the network.	Complex planning; advanced computerized tools and software packages; hundreds or thousands activities.	A central master plan with separate plans for subprojects; advanced computerized tools; up to ten thousand activities.
Control and reporting	Simple, in-house control; reporting to management or main contractor.	Tight and formal control on technical, financial and schedule issues; reviews with customers and management.	Master or central control by program office; separate additional control for subprojects; many reports and meetings with contractors.
Documentation	Simple, mostly technical documents.	Many technical and managerial formal documents.	Mostly managerial documents at program office level; technical and managerial documents at lower level.
Management style, attitude, and concern	Mostly informal style; familylike atmosphere.	Formal and bureaucratic style; some informal relationship with subcontractors and customers; often political and interorganizational issues.	Formal, tight bureaucracy; high awareness to political, environmental, and social issues.

Project Pace: How Critical Is Your Time Frame?

On this scale projects will differ by urgency, or how much time is available, and what happens if the time goal is not met. We identified three different levels of urgency, or pace: regular, fast-competitive, and critical-blitz (Shenhar et al., 2002):

Regular projects are those efforts where time is not critical to immediate organizational success. They are typically initiated to achieve long-term or infrastructure goals. They may include some public projects, projects in a noncompetitive or nonprofit environment, organizational improvements, or technology build-up efforts. Unless specifically prioritized, such projects are managed in a relatively casual format and may often be delayed or pushed aside by more urgent assignments.

Fast-competitive projects are the most common projects carried out by industrial and profit-driven organizations. They are typically conceived to address market opportunities, create a strategic positioning, or form new business lines. Time-to-market is directly associated with competitiveness, and although missing the deadline may not be fatal, it could hurt profits and competitive positioning.

Fast-competitive projects must be managed strategically. Project managers should focus on meeting schedule, but also achieving profit goals and addressing market needs. Managing the time frame should be one of the main concerns.

Critical-blitz projects are the most urgent, most time-critical. Meeting schedule is critical to success, and project delay means failure. Such projects are often initiated during a crisis, or as a result of an unexpected event. Examples may be industrial crisis projects to respond to a surprising move by competition, military projects during wartime, or a natural disaster that needs immediate recovery. Famous cases include the Y2K problem or saving the lives of Apollo 13 crew.

To succeed, such projects must be managed completely differently from other forms. The work flow in these projects is very tight. It is performed almost around the clock, with nonstop interaction and continuous decision making. There is normally no time for detailed documentation or report writing. Project managers in critical-blitz projects must possess high autonomy; project organization must be "pure project," and all team members must report directly to the project leader. Top management must be continuously involved to support and monitor project progress.

Table 8.6 describes characteristics of different project pace levels.

Managing Highly Uncertain and Complex Programs

Perhaps the two most important dimensions of the NCTP model are system scope and technological uncertainty. According to our studies, they have the highest impact on differences that can be found among projects and project management styles (for more details, see Dvir et al., 1998). Of particular interest are those projects that involve high complexity together with high technological uncertainty. In fact, many defense and space programs fall in this category, as well as some commercial efforts such as the development of Boeing 777 or GM's electric vehicle. Project management in this case is even more complex, and additional concerns and techniques must be employed.

As we have seen earlier, higher uncertainty involves longer design periods, more design cycles, and later design freeze, as well as better technical skills and more frequent technical

TABLE 8.6. CHARACTERISTICS OF DIFFERENT PROJECT PACE LEVELS.

Pace Level	Regular	Fast-Competitive	Blitz-Critical
Definition	Time not critical tto organizational success.	Time-to-market is a competitive advantage and has an impact on business success.	Time is critical for project success. Delays mean project failure.
Examples	Public works, government initiative, internal projects.	Business related projects, new product introduction.	Crisis situations, war, fast response to natural disasters, fast response to business related surprises.
Organization	Matrix or functional.	Matrix, teams, subcontractors.	Pure project, special task force.
Personnel		Qualified to the job.	Specifically picked.
Focus	No particular focus.	Strategically focused on time to market.	Swift solution of the crisis.
Procedures	No specific attention.	Structured procedures.	Shortened, simple, nonbureaucratic.
Top management involvement	Management by exception.	Go ahead at stages.	Highly involved and constantly supportive.

decisions. Similarly, higher complexity (scope) involves increase in formality and documentation. When both dimensions are at their higher levels, an important addition to classical project management is the discipline of *systems engineering management*. Developed previously in the defense environment, systems engineering management is a multidisciplinary discipline dedicated to controlling requirements, specifications, design, development, building, testing, manufacturing, and operation so that all elements are integrated to provide an optimum, overall system. It requires a "holistic view"—the design of the whole, rather than the design of the parts—and it involve identifying and understanding the need; creating the system concept and architecture; and combining economic, societal, environmental, and political issues into the design of the system.

The problem of system integration mentioned earlier is typically critical in high-uncertainty, high-complexity projects. Subsystem qualification does not guarantee total system performance. Considerable time and effort must be allocated to solving the integration issue. Numerous problems of interface, fit, and mutual effects must be resolved before the entire systems is qualified. And, obviously, configuration management becomes extremely important, as any change in any part may impact numerous other components and subsystems.

Finally, although all projects involve a certain level of risk, higher-scope and higher-tech projects are more risky and are more prone to problems. Thus, they need more systematic risk management. The objective of risk management is not to eliminate risks, but to manage them across the project so as to avoid investing excessive resources in the solution of a given risk while neglecting others. Risk management is conducted in a rigorous and systematic way to identify, analyze, and mitigate all sources of program risk. Figure 8.4

FIGURE 8.4. TECHNOLOGICAL UNCERTAINTY AND SYSTEM SCOPE IMPACT ON PROJECT MANAGEMENT.

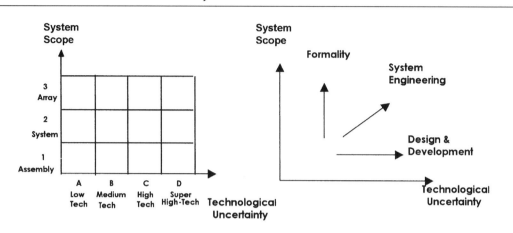

describes the two dimensions of technological uncertainty and system scope and the impact of a simultaneous increase of uncertainty and scope. Neglecting the impact of high uncertainty or high complexity may result in difficulties, delays, and even project failure (Dvir et al., 2003).

The Work Package Framework

This framework is particularly useful for team members and subcontractors when dealing with individual work packages. The work breakdown structure provides a tool to identify project tasks, allocate time and resources, create the product cost structure, develop a network-diagramming schedule (PERT and critical path), and assign responsibility for the execution and monitoring of task completion. The lowest level in the WBS is a work package, which may serve a basis for identifying project risks and deciding how to avoid or reduce them.

Our studies show that conventional treatment of all work packages as the same is unrealistic and must be replaced by a more adaptive approach. We use two dimensions to distinguish among packages: the type of task outcome and the type of activity or work that needs to be done to achieve it. Task outcomes may be tangible or intangible, and activity can be distinguished as requiring craft or intellectual work.

Task Outcome: Product of Work Package

Tangible outcomes produce a physical artifact (hardware). Any piece of equipment, such as a computer keyboard or a car's steering wheel, is a tangible product. Tangible products must be physically assembled and reproduced. Product and process design must be integrated to create a manufacturing process and is subject to quality control, assembly operations, and cost of production.

Intangible outcomes, in contrast, do not produce an artifact. Intangible products produce new information, which can be stored on different physical media such as a CD-ROMs. Software code, manuals, books, newspapers, blueprints, or movies are intangible products. Reproducing them in high quantities is easy, instant, and cheap and does not require dedicated production lines.

Type of Activity on Work Package

Craft activity involves repetitive efforts that have been done previously. Work outcomes are predictable, and duration is accurately anticipated and subjected to the classical learning curve. Craft examples include machine shop work, painting jobs, watch assembly, or car servicing.

Intellectual activity requires creative effort. Since such activities have not been done before, they are exploratory in nature and require new ideas and imagination. Artwork is obviously intellectual, but so are activities such as developing a new technology, new designs, or "building a better mousetrap."

As shown in Table 8.7, the four cells of our framework suggest different strategies for different work packages.

TABLE 8.7. ADJUSTING THE RISK OF WORK PACKAGES IN WORK BREAKDOWN STRUCTURES.

	Intangible	Tangible
Intellectual	New software code. Most effort in exploratory and creative work. Less risky than intellect tangible. Does not need process building.	New kind of hardware. Hardware never done before. Needs creativity, development, iterations, and testing. Must also plan and build new process. lowest accuracy in resources planning. Needs contingency resources and backup plans. Highest risk in product and process.
Craft	Writing routine plans or procedures. Minimal risk to produce routine text. No new ideas are required.	Building or producing well-known types of hardware. Repetitive tasks. Good estimation of work duration and other resources. Needs costly but predictable process building resources. Production quality becomes the main concern.

Using the Frameworks in Practice

How can managers and organizations benefit from adapting the conceptual framework presented here? An explicit, clear identification of the project type prior to execution should provide a basis for a proper adaptation of managerial attitudes and management style, for the selection of project managers and project team members, for establishing the proper project organization, and for a better choice of managerial tools. For example, identifying a project as high-tech will explicitly have to lead to a highly flexible style, to longer development periods, and to more design cycles. Dealing with a super-high-tech project will entail, in addition, the establishment of an intermediate program to be used as a testing bed for new technologies. Similarly, system projects will require the establishment of an efficient subcontracting procedure, as well as adequate time for system integration.

The framework suggested here might also be useful for identifying technical skill development needs, managerial development needs, and management training needs. Individuals would often gain compelling capabilities and expertise in managing specific kinds of projects. Moving into different, sometimes remote kinds of projects requires adopting a different style and the development of additional skills. Understanding the strategic, as well as the operational, differences between projects may help avoid potential errors and may considerably shorten the learning process. For example, moving into high- and especially into super-high-tech projects requires exceptional technical skills as well as the capability to assess potential value and risk involved in new or even not-yet-developed technology. Moving from assembly to system projects requires additionally a wealth of administrative skills. When managers are dealing with the system level for the first time, they need to develop the "system view"—being able to see the system as a whole and understand the effect of its separate components on the entire system. System project managers must be mature enough and able to utilize the accumulation of technical intuition from their original field to interdisciplinary problems of systems. When moving into the array level, managers must learn to back off from the technical matters and develop instead a broader view of the industry and its players. They must learn to deal with legal, environmental, and political issues, usually not addressed by managers of lower scope levels.

Summary

While the frameworks presented in this chapter may not be conclusive, they provide a good basis for organizations to distinguish among their project types. As mentioned, sometimes organizations will need to find their own way for classification. For example, the dimensions of uncertainty may identify other forms of uncertainty—political, economical, or environmental; and the dimension of complexity may be distinguished according to geography and spread of the project team—either in-house, across different organizations, or across international boundaries.

In conclusion, companies must realize that project management is not universal and adapting project management styles is critical to project success. Each company must identify its specific typical project types and select the right tools, procedures, and people for the specific project. The potential of improved project management is much too large to be neglected in this dynamic age of increased competition.

Acknowledgments

The authors are grateful for the help and support that the following institutions have provided to these studies over the years:

Israeli Institute for Business Research at Tel Aviv University; The Directorate for the Development of Armament and Production Infrastructure at the Israeli MOD; Center for the Development of Technological Leadership at the University of Minnesota; Center for Technology Management Research at Stevens Institute of Technology; and the National Science Foundation and its Management of Technological Innovation Program.

References

Ahituv, N., and S. Neumann. 1984. A flexible approach to information system development. *MIS Quarterly* (June): 69–78.

Balachandra, R., and J. H. Friar. 1997. Factors for success n R&D project and new product innovation: A contextual framework. *IEEE* Transactions on Engineering Management 44:276–287.

Boulding, K. 1956. General systems theory: The skeleton of science. *Management Science* (April): 197–208.

Brown, S. L., and K. M. Eisenhardt. 1997. The art of continuous change: Linking complexity theory and time-paced evolution in relentlessly shifting organizations. *Administrative Science Quarterly* 42: 1–34.

Burns, T., and G. Stalker. 1961. *The management of innovation.* London: Tavistock.

Cash, J. I. Jr., W. F. McFarlan, and J. L. McKenney. 1988. *Corporate information systems management.* Homewood, IL: Irwin.

Drazin, R., and A. H., van de Ven. 1985. Alternative forms of fit in contingency theory. *Administrative Science Quarterly* 30:514–539.

Dvir, D., S. Lipovetskey, A. J. Shenhar, and A. Tishler. 1998. In search of project classification: A non-uniform mapping of project success factors. *Research Policy* 27:915–935.

Dvir, D., A. J. Shenhar, and S. Alkaher. 2003. From a single discipline project to a multidisciplinary System: Adapting the right style to the right project. *System Engineering* 6(3):123–134.

Eisenhardt, K. M. 1989. Building theories from case study research. *Academy of Management Review* 14: 532–550.

Eisenhardt, K. M., and B. N. Tabrizi. 1995. Accelerating adaptive processes: Product innovation in the global computer industry. *Administrative Science Quarterly* 40:84–110.

Lawrence, P. R., and J. W. Lorch. 1967. *Organization and environment: Managing differentiation and integration,* Boston: Graduate School of Business Administration, Harvard University.

Lewis, W., and A. Samuel. 1989. *Fundamentals of engineering design.* Englewood Cliffs, NJ: Prentice Hall.

Morris, P. W. G. 1997. *The management of projects.* London: Thomas Telford.

Pahl, G., and W. Beitz. 1984. *Engineering design.* New York: Springer-Verlag.

Pearson, A. W. 1990. Innovation strategy. *Technovation* 10(3):185–192.

Pennings, J. M. 1992. Structural contingency theory: A reappraisal. *Research in Organizational Behavior* 14:267–309.

Pinto, J. K., and J. G. Covin. 1989. Critical factors in project implementation: A comparison of construction and R&D projects. *Technovation* 9:49–62.

Rechtin, E. 1991. *Systems architecting.* Englewood Cliffs, NJ: Prentice Hall.

Shenhar, A. J. 1991. Project management style and technological uncertainty: From low- to high-tech. *Project Management Journal* 22(4):11–14, 47.

———. 1992. Project management style and the space shuttle program: A retrospective look. *Project Management Journal* 23(1):32–37.

———. 1993. From low- to high-tech project management. *R&D Management* 23(3):199–214.

———. 1998. From theory to practice: Toward a typology of project management styles. *IEEE Transactions on Engineering Management* 41(1):33–48.

———. 2001a. One size does not fit all projects: Exploring classical contingency domains. *Management Science* 47(3):394–414.

———. 2001b. Contingent management in temporary organizations: The comparative analysis of projects. *Journal of High Technology Management Research* 12:230–271.

Shenhar, A. J., and Z. Bonen. 1997. A new taxonomy of systems: Toward an adaptive systems engineering framework. *IEEE Transactions on Systems, Man, and Cybernetics* 27(2):137–145.

Shenhar, A. J., and D. Dvir. 1996. Toward a typological theory of project management. *Research Policy* 25:607–632.

Shenhar, A. J., D. Dov, T. Lechler, and M. Poli. 2002. One size does not fit all: True for projects, true for frameworks. PMI Research Conference, Seattle, Washington.

Song, M. X., W. E. Souder, and B. Dyer. 1997. A casual model of the impact of skills, synergy, and design sensitivity on new product performance. *Journal of Innovation Management,* 14:88–101.

Souder, W. E., and M. X. Song. 1997. Contingent product design and marketing strategies influencing new product success and failure in U.S. and Japanese electronics firms. *Journal of Product Innovation Management* 14:21–34.

Steele, L. W. 1975. *Innovation in big business.* New York: Elsevier Publishing Company.

Van Gigch, J. P. 1978. *Applied general systems theory.* 2nd ed. New York: Harper & Row.

Wheelwright, S. C., and K. B. Clark. 1992. *Revolutionizing product development.* New York: Free Press.

Yap, C. M., and W. E. Souder. 1994. Factors influencing new product success and failure in small entrepreneurial high-technology electronics firms. *Journal of Product Innovation Management* 11:418–432.

CHAPTER NINE

VALUE MANAGEMENT:

A Group Decision-Making Process to Achieve Stakeholders' Needs and Expectations in the Most Resource-Effective Ways

Michel Thiry

Value management (VM) is not a "new fad," but a proven methodology formally developed in the late 1940s that evolved from "a problem-solving system to deliver products with appropriate performance and cost" (Miles, 1972) to "a style of management . . . with the aim of maximizing the overall performance of an organization" (BSI, 2000).

From Value Analysis to Value Management

Historical Background

From the late 1940s to the early 1960s, Lawrence D. Miles, who is today considered the father of value methodologies, developed the concept of value analysis at General Electric. He specifically identified three elements that are at the core of all value methodologies: the concept of function, the multidisciplinary team workshops, and a structured "job plan." During that period, value analysis (VA) and value engineering (VE), considered synonymous, were mainly applied to manufactured products.

In the early 1960s, Charles W. Bytheway (1965) developed the concept of the Function Analysis Systems Technique (FAST) diagram, and VA and VE started to be applied in construction and new product development. In the early 1980s, the Society of American Value Engineers developed the 40-hour (5-day) workshop in the United States, as part of their certification system. (A few issues limit the usefulness of the 40-hour workshop: it is generally carried out by external experts, which greatly reduces buy-in from the project team and implementation of proposals, and it is mostly focused on cost reduction, which limits its use in early strategic situations.) In the late 1980s and early 1990s, in Europe and

Canada, a number of practitioners started to apply VA and VE in earlier stages of projects and to integrate them into the project process; it is also the period when the first applications to organizational processes were attempted.

Although the term value management was first used in the 1980s, it was only in the late 1990s that value management emerged as a discipline distinct from VA/VE, drawing on management techniques and fully integrating it in the project life cycle as a "collaborative group-learning approach" (Barton, 2000). Today, VM is used in a number of new fields, like strategic planning, process reengineering, organizational management, change management, concurrent engineering, and others; it is also integrated with known processes like organizational effectiveness, quality management, design to objectives (DTO), and risk management.

Associations and Standards

In late 1958, a first group of value "engineers" associated under the acronym of SAVE (Society of American Value Engineers), now renamed SAVE International. Soon after SAVE was founded, VE associations started to spread around the world. Value societies were founded in Japan in 1965, the United Kingdom in 1966, Germany in 1974, France in 1978, and a number of other countries since; today more than 20 countries are represented by value associations, among them the United States and Canada, eight EU countries that have formed the European Governing Board of the VM Training and Certification System (EGB), Hungary, the Czech Republic, Kuwait, Brazil, Japan, Hong Kong, South Korea, Australia and New Zealand, India, and South Africa.

The U.S. Department of Defense established the first governmental value program in 1954. The method was to be applied at the engineering stage, which brought about a change in name from value analysis to *value engineering*. In the early 1970s, GSA asked SAVE to develop a certification program for value practitioners. The status of Certified Value Specialist (CVS) was established by SAVE as a standard (SAVE, 1993), recognizing competence in the field of value engineering. Today, many such programs exist around the world; specifically a European System of Training and Certification in Value Management was set up in 1997 as a result of discussions between eight European National Value Associations.

Germany developed the first value standard: DIN 69 910 on "Wertanalyse" was created in 1973. From 1985 to the early 1990s, AFNOR (French Standards Body) developed a number of value analysis standards in France. In 1987, the Bureau of Indian Standards set up standard IS:11810-1986 on VE. Australia and New Zealand published Standard AS/NZS 4183:1994 "Value Management"—*as an analytical process*—in 1994. The American Society for Testing Materials (ASTM, 1995) released a "VE in Construction" standard in December of 1995 and the European Committee for Standardization (CEN) issued the first standard on value management as a management approach in 2000 (BSI, 2000).

Current Definitions

As early as 1731, Daniel Bernoulli, presenting a paper at the Imperial Academy of Sciences in St. Petersburg, stated that "the value of an item must not be based on its price, but rather on the utility which it yields" (Bernstein, 1996). This concept of the relationship between value and function is at the heart of all value disciplines.

In 2003 and 2004, SAVE International and the EGB agreed to develop a framework for certification reciprocity. As part of this framework, definitions were developed for common value terms; they are recognized as a common basis for value practice. The definitions are as follows

- "*Value* has been defined as a ratio between 'Quality & Cost', 'Function & Cost', 'Worth & Cost', 'Performance & Resources', 'Satisfaction of needs & Use of resources' and 'Benefits & Investment'. All those ratios and others are acceptable, as long as, on one side, lies the satisfaction of an explicit or implicit need and on the other, the resources invested to achieve it." In the United States and U.S.-influenced countries like English Canada, the Middle East, Japan, the focus is on cost, whereas Europe and countries like Australia, French Canada, and Hong Kong have developed a broader view of resources.
- "*Value Management* (VM) consists of the combined application of value methodologies and other methodologies at organizational level (from strategic to operational) in order to improve organizational effectiveness." (In the U.S.-influence zone it is an umbrella term for VA and VE, whereas in Europe and other countries it is considered a management approach, as described in EN-12973:2000 European Standard).
- "*Value Analysis* (VA) and *Value Engineering* (VE) are specific value methodologies aimed at improving existing products or developing new products. Products can include both goods and services."
- "*Value methodologies* [no acronym] include all processes, tools and techniques derived from the work of Lawrence Miles. They specifically include the concept of functions and function analysis (and its derivatives), the concept of cross-functional teamwork and the concept of a structured process based on creative thinking (alternative left/right brain use).

  "There are a number of specific methodologies that have been developed worldwide on the basis of these, including, but not limited to: FAST Diagramming (US and Canada), 40 hour workshop with job plan (US & Canada), Split workshops (UK), Function Tree (Europe & Canada), Function Cost (US & Europe), Functional Performance Specification (France, French Canada & Europe), Soft VM (Australia, UK) etc. The use of one methodology or another should not limit the value practitioner, whose first aim is to improve value within the limits of their intervention."

There are three key concepts that, combined, underlie all value methodologies:

- The *function*, which is the expression of needs in terms of purpose, independent from any solution
- The *cross-functional team*, which enables a broad view and an increased knowledge of a situation
- The *structured process*, based on creative thinking; the alternate use of creativity and analysis, or lateral and vertical thinking (de Bono, 1990)

The Concept of Value

As outlined previously, there are many definitions of value. For the purpose of this chapter, I will concentrate of the one developed for the European Standard: "Value may be described, in the context of VM, as the relationship between the satisfaction of need and the resources used in achieving that satisfaction". Although there are other definitions of value, this one is the most appropriate to the practice of VM.

The European Standard (BSI, 2000) uses the representation shown in Figure 9.1, with the following note: "The symbol α signifies that the relationship between the satisfaction of need and the resources is only a representation. They must be traded off one against the other in order to obtain the most beneficial balance" (BSI, 2000).

Using a more specifically strategic approach, I have developed a more elaborate diagram for value, shown in Figure 9.2, where the Benefits Variance (BV) = Offered Benefits (OB) − Expected Benefits (EB) and BV must be ≥ 0. The assumption is that *Benefits* is directly related to the *Satisfaction of needs*. In addition, Resource Variance (RV) = Available Resources (AR)–Required Resources (RR) and RV must be ≥ 0.

This means that the value manager is not only concerned with balancing resources with satisfaction of needs (benefits) but also with making sure that both for benefits and resources, capacity—including capability—is matched with intent.

Distinctions between Value Analysis, Value Engineering, and Value Management

Value analysis was the term first used by Miles to define the process he developed; it was later changed to value engineering when he started working with the military. Today there is a consensus to define VA and VE as specific methodologies aimed at improving or developing better products. A common distinction is to consider VA as the value-based analysis of existing products to improve them and VE as the application of value techniques to develop new products. Officially, VE and VA are considered synonymous in both the SAVE (1997) Standard and in the European Standard (BSI, 2000). In their 1997 Standard, SAVE uses the term value methodology to encompass both VE and VA and defines it as "the

FIGURE 9.1. THE EUROPEAN STANDARD DIAGRAM DESCRIBING VALUE.

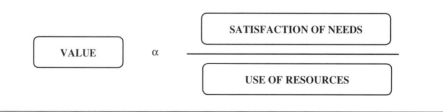

FIGURE 9.2. THE VALUE RATIO.

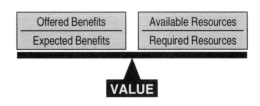

systematic application of recognized techniques which identify the functions of the product or service, establish the worth of those functions, and provide the necessary functions to meet the required performance at the lowest overall cost".

SAVE International has traditionally used the term value management to define the management of value proposals generated through the use of VE and VA (SAVE, 1997). Considering the recent reciprocity agreement between SAVE International and the EGB, as well as recent publications (Green, 1995; Kaufman, 1997; Barton, 2000; Brun and Constantineau, 2001; Woodhead and Downs, 2001). We can now consider that the definition of value management given previously and detailed in the European Standard (BSI, 2000) is widely accepted. In Australia (Barton, 2000) and in the United Kingdom (Green, 1995), researchers, using developments in systems engineering (Checkland and Scholes, 1994), have distinguished "soft" VM from "hard" VM; the former being applied at strategic level or in the early stages of projects to identify stakeholders and define needs and expectations, and the latter, based on VA and VE techniques, being applied in later stages of projects, when customers and problems can be clearly identified, to improve delivery and outcomes.

Value Management Today: A Style of Management

At a management level, VM is based on three root principles:

1. A focus on objectives and targets: stakeholders' needs and critical success factors
2. A focus on the function: purpose of the product, system, or service
3. A continuous awareness of value: measures for improvement and innovativeness

Three key concepts underlie it:

(a) *A transverse approach, translated in practice by cross-functional teamwork.* The VM methodology is based on a multidisciplinary team approach in a workshop environment; it enables the team to broaden perspective, assess every angle, share opinions, and reach consensus.

(b) *A structured decision process leading to better business decisions.* The VM process is a systematic approach, based on creative thinking, which uses time in the most efficient way to ensure that the scope of the VM study is covered in the best sequence.

(c) *The use of functions, increasing effectiveness and enhancing competitiveness.* VM relies on functions, rather than on predefined solutions, to identify the expected outcome of a project, thus allowing evaluation of a broader range of possible options.

A Management Approach

The application of VM at the organizational level is a source of competitive advantage. The combined use of a cross-functional group process, the concept of functions and of creative thinking principles combined to foster learning and enhance innovation, creates an undeniable combination of differentiation and implicit knowledge, leading toward competitive advantage.

Organizational decisions are often characterized by a high degree of uncertainty (lack of information), ambiguity (conflicting information), and stress (lack of time). Group decision making (GDM) enhances the ratio of facts versus assumptions, the quality (truthfulness) of information, consensus and objective sharing, and group commitment to decisions (also a relief of stress). VM is a GDM process that through sensemaking (Thiry, 2001) reduces ambiguity and enhances quality of information, and builds consensus and buy-in from participants in the process. At the elaboration phase, it uses scope definition, feasibility, and risk management concepts to increase the level of information and reduce uncertainty, as well as the possibility to develop more resource-effective options. If the players are well chosen among the key stakeholders, VM will increase decision support and implementation success through its participative group process.

The concept of functions is central to all value methodologies; applied at the strategic level, the traditional concept of functions can be defined as expected benefits in general and critical success factors (CSFs) in particular. CSFs have been defined in many studies as key issues to be addressed in every project, in which case they are generic and apply to all projects (Pinto and Rouhiainen, 2001). In a VM perspective, CSFs are specific expected benefits that are identified as critical for success and prioritized for each project by the key stakeholders, which are qualitative; these are defined in quantitative measurable terms through key performance indicators (KPIs).

Recently, an emergent perspective has linked VM to a learning paradigm (Thiry, 2000, 2002). This concept aligns VM with developments in organizational effectiveness like balanced scorecard (Kaplan and Norton, 1986), the EFQM Excellence Model (EFQM, 2000), and ISO-9000:2000 (ISO, 2000). When used iteratively in a systemic perspective, at program level and at project gateways, this approach ensures that strategic benefits will not only be explicitly defined at a strategic level but also effectively measured and delivered.

All the above would not be possible without a robust and structured process. Each value association around the world has developed their own "job plan," which can vary from five (SAVE, 1997) to seven (AFNOR, 1985) or even ten (BIS, 1987) steps. All those "plans" originate from the concepts developed by Miles for VA. They are all based on creative thinking, which has been detailed by de Bono (1990) as the alternate use of lateral and vertical thinking. Miles described a standard problem-solving process, where the decision-making authority lay outside the process, to which he added cross-disciplinary participation and function identification.

Today, VM is more associated with a decision-making process, where the decision makers are actively participating in the process and have authority over the resources required to implement the decision. In complex situations and environments, standard problem-solving or decision-making techniques are not applicable to define the situation; one needs to work with soft systems methodologies (Checkland and Scholes, 1994; Green, 1994; Neal, 1995) or more constructivist approaches (Weick, 1995; Guba and Lincoln, 1989; Quinn, 1996) like sensemaking.

The VM Process

For the purpose of this chapter, we shall treat VM as a strategic decision-making and control process in the context of program and project management. Figure 9.3 represents this view.

The VM process comprises the following subprocesses, which are typically carried out as facilitated workshops or meetings where the key stakeholders (at least) participate actively. Some tasks can be carried on individually, but research (Vennix, 1996; Woodhead and Downs, 2001; Fong, 2002) has demonstrated that facilitation is a key aspect of successful

FIGURE 9.3. VALUE MANAGEMENT IN CONTEXT.

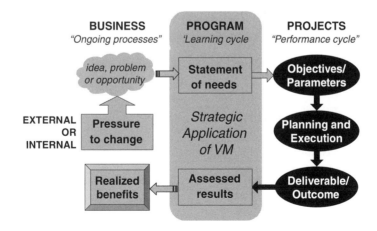

group decision processes in general and VM in particular. In this section, the "team" will mean those participating in the VM process through workshops or individual work.

- *Sensemaking*, which includes function(al) analysis and can use a variety of techniques like scenario planning, soft systems analysis, gap analysis, and others, is used to understand the situation and come to a shared agreement about the critical success factors (qualitative-level expected benefits) and key performance indicators (quantitative measures).
- *Ideation* is the creative generation of alternatives that enables the process to be truly innovative.
- *Elaboration* consists of the evaluation of alternatives in terms of their achievability and contribution to expected benefits, and their combination/modification to develop viable options. The definition given by Spradlin (1997) will be used for this chapter: "An option is an alternative that permits a future decision following revelation of information. All options are alternatives, but not all alternatives are options."
- *Choice* is the action of selecting/prioritizing the best options, in regards of the critical success factors.
- *Mastery* is a formative evaluation and control process based on improvement rather than on a baseline.

Focus on Results

The main success factor of value methodologies has always been their focus on tangible results. It starts with the analysis of stakeholders' needs and expectations and their translation into measurable objectives, which are then addressed to identify the most profitable options. Following the identification of the best options, a "formative evaluation" (Guba and Lincoln, 1989) process enables the team to deliver results that are in line with the expected benefits. The key aspect of VM is the direct link that is established between needs and results, through functions.

A case study follows, which will be used to demonstrate application of the VM techniques outlined in this section.

Case Study: Outline of Situation

Consider a medium-size company that develops and implements projects for external clients. The company has been in this business since 1985 and has built a good reputation with its clients.

The market is growing, and in the last year the company has hired 50 new personnel, of which 15 are project managers. About 20 percent of the personnel are project managers; the others are technicians, operations and support staff, and product developers. There are five program managers.

Recently, a number of clients have complained about project performance and it has come to the ears of members of the board. A significant number of projects are either running late or over budget.

The MD calls one of the program managers and asks her to take care of the problem. The mandate calls for quick results on the most significant projects and general improvement of the situation within six months.

Among the new people that have been hired, many have good experience but have not yet fully integrated in the company; others have little practical experience, although they show good potential.

The company has always relied on a few experienced PMs to "run the show," but those are the ones that are also the most resistant to change; they know their business and do not accept criticism of their methods easily: "I have always delivered what I was asked; don't come and tell me what to do."

For a few months now, the HR Department has talked about induction courses and PM courses. The Quality Management Department has put in a budget for the standardization of PM processes and procedures; it would involve hiring an external consultant.

Most people work remotely at client sites.

Sensemaking

The first step in the VM process is to understand the situation that has created the need for change and, from that understanding, identify tangible benefits expected by the key stakeholders—the *functions* that the program or project must provide. The traditional *information* and *function analysis* phases of VA/VE are not adequate to deal with complex situations; "[. . .] whereas in a well-defined situation, with cohesive groups it may be possible to shorten the sensemaking process to a simple 'information phase', in complex, ambiguous, multileveled situations it is necessary to allow and foster a sensemaking interaction . . ." (Thiry, 2001).

A well-managed sensemaking process enables participants to construct a shared view of a complex situation and model their expected benefits through a function diagram. This model, which has been called function breakdown structure (FBS) (Thiry, 1997) enables the group to define CSFs and KPIs, which are the key to the successful achievement of expected benefits.

1. The first step of sensemaking in VM is to perform a *stakeholder analysis*, which encompasses the identification of the stakeholders, their classification, and their ranking.
2. The second step is to carry out a *functional analysis*, which consists of determining stakeholders' needs and expectations, translating them into expected benefits using a verb-noun semantic, identifying any additional benefits required, and organizing all these into an FBS.
3. The next step lies in the identification of those expected benefits that are *critical success factors* of the program or project and their prioritization, which will support decision making throughout the delivery process.

4. Finally, those functions that have been identified as CSFs are *characterized* (BSI, 2000) through the definition of key performance indicators (KPIs), using the concept of a criterion of measure, an expected level of performance, and an accepted range of flexibility. (This method was standardized by the French Value Analysis Association and is now included in the VM Standard.) The definition of KPIs enables the stakeholders to move from qualitative to quantitative measures of success and to be able to assess benefits on clearly quantifiable terms.

Stakeholder Analysis

To define needs and expectations of stakeholders, the team must first identify the stakeholders. They will typically use intuition and historical data to list a number of possible stakeholders without attempting to categorize them or classify them—a lateral thinking process (de Bono, 1990). Once stakeholders have been identified, the team can group them in a hierarchical structure by creating groups and subgroups, not unlike a WBS; it is usually represented as a *Mind Map™* (Buzan, 1974). This is a vertical thinking process, the purpose of which is to complete the list, understand it, and simplify it; *it is not a hierarchical representation of the structure of the organization.*

The second step is to categorize the stakeholders—a vertical thinking process—to measure their potential influence on the program or project process and their outcome, in order to identify the key, or significant, stakeholders. There are many ways to do this; they can be classified by power level (preponderant to affected party), area of interest (financial, technical, regulatory, etc.), or structural layer (regardless of direct influence). A simple, effective way to quickly map the stakeholders is to develop an influence diagram. Based on the level of power, of all types, they could exercise on the program or project and their level of interest (whichever the area of interest), which can be positive or negative.

FIGURE 9.4. THE STAKEHOLDER INFLUENCE DIAGRAM.

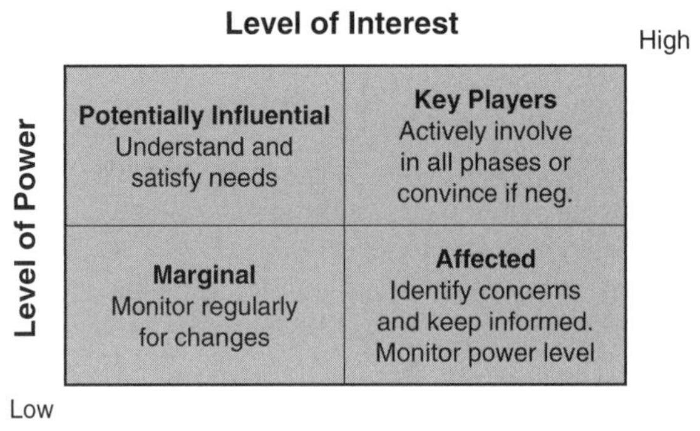

The influence diagram can be divided into four major areas (as shown in Figure 9.4), or each axis can be graded from 1 to 10 to create a more detailed picture of stakeholders' influence. Level of interest could also be divided into negative and positive, each with a gradation. When doing this, though, the team should not forget that a high level of interest, whether positive or negative, should be seen as similarly influential, but with a different potential impact.

Case Study: Stakeholder Influence Grid

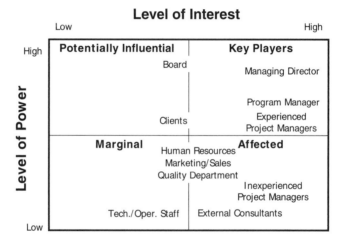

Function(al) Analysis

The PMBOK Guide (PMI, 2000, p.18) states that "finding appropriate resolution to [stakeholders] differences can be one of the major challenges of project management." The VM Standard (BSI, 2000) claims that "stakeholders . . . may all hold differing views of what represents value. The aim of VM is to reconcile these differences. . . ." To reconcile the stakeholders' differences, the first step is to identify stakeholders' needs and expectations; traditionally, project management has associated needs with requirements and expectations are considered undefined requirements. More precisely, Dorothy Kirk has identified expectations as "beliefs or assumptions about the future . . . usually set unintentionally" (Kirk, 2000). Conversely, VM has defined needs as "what is necessary for, or desired by the user. A need can be declared or undeclared, it can be an existing or potential one" (BSI, 1997). For VM, needs and expectations should be considered as one and the same, but it identifies needs and wants in order to distinguish what is absolutely necessary from what is not.

Identification of Needs and Expected Benefits. Although identification of needs is technically part of functional analysis, stakeholders' analysis and functional analysis are usually a seamless process. Before the workshop, the team will be required to gather information about the subject to be addressed through expert judgment and historical information. The team may also be required to carry out interviews of stakeholders that have been identified as key and cannot, or will not, attend.

Functional analysis requires clarifying expectations, rendering these explicit by asking questions to make stakeholders express them openly (Barton, 2000), and making any remaining assumptions agreed and well documented, so stakeholders can examine and clarify them. Once expectations are explicit, they need to be managed, which means comparing available versus required resources to fulfill them and resolving any gaps. This can be achieved only by communicating openly and early in the program or project. Finally, measurable criteria will be identified to ensure that the achievement of these expectations can be followed through.

Generally, needs identification is carried out using intuitive techniques like brainstorming or simple discussion. The European VM Standard (BSI, 2000) defines a more thorough method called Method of Interaction with the External Environment, or more simply, Interactors' Analysis. It consists of identifying the different interactive agents of the environment of the program or project (stakeholders as well as physical parameters and constraints) and identifying both direct expectations and interactions regarding the product or process, as well as functions created between interactors through the product or process. Although most functions can be identified using intuitive methods, it is worthwhile, if resources are available, to complete the study with an interactors' analysis.

Example:

> In our case study, a direct expectation for the Managing Director toward the improvement of project management delivery may be to standardize PM processes; an interaction function may be the need to choose between different standards, which will be identified through the use of interactive agents like existing PM standards.

When identifying stakeholders' needs, one should consider both direct, or tangible, needs—usually translated into hard benefits (economic, technical, operational)—and indirect, or intangible, needs—usually translated into soft benefits (power, politics, communications). In addition, one should also identify contradictions and assess their consequences, prioritize values, and manage trade-offs. This process should be iterated during decision implementation.

Case Study: Needs/Expected Benefits

Board:

- *Maintain/increase revenue*
- *Increase market share*
- *Stop clients' complaints*

Managing Director:

- *Demonstrate control*
- *Deliver within set parameters*
- *Maintain profits*

Clients:

- *Deliver within agreed parameters*
- *Maintain quality*
- *Improve current processes*

Experienced Project Managers:

- *Maintain freedom*
- *Get clear objectives/deliverables*
- *Understand company culture (new)*

Inexperienced Project Managers:

- *Understand company culture*
- *Get clear instructions (processes and procedures)*
- *Know where to find information (mentor?)*

Organization of Expected Benefits: The Function Breakdown Structure (FBS). Once benefits have been identified, it is time to start organizing them. This organization is accomplished through a hierarchical structure, based on needs, which has been labeled function breakdown structure for its similarity to the WBS (except it is, by convention, oriented from left to right, instead of top-down). In the United States, diagrams based on the same basic concepts are called FAST diagrams (Function Analysis Systems Technique). FAST diagrams have stricter rules than FBS and are often too constraining for program or project preinitiation situations; they are well adapted to project-product improvement or development. In Europe true FAST is little practiced and VM practitioners use what they call function trees or functional diagrams, which are similar to the FBS. The FBS process and the development of high-level WBS for projects are an iterative process, and the FBS must be reviewed regularly to continue representing the stakeholders' views.

The foundation rule of any function diagram is to create a hierarchy of functions that goes from the more abstract needs (the why) to concrete actions or tasks/activities (the how). The objective for the FBS is to model the expected benefits of the stakeholders in a way that represents their shared view of the situation. At the higher levels, these will be labeled vision, mission, objectives, and goals; at the lower levels, they will be called functions, tasks, and activities. If used to define programs or projects, the FBS should be a seamless continuum from the objectives and goals to the actions (tasks and activities) required to satisfy that purpose; at that level it could be labeled a task-oriented WBS and therefore clearly links project definitions to the stakeholders' expected benefits in terms of their contribution. Like a WBS, it helps define the scope of the needs, by identifying gaps and verifying completeness and by eliminating unnecessary functions. It also shows specific relationships by identifying the main purpose, critical success factors, and prioritizing needs and wants. It is an effective way to model complex situations objectively, when ambiguity is high and agreement low.

To build the FBS, the team will start with a minimal number of functions and relate them to each other using a how?-why? logic, as shown in Figure 9.5, to develop the basic structure. More functions will then be added in relationship with those already in place. Eventually, the completeness of the FBS will be verified by adding a when? relationship vertically; the when? constitutes a sequence of functions/tasks that, if achieved, will fulfill the higher-level function/benefit.

Identification of Critical Success Factors (CSF). Once the team is relatively satisfied that the FBS represents their shared view of the situation and their expected benefits, it must identify the CSFs. CSFs must be of a high enough level to be manageable and of a low enough level to be easily measured; they are generally qualitative. There are few rules as to what constitutes a CSF, except that the team identifies them as significant measures of the program or project's success. CSFs can be identified at different levels of the same FBS but should cover the complete range of needs. This means that if the team decides to go down one level, they must take all the subsidiary functions of that function; also, they cannot identify CSFs in the same branch at two different levels. Breaking these rules would create a gap or a redundancy.

FIGURE 9.5. THE HOW?-WHY? LOGIC.

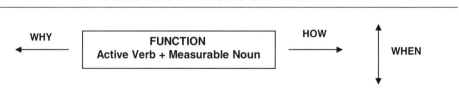

Case Study: Function Breakdown Structure with CSFs

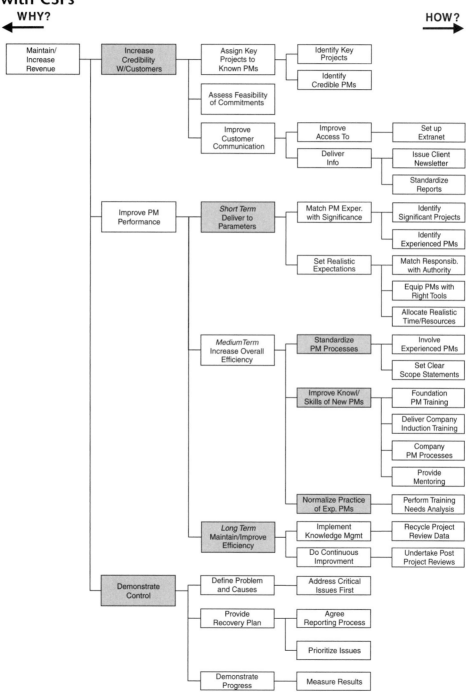

Note: High-level goal of program on left; critical success factors in gray; specific actions on right.

Eight to twelve CSFs is generally a good manageable number for most programs and projects, although large programs may require more. Once the CSFs have been identified and agreed by the team, they will become the baseline value criteria for the evaluation of all options in decision making and change management in the program or project. To increase the effectiveness of this process, CSFs need to be prioritized. Paired comparison is a fast and objective way to achieve this. Generally, 5 points will be allocated between each pair of CSFs. Combinations can therefore be 5-0, 4-1, 3-2, and vice versa, as shown in Figure 9.6. Once this is done, the scores will be brought back onto a 100 percent scale. Prioritization should be reviewed regularly to confirm ongoing validity, mostly in programs that are more long term and can therefore be more susceptible to changes in priorities.

Case Study: Weighting of CSFs using paired comparison:

A—Increase credibility with customers
B—Deliver to parameters (Short term)
C—Standardize PM processes
D—Improve knowledge and skills of new PMs
E—Normalize practice of experienced PMs
F—Maintain/improve efficiency
G—Demonstrate control

	A	B	C	D	E	F	G	Total
A		3	5	4	4	4	3	28
B	2		4	3	3	4	4	24
C	0	1		2	2	2	1	11
D	1	2	3		3	2	2	17
E	1	2	3	2		1	1	14
F	1	1	3	3	4		3	15
G	2	1	4	3	4	2		16

RANKING OF CSFS	SCORE	WEIGHT
1. Increase credibility with customers.	28	22
2. Deliver to parameters (short term).	24	19
3. Improve knowledge and skills of new PMs.	17	14
4. Demonstrate control.	16	13
5. Normalize practice of experienced PMs.	15	12
6. Maintain/improve efficiency.	14	11
7. Standardize PM processes.	11	09
Total:	125	100%

Characterization of Critical Success Factors. Although traditional VM calls for characterization of all functions of lower level, experience has shown that it is more resource-effective in programs and projects to characterize only the CSFs. Characterization, as explained earlier, requires identifying key performance indicators. The French have developed and standardized a characterization process that works very well in programs and projects (AFNOR, 1991; Thiry, 1997). It basically consists of three steps:

1. For each CSF, identify one or more measurable criteria.
2. Set the expected level of performance.
3. Identify an acceptable range or tolerance.

This process is repeated for each CSF in order to create a baseline related to the needs of the stakeholders.

Example: Deliver to Parameters (Short Term)

Criteria could be as follows:

- Actual time/cost vs. baseline in percentage of difference
- Time to achieve objective in weeks

Level could be as follows:

- 5% maximum
- 8 weeks

Flexibility could be:

- −10% or +5%
- +2 weeks

In general, sensemaking processes are performed in a team environment. It is possible, if stakeholders cannot commit the necessary time, to perform some of the tasks in smaller groups, as long as the results are offered to the whole team for discussion and final approval.

Ideation

VM practitioners use creativity techniques in a number of VM processes, but if there is one area where creativity becomes essential, it is the generation of alternatives. Although the functional analysis has already generated a number of actions (how?), ideation will broaden the scope of possibilities from which to develop options and therefore increase the quality of those options from which the team will choose solutions. It will also generate a bank of possible solutions that could be exploited and developed further if circumstances or stakeholders' expectations changed, thereby saving important redevelopment time and resources. Ideation is the traditional *creativity phase* of VA/VE; it consists of identifying as many alternatives as possible for the fulfillment of one or more benefits. It is an area where lateral thinking cannot be mixed with vertical thinking. The team will start with CSFs, or alternatively, low-level functions, and creatively identify possible ways of achieving them.

There are a number of techniques available to the team to generate alternatives. The best known are brainstorming (Osborn: where the subject is clearly defined, or Gordon: where the subject is known only to the facilitator), forced comparison, synectics, interviews, stepladder technique (Rogelberg et al., 1992), or Delphi technique. In a program or project context, forced comparison and synectics are usually not focused enough and competent facilitators in those techniques are not easily available; Delphi is time-consuming and generally costly; interviews are a good alternative, but robustness of alternatives given by the group process is lost. Osborn technique brainstorm, with an experienced facilitator, is an easily applicable and valuable technique for small groups (five to eight), and stepladder would be the best choice where large groups are concerned, as it combines the quality of exchanges of the group process with the participation and creativity of every individual. As there is an evolution toward more stakeholder involvement, VM is slowly moving from using traditional brainstorm for creativity toward techniques akin of stepladder.

In the *Osborn brainstorm technique*, the group must first agree on the definition of the problem and make a clear statement of it; this is usually achieved through the functional analysis. The group, under the guidance of a facilitator, then identifies as many alternatives as possible to resolve the problem; or in this case, the team brainstorms on the CSFs or lower-level functions to offer the expected benefits. There are typically ten rules for an effective brainstorming session:

1. Write down all ideas and comments.
2. Target quantity rather than quality.
3. Exclude criticism; assume that each idea will work.
4. Hold judgment until the evaluation phase.
5. Eliminate the word impossible from your vocabulary.
6. Let your imagination roam free (the craziest ideas are often the most important).
7. Use piggybacking (build on other ideas and comments).
8. Cross-fertilize ideas (associate or modify ideas and comments).
9. Let everybody talk.
10. Do not interrupt!

Research has demonstrated that there are some problems associated with traditional brainstorming, like social inhibition, created by shyness or dominant individuals; social loafing, which distracts from objectives; and production blocking, which increases inhibition of individuals in a group situation (Rogelberg et al., 1992; West, 1994).

The *stepladder technique* directly addresses some of the problems identified for brainstorming and particularly a strong criticism of group brainstorming, that "quantity and often quality of ideas produces by individuals working separately are consistently superior to those produced by a group working together." (West, 1994). The principle of stepladder is that each individual works independently on the problem before joining the group, which also fosters transformational learning through *critical reflection* (Cranton, 1996). Ideas are then shared with the group in turn as each new individual joins (to avoid being influenced), and discussion can start only when all ideas have been expressed. The first part of the discussion should follow the ten rules of brainstorming, as the objective is to develop group alternatives. Only when the group is satisfied that they have explored all possible alternatives will they switch into vertical thinking mode.

A short form of the stepladder technique consists of having individuals work first independently and then discuss their ideas in pairs and in fours before sharing all ideas with the group. In this case there is a series of short, lateral–vertical thinking processes, with the first part of every discussion lateral, before the final evaluation and development of options.

In Australia, Roy Barton (2000) used a variation of stepladder for large VM workshops (20 to 30 or more participants) at strategic planning level for the New South Wales Public Sector. He divided the group in smaller working units of five to seven people, who work independently from each other before sharing findings with the whole group. A survey of 200 participants in more than 20 of his workshops showed that 90 people felt their views were adequately considered, and in more general terms, 76 percent felt that VM was very important to their project.

Case Study: Alternatives for Three Top CSFs

CSF 1: Increase Credibility with Customers.

- *Assign key projects to known/best PMs.*
- Improve customer communications.
- *Assess feasibility of commitments.*
- Communicate success.
- *Deliver within parameters.*
- Acknowledge past failures to customers.
- Involve customers in problem identification and solving.
- Create focus groups.
- Give performance guaranties and metrics.
- *Review customer account history to ensure good fit.*
- Involve Business Unit Managers with clients.
- *Realign dates and estimates to be realistic.*
- *Validate sales commitments before commitment to customer.*

CSF 2: Deliver to Parameters

- *Match PM experience with significant projects.*
- *Set realistic expectations.*

- *Provide strong incentives to deliver.*
- Renegotiate expectations with customers, based on achievability.
- *Provide PMs with required tools.*
- *Create team incentives to meet objectives.*
- Implement overtime incentives.
- *Re-scope deliverables in regards of resources.*

CSF 3: Improve Knowledge and Skills of New PMs

- Provide general training.
- Send new PMs to PMI seminars.
- *Create a mentoring program.*
- *Share lessons learned.*
- *Start new PMs on small, less risky projects.*
- Have them develop their own teaching material on PM methodology.
- Provide regular feedback.
- *Reward good performance.*
- Spot recognition program.
- *Set-up incentive plans.*

Alternatives in italic are part of chosen option (see next section).

In a project of program context, it is important to keep the team focused during ideation; although creativity requires the group to accept all ideas, it is also easy to lose sight of the initial objective. While the loose-rein approach is valid for purely creative processes, the management of programs and projects require some focus because of the limited resources available. An experienced facilitator should be able to manage the process to allow enough freedom to foster creativity without losing sight of the objectives. It is also noticeable that fewer ideas will be generated at higher levels (strategic or tactical) than at lower levels (technical or operational) because they are more abstract and cover more ground.

Elaboration

This phase of the VM process combines two traditional VA/VE phases: *evaluation* and *development*. It is the assessment of those alternatives that have been generated in the previous phase and the development of viable and profitable options. It uses *vertical thinking* concepts like feasibility, cost-benefit analysis, weighted matrices, risk analysis, estimating, and so on.

The first step of elaboration is to eliminate all nonviable alternatives; alternatives could be deemed nonviable because they are clearly unachievable or deliver none of the expected benefits, or more simply, because they will not be accepted by the stakeholders. Other alternatives will be clearly viable because of their intrinsic value in terms of achievability and benefits, or their wide acceptance by stakeholders. Some alternatives will fall into a gray zone where acceptance or viability will not be clearly established and rejection is not clearly an option; these alternatives will be looked at more closely to establish if they can be combined or improved to make them clearly viable.

The second step is to combine, modify, or develop the alternatives further to generate options. The development of options includes the gathering of additional data and a degree of analysis that enables the team to make a well-documented decision and persuade sponsors to support them. A practical way to achieve this is to identify "champions" for each potential option and to ask these champions to develop their options up to a point where they feel comfortable to make a recommendation that outlines benefits, states advantages and disadvantages (risks), and draws contingency plans.

Case Study: Elaboration and Comparison of 3 Options

Option 1: *Develop a Comprehensive Staffing Plan*

- Addresses both short term performance objectives and long term strategic goals.
- Cost invested brings return.
- Plan addresses both achievability and benefits.

Option 2: *Develop a Performance Incentive Plan*

- Could be costly (escalation and equity).
- Does not address right PM for right project issue.
- May not increase credibility.

Option 3: *Develop a Customer Communication Plan*

- Does not really address overall problem.
- Could backfire if delivery does not follow.
- Only addresses short term.

The third step is to compare a number of options between them based on their relative capacity to deliver CSFs, in order to prioritize resource expenditure. Decision factors may consider aspects like available resources, quick wins, ease of implementation, or, simply, benefits.

The benefits variance (BV) concept ranks and weighs CSFs—using paired comparison, for example. The team then determines expected benefits (EBs) by defining a minimum acceptable score (note that, if weighting adds up to 100, as shown in example, and scores are set on a scale of 0 to 10, the maximum possible score will be 1,000). Options are evaluated against each of those CSFs to generate, for each, a combined score that corresponds to their offered benefits (OBs). Any option that is below the minimum score is rejected; others are ranked. Thus, an order of priority is determined.

Case Study: Evaluation of Options

WEIGHTED MATRIX SCORING OF OPTIONS ON BENEFITS VARIANCE.

Options CSF/Weight	Increase credibility with customers		Deliver to parameters (Short term)		Standardize PM processes		Improve knowledge and skills of new PMs		Normalize practice of experienced PMs		Maintain/improve efficiency		Demonstrate control		Total
	A	22	B	19	C	14	D	13	E	12	F	11	G	9	Total
1—Develop a comprehensive staffing plan	7	154	6	114	7	98	9	117	9	108	7	77	9	81	**749**
2—Develop a performance incentive plan	3	66	7	133	2	28	1	13	5	60	8	88	5	45	**433**
3—Develop a customer communication plan	6	132	2	38	0	0	0	0	0	0	0	0	7	63	**233**
Maximum Possible Score	10	220	10	190	10	140	10	130	10	120	10	110	10	90	**1000**
Minimum Acceptable Score	7	154	7	133	6	84	6	78	5	60	5	55	4	36	**600**

Resource variance (RV) is estimated, considering a concept of achievability based on the organization's capabilities, or available resources (ARs) against program or project requirements, or required resources (RRs), in terms of:

- *Financial factors*: Capital cost, cash flow, life cycle costs
- *Parameters and constraints*: Size/scope, cost, type of work, etc.
- *Human resources*: Expertise, spread, external versus internal, etc.
- *People factors*: Availability, competence (skills and expertise), customer perception
- *Complexity*: Innovativeness, interdependencies, stakeholders, etc.

To be accepted, an option must be clearly achievable and RV must be superior to the required minimum.

If a number of options are contemplated, the RV result is added to the BV and the combined score is used to prioritize resources against benefits; it is possible to establish a minimum acceptable score if achievability elements are quantified. This means that the value manager is not only concerned with balancing resources with satisfaction of needs (benefits) but also with making sure that both for benefits and resources, capacity is matched with intent.

Once this is done, alternatives that have been judged viable are offered for final decision/prioritization. To complete the process, each major option will be assessed against the KPIs and risks will be identified and analyzed; then risk responses should be included in the final options.

Choice

The decision itself is the last step of the process. Traditionally, in VA/VE this was the *recommendation* phase; in VM, the team is expected to "own" the power over resources and therefore make the actual decision rather than recommend options. This is a major change with traditional value methodologies, as the value practitioner has, until recently, mostly acted as an external consultant and did not have decision-making power. Under more recent developments, where VM is considered a group decision support process, decision making is part of the VM process, since the decision makers are on the team.

A number of authors (Vennix, 1996; Checkland and Scholes, 1994; Guba and Lincoln, 1989; Waring, 1989) point out that in complex situations, decision making is subjective and intuitive rather than objective and rational. These findings outline the importance of aiming toward consensus, which, as a number of research studies have demonstrated (Vennix, 1996), will foster support and commitment to the implementation of the decision and increase its intrinsic quality. Group decision making is influenced by the significance of issues for the group, the shared understanding among group members, and participants' representation in the process (Eden, 1992; Deetz, 1995). VM, through its sensemaking, ideation, and elaboration processes, fosters all those elements and makes consensus easier to achieve. Still, there is a gradation in decision making from the most to the least consensual: consensus, compromise following discussion, vote and consensual discussion, vote following discussion, simple vote, leader decision following discussion, and finally, leader decision. If there is no other choice than to have a leader decision, the team must still be given a fair opportunity to discuss and try and reach consensus before the final decision is made.

Interestingly, a study by Holloman and Hendrick in 1972 demonstrated that consensus after majority vote is the most effective method in terms of time, quality, and satisfaction with the decision (Vennix, 1996). In this case the team takes a majority vote to choose the options and subsequently tries to arrive at a consensual decision. Experience shows that, when using VM, consensus is fairly easy to achieve, as it is a logical conclusion to the process, and when vote needs to be taken, there is usually little discussion or conflict.

The following case study example shows a vote with discussion, where a vote was taken to choose option 1 (unanimously). Following agreement, a discussion was initiated to decide which elements it should include. The final choice is slightly different from the initial option presented for evaluation and now includes incentives, which were part of option 2.

Case Study: Choice

Option 1 Components:

- Assess competency levels of all staff.
- Assess PM relationship/experience with different customers.
- Empower PMs, give authority to make decisions.
- Identify required tools and provide them.
- Tie incentives to restructuring.
- Tie incentives to company goals.

- Gain comprehensive knowledge of workload versus demand.
- Assign lesser tasks to new PMs.
- Team up new PMs with experienced PMs across organization.

Mastery—Benefits Management

If VM is to be considered a style of management, gatekeeping becomes an essential part of the process to ensure that value is delivered. With power over resource prioritization, VM can really achieve management of value over time if it is used for program appraisal and at project "gateways," which correspond to milestones. These gateways generally relate to deliverables, allowing stakeholders to monitor benefits methodically. As the program or project progresses toward its outcome, the focus of VM will evolve from strategic to tactical and technical/operational level, and expected benefits and context may evolve along the way. Hence, VM must be an iterative process.

Change management is an essential part of the management of value; the VM process is also applied to change, especially regarding evaluation of results and integration with overall needs and expectations. When VM is applied to the change control process, it ensures that the "real" issues are addressed and that changes are made in line with the CSFs and other expected benefits. This leads us to the concept of benefits management, which has recently been identified as a significant aspect of the management of programs. VM provides a clear link between identified expected benefits (functions) at different levels of the organization and results. As an iterative process, VM regularly reassesses stakeholders' needs and expectations and alerts program and project managers early enough to identify the most resource-effective alternatives and evaluate them on a rational basis.

There are a few elements that must be put in place to support this iterative process:

- Gateways for approval of deliverables need to be defined (typically part of program management) and a VM process (sensemaking–ideation–elaboration) used to review the deliverables.
- Regular reviews of stakeholders' needs and expectations should be planned, specifically at project gateways and program appraisals; CSFs and other expected benefits adjusted in consequence.
- Value criteria (CSFs and KPIs) are to be the basis for change request evaluation; again the VM process should be applied to the management of change.

To achieve this, the value team must develop concepts of responsive and "formative" evaluation, as opposed to strictly baseline and "summative" evaluation. Whereas summative evaluation is based on preset standards (the baseline), where the actual situation is compared with these standards, formative evaluation is the iterated negotiation of program or project evaluation criteria and priorities, more appropriate to complex situations. VM, applied to program appraisal, project gateways, and more generally, change management enables the team to structure this process around a robust framework.

Organizations that apply the preceding principles often closely link the VM team with finance or resource management and follow projects on an ongoing basis. Project initiation is based on VM; reviews are carried out at gateways by the VM team and include the project team. Outputs of this process include resource reprioritization, contingencies and risk response management (reallocation of unused funds or resources to other projects on an ongoing basis), revalidation of CSFs, and other success factors to improve value.

Summary

The PMBOK Guide (PMI, 2000) states: "Finding appropriate resolution to [stakeholders] differences can be one of the major challenges of project management." On the other hand, the European Value Management Standard (BSI, 2000) says: "Stakeholders . . . may all hold differing views of what represents value. The aim of VM is to reconcile these differences and enable an organisation to achieve the greatest progress towards its stated goals with the use of minimum resources." This chapter has defined the methodology, tools, and techniques that enable the achievement of the latter statement.

To be most effective, value management must be linked to strategy, success factors, programs, and prioritization, as well as to change management. VM is a group decision management (GDM) process, which enables groups of stakeholders to make sensible and well-grounded strategic decisions, based on needs; define and prioritize expected benefits; and quantify them. It also enables program managers to select and prioritize projects and other actions, based on the expected benefits that have been defined at a strategic level and the most effective use of resources. Value management, if it is applied at the project gateways and program appraisal phase, becomes a change management methodology that ensures a choice of options directly related to the expected benefits.

Over its history of more than 50 years, value methodologies have spread across the world, and although VM, VE, and VA are widely practiced, there are still differences in practice depending on regions and value association allegiance. However, although methods and techniques may be different from region to region, the same basic principles apply everywhere. The grouping of European value associations under a common Standard and Training and Certification System, and the recent discussions toward certification agreement between SAVE International (United States), the EGB (European Governing Board), and preliminary talks between the EGB and the AIVM (Australia), are all signs of the desire to unify the knowledge and practice of VM. The SAVE-EGB Certification agreement proposal identifies, among others, the two following medium-term objectives:

- Develop universally recognized value knowledge and practice standards.
- Extend common understanding and respect for regional differences in the concepts and practice of value methodologies.

In conclusion, VM is the method of choice to deal with the ambiguity of stakeholders' needs and expectations and the complexity of changing business environment at program level and project initiation. It will bring structure and objectivity to what has often been a highly subjective and intuitive process and provide a framework for decision making throughout the delivery process. The VM process requires involvement of the whole program/project team, at different levels and times, but to be effective, decision makers—with authority over resources—must be involved at all stages of the process.

References

AFNOR, Commission de normalisation. 1985. *Analyse de la valeur, recommandations pour sa mise en oeuvre,* norme NF X 50-153, AFNOR, Paris.

———. 1991. *Analyse de la valeur, Analyse fonctionnelle, Expression fonctionnelle du besoin et cahier des charges fonctionnel,* norme NF X 50-151, AFNOR, Paris.

ASTM. 1995. Subcommittee E-06.81 on Building Economics. *Standard Practice for Performing Value Analysis (VA) of Buildings and Building Systems.* Standard Designation: E 1699-95. Philadelphia: American Society for Testing Materials.

Barton, R. 2000. Soft value management methodology for use in project initiation: A learning journey. *Journal of Construction Research* 2(1):109–122.

BSI. 2000. *Value management.* Standard BS EN 12973:2000. European Committee for Standardization (CEN) Technical Committee CEN/TC 279–British Standards Institute Technical Committee DS/1, Chelsea, UK.

———. 1997. *Value management, value analysis, functional analysis vocabulary.* Standard BS EN 1325-1:1997. European Committee for Standardization (CEN) Technical Committee CEN/TC 279–British Standards Institute (BSI) Technical Committee DS/1, Chelsea, UK.

Brun, G., and F. Constantineau. 2001. *Le management par la valeur: Un nouveau style de management.* AFNOR, Paris, FR.

BIS. 1987. *Guidelines to Establish a Value Engineering Activity.* Management and Productivity Sectional-Committee, EC 9., IS: 11810-1986. New Delhi: Bureau of Indian Standards.

Buzan, T. 1974. *Use your head.* London: BBC Consumer Publishing.

Bytheway, C. W. 1965. FAST diagramming. *SAVE Proceedings.* SAVE, Northbrook, IL.

de Bono, E. 1990. *Lateral thinking: A textbook of creativity.* 3rd ed. Harmondsworth, UK Penguin.

Checkland, P., and J. Scholes. 1994. *Soft systems methodology in action.* Chichester, UK: Wiley.

Cranton, P. 1996. *Professional development as transformative learning.* San Francisco: Jossey-Bass Publishers.

Deetz. S. 1995. *Transforming communication, transforming business.* Creskill, NJ: Hampton Press.

EFQM 2000. *The EFQM Excellence Model.* Brussels: European Foundation for Quality Management. www.efqm.org/.

Fong, P. S. W. 2002. Effective facilitation in value management workshops. *Proceedings of the 5th PMI-Europe Conference.* Cannes, June.

Green. S. D. 1994. Beyond value engineering: SMART value management for building projects. *International Journal of Project Management* 12(1):49–56.

Guba, E.G., and Y.S. Lincoln. 1989.. *Fourth generation evaluation.* Newbury Park, CA: Sage Publications.

ISO 2000. *ISO-9000:2000 Quality management systems.* Geneva: International Organization for Standardization.

Kaplan, R. S., and D. P. Norton. 1996. Using the balanced scorecard as a strategic management system. *Harvard Business Review.* (January–February): 75–85

Kaufman, J. 1997. *Value management: Creating competitive advantage.* Menlo Park, CA: Crisp Publications Inc.

Kirk, D. 2000. Managing expectations. *PM Network.* (August).

Miles, L. D. 1972. *Techniques of value analysis and engineering.* 3rd ed. New York, McGraw-Hill.

Neal. R. A. 1995. Project definition: The soft systems approach. *International Journal of Project Management* 13(1):5–9.

Project Management Institute. 2000. *A guide to the Project Management Body of Knowledge.* Newtown Square, PA: Project Management Institute.

Quinn, J. J. 1996.. The role of "good conversation" in strategic control. *Journal of Management Studies* 33(3):381–394.

Rogelberg, S. G., J. L. Barnes-Farrell, and C. A. Lower. 1992. The stepladder technique: An alternative group structure facilitating effective group decision-making. *Journal of Applied Psychology* 77: 730–737.

SAVE. 1993. *Certification examination study guide.* Northbrook, IL: Society of American Value Engineers.

———. 1997. *Value methodology standard.* Northbrook, IL: Society of American Value Engineers.

Spradlin T. 1997., A lexicon of decision making. *Decision Analysis Society Web site:* http://faculty.fuqua.duke.edu/daweb/lexicon.htm.

Thiry, M. 1997. *Value management practice.* Sylva, NC: Project Management Institute..

———. 2000. A learning loop for successful program management. *Proceedings of the 31st PMI Seminars and Symposium.* Newtown Square, PA: Project Management Institute.

———. 2001. Sensemaking in value management practice. *International Journal of Project Management.* Oxford, UK: Elseveir Science.

———. 2002. Combining value and project management into an effective programme management model. *International Journal of Project Management.* Oxford, UK: Elseveir Science. Also appears in 2001 *Proceedings of the 4th Annual PMI-Europe Conference.* London.

Vennix, J. A. M. 1996. *Group model building: Facilitating team learning using system dynamics.* Chichester, UK: Wiley.

Waring, A. 1989. *Systems methods for managers: A practical guide.* Oxford, UK: Blackwell Science.

Weick, K. E. 1995. *Sensemaking in organizations.* London: Sage Publications.

West, M. A. 1994. *Effective teamwork.* Leicester, UK: The British Psychological Society.

Woodhead, R., and C. Downs. 2001. *Value management: Improving capabilities.* London: Thomas Telford.

CHAPTER TEN

PROJECT SUCCESS

Terry Cooke-Davies

Few topics are more central to the art and science of managing projects than project success. It would seem to be self-evident that every person involved in the management of a project will be striving to make it successful. In the world of the twenty-first century, "success," like its close relative "winning," seems to be an unquestioned "good." So surely there can be nothing too difficult about measuring project success?

Unfortunately, behind this rather obvious-sounding question, there lies a seething mass of complex assumptions and interrelated concepts that have led one author almost despairingly to ask, "Measuring success—can it really be done, and if carried out, what purpose does it serve?" (De Wit, 1988). Difficulties abound for many reasons: the different viewpoints, interests, and expectations of groups of stakeholders involved in any given project; the subjective nature of perceptions of "success"; the tendency of perceptions to evolve over extended periods of time; the difficulty of assessing complex phenomena using simple metrics— the list is a lengthy one. On closer examination, project "success" turns out to be a rather slippery subject.

And yet the need remains. Every project is undertaken to accomplish some purpose, and it is both natural and right to seek to assess the extent to which that purpose has been achieved. Equally, if the art and science of project management is to advance, then practices that lead to success are to be encouraged over those that lead to failure. Indeed, these two aspects of the need to understand project success each lead to a different aspect of the topic that will be covered more fully in this chapter. Success criteria are the measures against which the success or failure of a project will be judged, and success factors are those inputs to the management system that lead directly to the success of the project. Each is important, but the two should not be confused.

A Brief Survey of the Literature on Project Success

Many of the practitioner-focused textbooks on project management define project success criteria in terms of the time, cost, and product performance (expressed as quality, or scope, or conformance to requirements) compared to the plan. Indeed, this is so widely accepted that one popular book aimed at practitioners is subtitled "How to plan, manage, and deliver projects on time and within budget" (Wysocki, Beck, and Crane, 1995). As a headline, it commands attention, although in the body of their book, the same authors acknowledge the need to define success criteria more completely during the early stages of project definition. The difference of emphasis, however, serves to highlight a distinction that is well expressed by De Wit (1988), who differentiates between the success of project management (for which measures of time, cost, and quality might be broadly appropriate) and the success of the project, which will depend on a wider range of measures. This distinction is important, although often ignored.

The importance of the distinction is emphasized by Munns and Bjeirmi (1996), who draw attention to the short-term goals of the project manager (in delivering the required product or service to schedule and within budget) as opposed to the long-term goals of the project (to deliver the promised business benefits). Kerzner makes a similar distinction between "successful projects" and "successfully managed projects." "Successful implementation of project management does not guarantee that individual projects will be successful . . . Companies excellent in project management still have their share of project failures. Should a company find that 100 percent of their projects are successful, then that company is simply not taking enough business risks" (Kerzner, 1998; p. 37).

De Wit, as it happens, is following Baker, Murphy, and Fisher's classic analysis of 650 completed aerospace, construction, and other projects (1974), which was subsequently developed further by the same authors (1988). They concluded (p. 903) that "if the project meets the technical performance specifications and/or mission to be performed, and if there is a high level of satisfaction concerning the project outcome among key people in the parent organization, key people in the client organization, key people on the project team and key users or clientele of the project effort, the project is considered an overall success." A definition that includes elements of both project management success (technical performance specifications; satisfaction of key people on the project team) and project success (meets mission to be performed; satisfaction in parent and client organization).

This tendency to blur the distinction is also followed in work subsequent to Baker, Murphy, and Fisher by authors writing both before and after De Wit's article. Morris and Hough (1987), for example, in their seminal work on major projects, make a convincing case for the popular perception that an excessively large number of "major projects" are perceived by the public to fail, and then argue on the basis of both a comprehensive survey of the literature and also eight meticulously conducted case studies for three or possibly four dimensions to project success criteria: project functionality, project management, contractors' commercial performance, and possibly, in the event that a project was canceled, whether the cancellation was made on a reasonable basis and the termination handled efficiently. Project functionality, as defined by Morris and Hough, includes an assessment of both project technical performance, which forms a part of "project management success," and other aspects of performance, which presages the much more recent language of benefits management.

More recently, a survey of 127 Israeli project managers (Shenhar, Levy, and Dvir, 1997) concluded that there are four dimensions to project success: project efficiency (broadly De Wit's "project management success"), impact on customer, business success, and preparing for the future. The latter three fall within De Wit's category of "project success," as well as being remarkably similar to Baker, Murphy, and Fisher's conclusions.

A backdrop to the discussion on success criteria is provided by an understanding of the different parties to the project that have a legitimate interest in its success or failure. Baker, Murphy, and Fisher (1988; pp. 903*ff.*) emphasize the importance of perceptions and name the "client" and the "parent" in addition to the project team. Morris and Hough (1987, pp. 194*ff.*) refer to "sponsors," contractors, owners, regulators, financiers, and governments, as well as citizens and environmentalists. DeWit (1988, pp. 167–168) reviews the breadth of possible project "stakeholders," as does Geddes (1990). Authors generally acknowledge that each stakeholder group can have different criteria for the success of the project, thereby introducing greater complexity to the subject.

The literature on project success factors is more extensive than that on success criteria (Crawford, 2001), although much of it is based on anecdotal evidence or studies with very small sample size: The state of current understanding can perhaps best be illustrated by considering three representative studies: Baker, Murphy, and Fisher's (1988) considered findings from their analysis of 650 aeronautical, construction and other projects; Pinto and Slevin's studies (1988b; 1988a) of answers provided by 418 project managers from various industries, and Lechler's survey (1998) of 448 projects in Germany. These three have been chosen as representative because of their large samples of empirical data, because they include projects from different industries, because they use complementary data analysis methods, and because they cover the past three decades, during which time 99 percent of all the articles published about project management have been written. (Kloppenborg and Opfer, 2000).

Baker, Murphy, and Fisher

Baker, Murphy, and Fisher adopted the definition of success that has already been cited. It includes a number of factors and the perceptions of success of different groups of stakeholders. Their conclusion is that there are 29 factors that strongly affect the perceived failure of projects; 24 factors that are necessary, but not sufficient, for perceived success; and 10 factors that are strongly linearly related to both perceived success and perceived failure (i.e., their presence tends to improve perceived success, while their absence contributes to perceived failure).

The output measure (whether the project was successful or not) was a simple categorization of projects into three success "bands," based on a multiple of the factors contributing to their definition of success, which has already been discussed.

The ten factors are as follows:

1. Goal commitment of project team
2. Accurate initial cost estimates
3. Adequate project team capability
4. Adequate funding to completion
5. Adequate planning and control techniques
6. Minimal start-up difficulties
7. Task (vs. social) orientation
8. Absence of bureaucracy
9. On-site project manager
10. Clearly established success criteria

Pinto and Slevin

Pinto and Slevin derived from Baker, Murphy, and Fisher an understanding of the factors that influence project success, and then developed from it a more explicit definition of the criteria for judging project success (see Figure 10.1).

They then assessed the opinions of 418 PMI members who responded to a questionnaire about which factors were critical to which elements of project success (just over half of them

FIGURE 10.1. PINTO AND SLEVIN'S MODEL OF PROJECT SUCCESS CRITERIA.

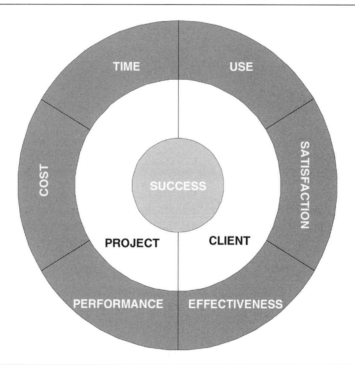

Source: Pinto and Slevin (1988a, p. 69). Project Management Institute, *Project Management Journal.* Project Management Institute, Inc., 1988. Copyright and all rights reserved. Material from this publication has been reproduced with the permission of PMI.

were project managers and nearly a third were members of project teams). They also related the results to the particular phase of the project's life cycle within which each of the factors were significant, using a simple four-phase model: conceptualization, planning, execution, and termination. Participants were instructed to "think of a project in which they were involved that was currently under way or recently completed. This project was to be their frame of reference while completing the questionnaire. The four-phase project life cycle . . . was included in the questionnaire, and was used to identify the current phase of each project" (Pinto and Slevin, 1988a; p. 70).

The results identified ten "critical success factors," which were then developed into an instrument to allow project managers to identify how successful they were being in managing their project. The ten factors are as follows:

1. *Project mission.* Initial clarity of goals and general direction.
2. *Top management support.* Willingness of top management to provide the necessary resources and authority/power for project success.
3. *Project schedule/plans.* A detailed specification of the individual action steps required for project implementation.
4. *Client consultation.* Communication, consultation, and active listening to all impacted parties.
5. *Personnel.* Recruitment, selection, and training of the necessary personnel for the project team.
6. *Technical tasks.* Availability of the required technology and expertise to accomplish the specific technical action steps.
7. *Client acceptance.* The act of "selling" the final product to its ultimate intended users.
8. *Monitoring and feedback.* Timely provision of comprehensive control information at each phase in the implementation process.
9. *Communication.* The provision of an appropriate network and necessary data to all key factors (*sic*) in the project implementation.
10. *Troubleshooting.* Ability to handle unexpected crises and deviations from plan.

Lechler

Lechler, in the most recent of the three empirical studies, also started from an analysis of the literature. His starting point was that "cause and effect" is rarely taken into consideration, but rather that the "critical success factors" are analyzed as separate, independent variables. He reviewed 44 studies, covering a total of more than 5,700 projects, and from them deduced that 11 discrete key success factors could be identified. Out of these, he chose the eight that were most frequently cited for his own empirical analysis.

Working from Pinto and Slevin's questionnaire, Lechler isolated 50 questions that corresponded to his chosen eight critical success factors, and distributed them to members of the German Project Management Society (Gesellschaft für Projektmanagement—GPM). Each respondent was sent two questionnaires and asked to complete one for a project that they considered to be successful and one for a project that they considered to be unsuccessful. They were invited to assess the project as successful if "all people involved" regarded the process (social success), the quality of the solution (effectiveness), and the adherence to time and cost objectives (efficiency) as overall positive. A total of 448 questionnaires were received and analyzed; 257 of them relating to "successful" projects and 191 to "unsuccessful" ones.

The first step in Lechler's analysis was to seek correlations between individual technical factors included in the questionnaire and overall project success. Only four factors were found to have significant correlations:

- The appropriate technology (equipment, training programs, etc.) has been selected for the project.
- Communication channels were defined before the start of the project.
- All proceeding methods and tools were used to support the project well.
- The project leader had the necessary authority (a composite of four different questions).

The second step in the analysis was to carry out a LISREL analysis (Linear Structural Relationships) for the eight critical success factors. This resulted in the path diagram shown in Figure 10.2.

Weightings were calculated for the various different paths of "causality," which, cumulatively, gave a value for r^2 of 0.59: 0.47 from the "people" factors (top management, project leader, and project team) and only an incremental 0.12 from the "activities" (participation, planning and control, and information and communication) and "barriers" (conflicts and changes of goals). Lechler indicates the importance that he attaches to this

FIGURE 10.2. LECHLER'S CAUSAL ANALYSIS.

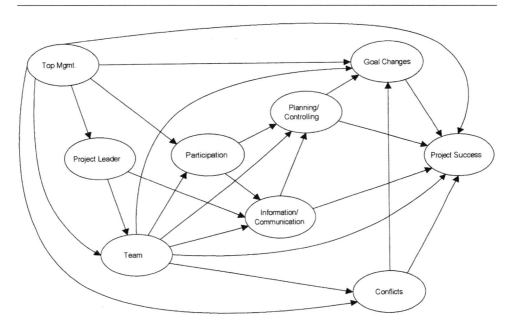

conclusion through his choice of title for the paper: "When it comes to project management, it's the people that matter."

The weighting given to each of the eight factors is shown in Table 10.1.

The Three Studies in Summary

The three studies show a certain commonality. Each of them emphasizes the importance of clear and doable project goals, of careful and accurate project planning, of adequate resources provided through top management support, and of what would today be referred to as stakeholder management. Perhaps this is not too surprising, since each of the latter two builds on the earlier work. What it has meant, however, is that these factors have become "accepted wisdom" within the world of project management practice. There are, moreover, some serious questions to be asked about how generally valid the results are for all types of projects under all circumstances. After all, each study ultimately employed a single "composite" criterion for success and based conclusions about the extent to which it was achieved on the answers provided by the same respondents who identified the role of different factors in contributing to that success. Further discussion is clearly called for.

Distinguishing Three "Levels" of Success

Regardless of what criteria are used to assess project success, and even with the broad agreement within the literature on the kinds of factors that are essential prerequisites to success, the fact must be faced that a disproportionately large number of projects are unsuccessful (Morris and Hough, 1987; O'Connor and Reinsborough, 1992; KPMG, 1997; Cooke-Davies, 2001). This suggests that there is something missing from the debate on project success, and continuous action research with more than 70 multinational or large national organizations in the United States, Europe, and the Asia-Pacific suggests that even the distinction between project success and project management success may be insufficient (Cooke-Davies, 2002a).

TABLE 10.1. WEIGHTINGS OF EIGHT FACTORS.

Factor	Direct	Indirect	Total
Top management	0.19	0.41	0.60
Project leader	—	0.18	0.18
Project team	0.16	0.36	0.52
Participation	—	0.10	0.10
Planning/controlling	0.16	0.01	0.17
Information/communication	0.12	0.06	0.18
Conflicts	−0.21	−0.08	−0.29
Goal changes	−0.20	—	−0.20

It has been argued elsewhere that the question of "which factors are critical to project success depends on answering three separate questions: 'What factors lead to project management success?' 'What factors lead to a successful project?' and 'What factors lead to consistently successful projects?'" (Cooke-Davies, 2002a, p185). The same article describes the relationship between business success and project success.

So what can be gained by regarding these three questions as pertaining to three different "levels"? Are there essential characteristics that can be used to distinguish each level from the other two? Or is this simply another conceptual framework to further bedevil a field of practice that already could be said to suffer from a surfeit of conceptual models along with a paucity of empirical data? The answer to these questions will emerge as each level is considered in turn, first, in the next section, with regard to success criteria and then subsequently with regard to success factors.

Three Levels of Success Criteria

1. Project Management Success—Was the Project Done Right? This is the measure of success that has dominated the practitioner-oriented literature on project management. In the folklore of the project manager, it is about managing time, cost, and quality. In reality, project objectives are rarely this simple. There will often be a business case to be borne in mind or a gross profit to be made; there may be health, safety, and environmental objectives to be accomplished; if the project is a technical one, or a "platform" new product development, there could be scientific or technical goals to reach. Nevertheless, the principle is simple: the endeavor is to deliver the project so that it meets the objectives within the constraints. If anything changes, which is likely given the inherent uncertainty that is involved in any new endeavor, techniques such as project risk management and project change control can be called into play as appropriate. As a guided missile seeks its target, adjusting its trajectory as appropriate along the way, so the project team seeks to achieve the project objectives. Is this then an appropriate level at which to measure the success of a project? There are three different kinds of arguments that suggest that it is.

First, modern project management has developed from a base of managing relatively "discrete" projects, each with its own organization and each established to accomplish specific purposes (Morris, 1994). The kind of success criteria that are broadly used as measures of "project management success" have not only been those most commonly applied in the history of project management (e.g., *A Guide to the Project Management Body of Knowledge*, PMI, 1996), but they also allow the project team as a coherent organizational unit to be accountable for its own performance, and the practice of aligning accountability with authority is one of the well-attested principles of good management practice.

Second, the underlying concept behind measuring success at this level is based on the well-understood principles of first-order cybernetics (Schwaninger, 1997) in much the same way that a thermostat or a guided missile operate. This is clearly appropriate for projects in which both the goals and the methods of achieving them are relatively clear at the outset (Turner and Cochrane, 1993). Third, the capture of data about the extent to which projects within the same enterprise are successful in terms of project management success enables the enterprise to compare and contrast the practices that are generally associated with successful projects with those associated with unsuccessful ones. This in turn provides the enterprise with valuable information about which project management practices are in need of improvement within project teams.

These are convincing arguments that support the case for continuing to measure project management success for many projects in many organizations. It is far from being the whole story, however, and for the second level of success, it is necessary to turn to the second of De Wit's levels—what he calls "project success."

2. Project Success—Was the Right Project Done? This level of project success is perhaps the one that is of most interest to the owner or sponsor of the project. It is, in a sense, a measure of "value for money" in its broadest sense. The assumption is that the project will be successful only if it successfully delivers the benefits that were envisaged by the people and organizations (i.e., the stakeholders) that agreed to undertake the project in the first place. In an attempt to isolate those core elements that are central to the way a project manager thinks about his or her work, a detailed analysis of the topics contained in six recent "bodies of knowledge" (Cooke-Davies, 2001, pp. 51 to 90 and Appendices P1 and P2) has shown that they can be clustered into 11 topic areas and related to each other in narrative fashion through a "systemigram" (see Figure 10.5). Viewed in this way, it becomes clear that "anticipated benefits" become the touchstone not only for formal "stage gate" reviews of projects but also for the continuous "informal assessment" of the likely success

FIGURE 10.3. THE INVOLVEMENT OF BOTH PROJECT MANAGEMENT AND OPERATIONS MANAGEMENT IN THE ACHIEVEMENT OF "PROJECT SUCCESS".

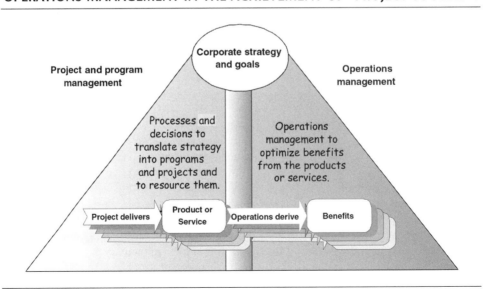

of projects carried out by owners, sponsors, or senior management and for influencing decisions about priorities and resource allocation.

Comparison of the 11 topic areas with previously published research about project success reveals a silence about "benefits" (Cooke-Davies, 2001; p. 90, Figure 7) perhaps because little has been written about benefits management or benefits realization until recently, and perhaps because the subject of benefits has been subsumed in the general discussion about project purpose or project goals. Nevertheless, there are three reasons why this is an appropriate level at which to measure the success of a project separately from the first level that was discussed, project management success.

First, as Figure 10.3 shows, benefits are not delivered or realized by the project manager and project team; they require the actions of operations management. This calls for a close cooperation between the project team on the one hand and the sponsor, customer, and/or user(s) on the other. Thus, the discussion of project success involves dialogue with a wider cross section of the organization than is appropriate or necessary for project management success. Second, delivering project success is necessarily more difficult than delivering project management success, because it inevitably involves "second-order control" (both goals and methods liable to change) and thus brings into play an additional set of corporate processes to those that are involved in delivering project management success. And third, the extent of project success is unlikely to become clear during the life of the project itself, whether success is measured quantitatively in terms of financial benefits or qualitatively in some less tangible form. For these three reasons, project success is itself a viable level at which to establish success criteria.

It is not being suggested here that project success is somehow a "better" level at which to establish success criteria. Both project success and project management success are important to any project. If a project achieves project success without project management success, there is the inevitable conclusion that even greater benefits could have been realized. On the other hand, if project management success is achieved without project success, then the owner or sponsor has failed to obtain the benefits that the project was designed to provide. And that brings us to the third level of success.

3. Consistent Project Success—Were the Right Projects Done Right, Time after Time? As the focus moves from project management success, through project success to consistent project success, a completely new set of criteria come into play, as adjudged by different groups of stakeholders. Projects are the means by which all organizations accomplish business change, as well as the means by which some organizations deliver profits to their shareholders. The consistency with which projects accomplish both project success and project management success is thus a matter of interest to every organization that is competing in markets for scarce resources, such as customers or finance.

At this level, a discussion of the criteria by which consistent project success is achieved is one that embraces the whole organization, and that will inevitably be influenced by its chosen strategy. For operations-driven organizations (such as financial services companies or mass manufacturers), consistent project success in such areas as effective overall IT expenditure and new product development can lead to competitive advantage. For project-based organizations (such as engineering contractors, defense suppliers, or turnkey IT systems providers), consistent project success can lead to profitable expansion. In either case, as the proportion of total work that is carried out in the form of projects increases, so consistent project success assumes an increasing strategic significance.

In recent years there has been a growing interest in project portfolio management for new product development (e.g., Cooper, Edgett, and Kleinschmidt, 2001), specifically for R&D (e.g., Matheson and Matheson, 1998) or generally for project spend in organizations (e.g., Artto, Martinsuo, and Aalto, 2001). But many organizations, particularly in traditional project-based industries, do not adopt a portfolio approach. For such organizations, as for all others, the effective and efficient use of scarce resources (particularly, but not only, people and money) remain of paramount importance. Thomas and Jugdev (2002) in their award-winning article on project management maturity models emphasize that long-range competitive advantage is enjoyed by those organizations that make the best use of their strategic assets (i.e., resources). Further, they conclude that maturity models are not in and of themselves sufficient to enable organizations to capitalize on their intangible assets, such as strength in project management, they do, however, go some way toward establishing the value of "project management maturity" as a further criterion of success at this third level.

A Word about Project Metrics

One practical implication of this discussion of three different levels of success criteria is that an organization would do well to monitor its performance using a "suite" of project metrics that incorporates all three levels of success, if it is serious about understanding and improving its success in the field of projects. As Figure 10.4 shows, each of the different levels of success is of interest to different levels of the corporate hierarchy, and each is visible after different amounts of time have elapsed relative to the project duration. No significant studies have as yet been published about the nature and extent of project metrics, although Atkinson (1999) and Lim and Mohamed (1999) each argue that the need for multilayered project success criteria is intimately linked to the need for more comprehensive metrics. Unpublished research (Egberding and Cooke-Davies, 2002), however, indicates that very few organizations are happy with the metrics that they use.

After considering which factors influence success at which level, a framework for a hierarchical suite of metrics will be suggested at the conclusion of this chapter.

Factors Contributing to Success at Each of the Three Levels

Although much has been written about project success factors, the distinction between different "levels of success" is a recent addition to the conceptual language of the project management research community. The predominant tenor of the discussions is to construct (or, at worst, to imply) some overall measure of "success" and then to establish by primary research, by secondary research, or by personal observation those factors that seem to correlate to success or to failure. The three chosen examples from the literature that were

FIGURE 10.4. MEASUREMENT OF PROJECT SUCCESS—ORGANIZATIONAL COMMITMENT AND TIME ELAPSED.

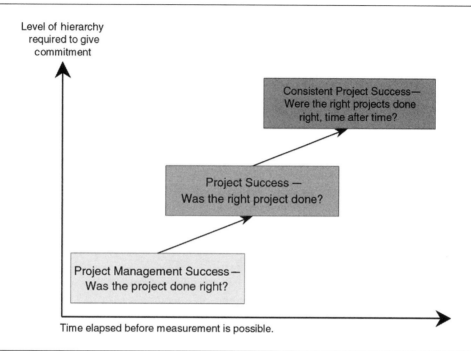

reviewed earlier in this chapter illustrate this point (Baker, Murphy, and Fisher, 1988; Pinto and Slevin, 1988a; Lechler, 1998). This is not the only criticism that can be leveled at the whole body of research into project success. Much of it uses survey techniques to collect answers from respondents both about the success or failure of individual projects and about the factors that contributed to that success or failure. It is thus better presented as research into the opinions of the project management community about success factors than as absolute success factors themselves. That is not to say that it is not useful—it is—but it is less useful than could be wished for.

Lest the pendulum be pulled too far in the opposite direction at this point, it is worth reflecting on the danger of what accountants call "spurious accuracy" in quantitative research into project success. Any assessment of project success will be carried out by specific stakeholders at specific times, and this will inevitably be influenced by many factors that are not directly related to the project itself. Business transformation or new product development projects, for example, may well be at the mercy of unforeseen and even unforeseeable developments that an assessor may or may not to take into account when judging success. And the longer the delay between project initiation and the point of assessment, the more difficult it becomes. It can be very difficult to distinguish between "luck" and "success" for any single project!

This is not a counsel for despair of ever producing any useful quantitative data—other "soft" science disciplines such as economics suffer from the same difficulties. But it does suggest the need to discern patterns or laws within large quantities of data, and thus as a prerequisite to create semantic frameworks that allow data to be compared on as near as possible a like-for-like basis.

Morris and Hough (1987), Belassi and Tukel (1996), and Crawford (2001, Appendices C and D) include excellent tabular listings of published research that between them account for 44 different research-based studies. Each of these three tables shows the breadth of conclusions that different researchers have reached concerning which factors are truly "critical" to success, although Crawford (who includes the Morris and Hough work in her own table, as well as all three studies described earlier) categorizes them into 24 groups of similar factors. Nevertheless, 24 is a very large number of "critical" factors, and if so many things are all equally important, it is also fair to conclude that nothing is especially important. What can the perplexed practitioner conclude from all this?

The first legitimate conclusion is that this is a genuinely difficult field of study that is bedeviled by at least three dimensions of difficulty. The first of these is the absence of generally accepted definitions for all the terms used to describe the subject, and it has already been noted what a slippery topic it is. Variations in language occur in different places: between researchers both as they frame the research questions and as they describe the results; between project managers and teams as they provide the data for analysis; between stakeholder groups with differing interests in the same project, and even between any given stakeholder group as its perceptions change over time; between organizations in their own internal project management guidance literature; and between industries and markets that each have their own distinct vocabulary (try talking to a research chemist in pharmaceutical R&D about "project scope management").

But if that were not enough, it is still only a part of the story. A second dimension of difficulty is the multifactorial variability of projects themselves, which makes comparisons between any two projects fraught with uncertainty. Projects are undertaken by unique temporary organizations, using unique combinations of resources (human and other) to undertake a unique, novel, and temporary endeavor that is faced with unique inherent uncertainty in order to deliver unique beneficial objects of change (Turner and Müller, 2002). On top of this, it may be the case that projects, like the weather or stock markets, are subject to "sensitive dependence on initial conditions" (Richardson, Lissack, and Roos, 2000). The third dimension of difficulty is the problem of developing robust research methods that need to encompass three worlds as varied as the physical, the social, and the personal, each of which plays an important part in the management of projects. Taken together, these three dimensions present a call to action to the project management research community that it needs to raise its game if it is to offer practical assistance to project management practitioners and the organizations that employ them.

The second conclusion is perhaps more helpful in terms of improving project management practice—there is no silver bullet with which instant success can be achieved. As in the majority of challenging human endeavors, achieving project success comes through a combination of factors, and an organization can be sure that it understands them only when it begins to see improvements in its level of consistent project success.

The third conclusion is more helpful still. Although success can be achieved only through a combination of factors, there is a relatively high degree of agreement on the kind of factors that are critical to project success. It is these that the remainder of this section will address.

But first, there is a need to map the existing research onto the three levels of success that have been defined in this article. This can be done with the aid of the project manager's worldview analysis (Cooke-Davies, 2001) that was mentioned earlier, in the discussion of the criteria for project success. Figure 10.5 shows the relationship of the 11 groups of topics expressed as a systemigram.

Of these, an examination of the detailed concepts incorporated into each allows seven of the groups to be associated with criteria for project management success as follows (numbered to correspond to those used in Figure 10.5):

2. *Project goals.* Establishing, specifying, and achieving the projects goals.
3. *Product or service.* Defining, specifying, ensuring, manufacturing, and delivering the product or service.
4. *Project work.* Identifying, structuring, planning, executing, and controlling the work to be carried out.

FIGURE 10.5. ELEMENTS OF A PROJECT MANAGER'S WORLDVIEW.

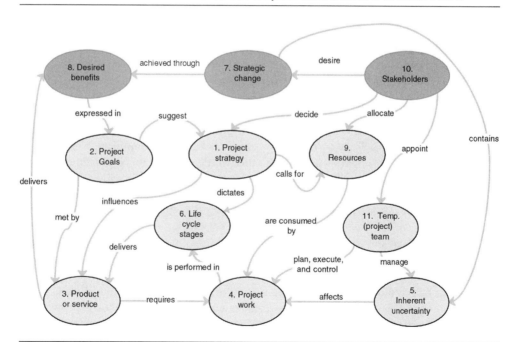

5. *Inherent uncertainty*. Managing the uncertainty that is inherent in the uniqueness of the project.
6. *Life cycle stages*. Practices relating to managing the stages that the project will need to pass through.
9. *Resources*. Allocating organizational resources to the project.
11. *Temporary team*. Creating, leading, and managing the temporary team that will initiate, plan, control, execute, and close the project

The remaining four (along with project goals, which acts as the hinge that links the two levels) can best be associated with project success as follows:

1. *Project strategy*. Establishing a strategic framework for the project.
2. *Project goals*. Establishing, specifying and achieving the projects goals.
7. *Strategic change*. Practices relating the project to the elements of business strategy to which it contributes.
8. *Benefits*. Defining, quantifying, and harvesting organizational benefits as a result of carrying out the project.
10. *Stakeholders*. Identifying and aligning the interests of the project stakeholders.

The detailed analysis underlying the worldview identified no groups of topics that could be associated directly with consistent project success, although individual topics that are contained at a lower level such as quality, culture, and organizational learning clearly contribute to this third level.

Project Management Success—How to Ensure That the Project Is Done Right?

The criteria for project management success, as has been seen, may include cost, time, quality, scope, commercial performance, technical achievements, or safety record. Although these can all be said to be indicators of project management success, the achievement of each of them is likely to depend on different factors, as one piece of recent research has indicated (Cooke-Davies, 2002a). In other words, if cost matters much more than time, then different factors are likely to be critical to the project team. Having said that, taking all the published research into account, six groups of factors can be identified as contributing to success at this level:

1. Achieve and maintain clarity about the goals of the project. Define the project in a way that clarifies both the goals of the project and the needs of stakeholders. Minimize changes to the goals once the project has started.
2. Select and assemble a capable project team of task-oriented individuals, led by a competent leader. Ensure that the team contains the right capabilities, is appropriately structured, communicates well, and has good processes for teamwork, problem solving, and decision making.

3. Ensure that the project is resourced adequately to the project scope and objectives. Mobilize top management support and ensure that there is adequate support from the organization and effective project administration.

4. Establish clarity at the outset about the technical performance required from the product and manage the scope of work tightly, using a mature change management process.

5. Plan meticulously, using well-established estimating procedures, and to a sufficient level of detail to allow effective monitoring and control. Maintain excellent metrics that relate the technical content of work done to the elapsed time and expenditure incurred.

6. Employ established risk management practices that are well understood by all project participants, including effective risk response development and control.

A summary mapping the origin of these factors onto the seven relevant topic groups from the project managers' worldview is shown in Table 10.2.

Project Success—How to Ensure That the Right Project Is Done?

Before the "success" of any individual project can be measured, the benefits that it is intended to deliver must be considered, and these can vary considerably as the following partial list of project types indicates:

- Successful business process reengineering projects (which have a notoriously low rate of achievement of their objectives) can lead directly to improved competitiveness.
- If the organization is essentially project-based (as is the case in many of the traditional project management environments such as engineering, defense, petrochemical exploration, construction, or IT/IS systems integration), then successful project performance translates directly into an improved bottom line.
- If the organization is operations-based, then successful projects to support or to improve operations (such as marketing projects, plant shutdowns, or production engineering projects) lead indirectly to improved bottom-line performance.
- Successful research projects and (in the case of some industries such as pharmaceuticals) development projects lead to a maximized return on R&D spend, leading directly to the creation of new streams of operating revenue.
- Successful development projects improve time-to-market and can enhance competitive position, product sales, or product margins.
- Successful IT/IS projects deliver improved financial benefits (either directly or indirectly) and/or reduced wastage from aborted projects (Standish, 1995).
- Successful projects to design, procure, and construct new capital assets can enhance time-to-market, return on investment, reduced operating costs, or some combination of all three.

In spite of these complexities, recent work (e.g., Cooke-Davies 2002a) on benefits- and stakeholder-management supports the main body of literature in suggesting that there are

TABLE 10.2. CRITICAL FACTORS FOR PROJECT MANAGEMENT SUCCESS.

Worldview	Level	Baker et al.	Pinto and Slevin	Lechler	Crawford	Cooke-Davies
1. Project goals	1	Goal commitment Good cost estimates Clear success criteria	Project mission	Goal changes	Project definition	Project scope management
2. Product or service	1				Technical performance	Performance management
3. Project work	1	Planning and control	Schedule/plans Monitoring and feedback	Planning/controlling	Planning Monitoring and control	
4. Inherent uncertainty	1		Trouble-shooting		Monitoring and controlling (risk)	Project risk management
5. Life cycle stages	1	Few start-up problems				
6. Resources	1	Adequate funding	Top mgmt support Technical tasks	Top management	Organizational support Administration	
7. Temporary team	1	Adequate capability Task orientation On-site project manager	Personnel	Project leader Project team Participation	Team selection Communication Leadership Team development Organization structure Task orientation Decision making and problem solving	

in fact fewer factors that are critical. There are four of them, including the one that is also critical to project management success:

1. Achieve and maintain clarity about the goals of the project. Define the project in a way that clarifies both the goals of the project and the needs of stakeholders. Minimize changes to the goals once the project has started.
2. Establish and maintain active commitment to the success of the project and its mission on the part of all significant stakeholder groups, such as sponsors, clients, owners, operations management, parent company, and so on. Establish effective communication and conflict resolution methods.
3. Develop and sustain effective processes during the project and after completion to deliver the anticipated benefits of the project and ensure that they are realized. Ensure that a close link is developed and maintained between anticipated benefits, the business case for the project, and the explicit project goals.
4. Develop a project strategy, or "trajectory" in the words of Miller and Hobbs (2000), that is appropriate to the unique environment and circumstances of the project. (Trajectory is a term that encompasses both strategy and life cycle model, and is derived from a detailed study of 60 megaprojects.)

A summary mapping the origin of these factors onto the seven relevant topic groups from the project managers' worldview is shown in Table 10.3.

Consistent Project Success—How to Ensure That the Right Projects Are Done Right, Time after Time?

This third level of success has received little attention in the literature to date. The factors that are identified in the list that follows are thus necessarily more speculative than those for either of the other levels. These three have been identified from a variety of elements of my own continuous action research described elsewhere (Cooke-Davies, 2001, 2002b; Egberding and Cooke-Davies, 2002).

1. An effective means of learning from experience on projects, that combines explicit knowledge with tacit knowledge in a way that encourages people to learn and to embed that learning into continuous improvement of project management processes and practices. Indeed, in a number of recent project management maturity models (e.g., Kerzner, 2001; Fahrenkrog et al., 2003), continuous improvement represents the fifth and highest stage of project management maturity in an organization.
2. Portfolio and program management processes that allow the enterprise to resource fully a suite of projects that are thoughtfully and dynamically matched to the corporate strategy and business objectives. These processes include the dynamic allocation of scarce resources to competing projects, in a way that serves the enterprise as a whole.
3. A suite of project, program, and portfolio metrics that provides direct line-of-sight feedback on current project performance, and anticipated future success, so that project,

TABLE 10.3. CRITICAL FACTORS FOR PROJECT MANAGEMENT SUCCESS.

Worldview	Level	Baker et al.	Pinto and Slevin	Lechler	Crawford	Others
1. Project strategy	2					Project trajectory*
2. Project goals	1	Goal commitment; project mission Good cost estimates Clear success criteria		Goal changes	Project definition	
3. Strategic change	2		Project mission		Strategic direction	
4. Benefits	2					Benefits realized† Business Case‡ Benefits delivery & management§
5. Stakeholders	2	No bureaucracy	Client consultation Client acceptance Communication	Conflicts Information/ communication	Stakeholder management	

*Miller and Hobbs, 2000
†KPMG; Thorp, 1998
‡Beale, 1991
§Cooke-Davies, 2002

TABLE 10.4. THE ELEMENT OF PROJECT SUCCESS.

Success "Level" Accountable	Typical Criteria for Success at This Level	Possible Factors Critical for Success at This Level	Organizational Level
Level 1: Project management success. "Was the project done right?"	Time Cost Quality Technical performance Scope Safety	1. Clear project goals 2. Well-selected, capable and effective project team 3. Adequate resourcing 4. Clarity about technical performance requirement 5. Effective planning and control 6. Good risk management	Project manager Project team
Level 2: Project success. "Was the right project done?"	Benefits realized Stakeholder satisfaction	1. Clear project goals 2. Stakeholder commitment and attitude 3. Effective benefits management and realization processes 4. Appropriate project strategy	Project sponsor "Client", "owner," or "operator" (recipient of benefits)
Level 3: Consistent project success. "Are the right projects done right, time after time?"	Overall success of all projects undertaken Shareholders (or equivalent) Overall level of project management success Top managers Productivity of key corporate resources. Directors of project management. Effectiveness in implementing business strategy.	1. Continuous improvement of business, project and support processes. 2. Efficient and effective portfolio, programme and resource management processes. 3. Comprehensive and focused suite of metrics covering all three levels. Business unit managers. Portfolio managers.	

portfolio, and corporate decisions can be aligned. Since corporations are increasingly recognizing the need for upstream measures of downstream financial success through the adoption of reporting against such devices as the balanced scorecard (Kaplan and Norton, 1996), it is essential for a similar set of metrics to be developed for project performance in those areas where a proven link exists between project success and corporate success. (See the chapter by Brandon.) For the project management community, it is also important to make the distinction between project success (which cannot be measured until after the project is completed) and project performance (which can be measured during the life of the project). No system of project metrics is complete without both sets of measures (performance and success) and a means of linking them so as to assess the accuracy with which performance predicts success.

Summary

Table 10.4 summarizes the points made in this chapter in tabular form. The table indicates clearly the different organizational levels that are involved in the assessment of project success and shows how each of the three levels is necessary but, on its own, not sufficient for any organization that is serious about achieving project success consistently. The table as a whole represents a framework for thinking and talking about project success—a framework such as is necessary to underpin any attempts to advance the art and science of project management.

References

Abdel-Hamid, T., and S. Madnick. 1991. *Software project dynamics: An integrated approach.* Upper Saddle River, NJ: Prentice Hall.

Artto, K. A., M. Martinsuo, and T. Aalto. 2001. *Project portfolio management: Strategic management through projects.* Helsinki, Finland: Project Management Association Finland.

Baccarini, D. 1999. The logical framework method for defining project success. *Project Management Journal* 30(4):25–32.

Baker, B. N., D. C. Murphy, and D. Fisher. 1974. Determinants of project success. NGR 22-03-028. National Aeronautics and Space Administration.

———. (1988). Factors affecting project success. In *Project Management Handbook.* ed. D. I. Cleland and W. R. King, 2nd ed. 902–919. New York: Wiley.

Belassi, W., and O. I. Tukel. 1996. A new framework for determining critical success/failure factors in projects. *International Journal of Project Management.* 4(3):141–151.

Construction Industry Institute. 1993. Cost-trust relationship. Austin, TX: Construction Industry Institute.

———. 1995. Quantitative effects of project change. Austin, TX: Construction Industry Institute.

Cooke-Davies, T. J. 2000. Discovering the principles of project management. *IRNOP IV,* Sydney: University of Technology, Sydney.

———. 2001. *Towards improved project management practice: Uncovering the evidence for effective practices through empirical research.* Available at www.dissertation.com.

————. 2002a. The "real" success factors on projects. *International Journal of Project Management* 20(3): 185–190.

————. 2002b. Establishing the link between project management practices and project success. *PMI Research Conference*. Newtown Square, PA: Project Management Institute.

Cooper, K. G. 1993. The rework cycle: benchmarks for the project manager. *Project Management Journal* XXIV (1).

Cooper, R. G. 2000. *Product leadership: Creating and launching superior new products*. Cambridge, MA: Perseus Books.

Cooper, R. G., S. J. Edgett, and E. J. Kleinschmidt. 2001. *Portfolio management for new products*. Cambridge, MA: Perseus.

Crawford, L. 2001. Project management competence: The value of standards. Henley-on-Thames: Henley Management College.

————. 2000. Profiling the competent project manager. *Proceedings of PMI Research Conference*. Newtown Square, PA: Project Management Institute.

Crawford, L., and P. Price. 1996. *Project team performance: A continuous improvement methodology*. Paris.

De Wit, A. 1988. Measurement of project success. *International Journal of Project Management* 6(3):164–70.

Duncan, W. R. 1996. *A guide to the Project Management Body of Knowledge*. Newtown Square: Project Management Institute.

Egberding, M., and T. J. Cooke-Davies. 2002. *GTN Metrics Survey: Preliminary report on Findings*. Available at www.humansystems.net.

Fahrenkrog, S., C. M. Baca, L. M. Kruszewski, and P. R. Wesman. 2003. Project Management Institute's Organizational Project Management Maturity Model (OPM3). *PMI Global Congress 2003—Europe*. Newtown Square, PA: Project Management Institute.

Freeman, M., and P. Beale. 1992. Measuring project success. *Project Management Journal* XXIII(1):8–17.

Geddes, M. 1990. Project leadership and the involvement of users in IT projects. *International Journal of Project Management* 8(4):214–216.

Haalien, T. M. 1994. Managing the cultural environment for better results. *Internet '94 12th World Congress*, Oslo.

Hayfield, F. 1979. Basic factors for a successful project. *Proceedings of 6th Internet Congress*. Garmish-Partenkirchen FRG: I.P.M.A. (formerly "Internet").

Jiang, J. J., G. Klein, and J. Balloun. 1996. Ranking of systems implementation success factors. *Project Management Journal* 27(4):50–55.

Jiang, J. J., G. Klein G. and H. Chen. 2001. The relative influence of IS project implementation policies and project leadership on eventual outcomes. *Project Management Journal* 32(3):49–55.

Kaplan, R. S., and D. P. Norton. 1996. *The balanced scorecard: Translating strategy into action*. Cambridge, MA: Harvard Business Press.

Kerzner, H. 2001. *Strategic planning for project management using a project management maturity model*. New York: Wiley.

————. 1998. *In search of excellence in project management. Successful practices in high performance organizations*. New York: Van Nostrand Reinhold.

Kharbanda, O. P., and J. K. Pinto. 1996. *What made Gertie gallop? Lessons from project failures*. New York: Van Nostrand Reinhold.

Kharbanda, O. P., and E. A. Stallworthy. 1983. *How to learn from project disasters*. Aldershot, UK: Gower.

————. 1986. *Successful projects with a moral for management*. Aldershot, UK: Gower.

Kloppenborg, T. J., and W. A. Opfer. 2000. Forty years of project management research: Trends, interpretations and predictions. *Proceedings of PMI Research Conference*. Newtown Square, PA: Project Management Institute.

Kotter, J. P., and J. L. Heskett. 1992. *Corporate culture and performance*. New York: Free Press.

KPMG. 1997. Profit-focused software package implementation. *Profit-Focused Software Package Implementation.* London: KPMG.

Laufer, A., and E. J. Hoffman. 2000. *Project management success stories: Lessons of project leaders.* New York: Wiley.

Lechler, T. 1998. When it comes to project management, it's the people that matter: An empirical analysis of project management in Germany. *IRNOP III. The Nature and Role of Projects in the Next 20 Years: Research Issues and Problems.* Calgary: University of Calgary.

Matheson, D., and J. Matheson. 1998. *The smart organization: Creating value through strategic R&D.* Boston: Harvard Business School Press.

Miller, R., and B. Hobbs. 2000. A framework for managing large complex projects: The results of a study of 60 projects. *Proceedings of PMI Research Conference.* Newtown Square, PA: Project Management Institute.

Morris, P. W. G. 1988. Managing project interfaces: Key points for project success. In *Project Management Handbook*, 2nd ed., ed. D. I. Cleland, and W. R. King, pp. 16–55. New York: Wiley.

———. 1994. *The management of projects.* London: Thomas Telford.

———. 2000. Benchmarking project management bodies of knowledge. *IRNOP IV*, Sydney: University of Technology in Sydney.

Morris, P. W. G., and G. H. Hough. 1987. *The anatomy of major projects. A study of the reality of project management.* Chichester: Wiley.

Munns, A. K. and B. F. Bjeirmi. 1996. The role of project management in achieving project success. *International Journal of Project Management* 14(2).

O'Connor, M. M., and L. Reinsborough. 1992. Quality projects in the 1990s: A review of past projects and future trends. *International Journal of Project Management* 10(2):107–14.

Pettersen, N. 1991. What do we know about the effective project manager? *International Journal of Project Management* 9(2).

Pinto, J. K. 1990. Project implementation profile: A tool to aid project tracking and control. *International Journal of Project Management* 8(3).

Pinto, J. K., and D. P. Slevin. 1988a. Critical success factors across the project life cycle. *Project Management Journal* 19(3):67–75.

———. 1988b. Project success: definitions and measurement techniques. *Project Management Journal* 19(1):67–72.

———. (1998). Critical success factors. 379–395. San Francisco: Jossey-Bass.

Richardson, K. A., M. R. Lissack, and J. Roos 2000. Towards coherent project management. *IRNOP IV*, Sydney: University of Technology in Sydney.

Robins, M. J. 1993. Effective project management in a matrix-management environment. *International Journal of Project Management* 11(1).

Schwaninger, M. 1997. Status and tendencies of management research: A systems oriented perspective. *In Multimethodology: Towards theory and practice and mixing and matching methodologies*, ed. J. Mingers and A. Gill. Chichester, UK: Wiley.

Shenhar, A. J., O. Levy and D. Dvir. 1997. Mapping the dimensions of project success. *Project Management Journal* 28(2):5–13.

Sommerville, J., and V. Langford. 1994. Multivariate influences on the people side of projects: stress and conflict. *International Journal of Project Management* 12(4).

Thamhain, H. J. 1989. Validating technical project plans. *Project Management Journal* 20(4):43–50.

Thomas, J. and Jugdev, K. 2002. Project management maturity models: The silver bullets of competitive advantage? *Project Management Journal* 33(4):4–14.

Thorp, J. 1998. *The information paradox: Realizing the business benefits of information technology.* New York: McGraw-Hill.

Turner, J. R., and R. A. Cochrane. 1993. Goals-and-methods matrix: Coping with projects with ill defined goals and/or methods of achieving them. *International Journal of Project Management* 11(2):93–102.

Turner, J. R., and R. Müller. 2002. On the nature of the project as a temporary organization. *Proceedings of IRNOP V. Fifth International Conference of the International Network of Organizing by Projects.* Rotterdam: Erasmus University.

Wateridge, J. 1998. How can IS/IT projects be measured for success? *International Journal of Project Management* 16(1):59–63.

———. 1995. IT projects: A basis for success. *International Journal of Project Management* 13(3):169–72.

Wysocki R. K., R. Beck, Jr., and D. B. Crane. 1995. *Effective project management.* New York: Wiley.

CHAPTER ELEVEN

MANAGEMENT OF THE PROJECT-ORIENTED COMPANY

Roland Gareis

The general topic of the World Congress of the IPMA (International Project Management Association) in June 1990 in Vienna, Austria, was "Management by Projects." Since then the vision to cope with the complexity and the dynamics of companies by projects has become a reality. Management by projects is today the most appropriate organizational practice in many industries. Research results about the specific strategies, structures, and cultures of the "project-oriented company" (POC) have been published, even the project orientation of regions and nations have been assessed and benchmarked.

This chapter introduces the POC as a social construct and describes models for the organizational differentiation and integration in the POC, such as projects, programs, expert pools, the PM office, and the project portfolio group. Further, a *POC Maturity Model*, based on the specific business processes of POCs, is presented, which can be applied to assess and to benchmark the competences of POCs.

It is not intended to describe the specific business processes of the POC in detail. This is done in the literature on project and program management. The intention is rather to elaborate on the new perceptions of projects and programs as temporary organizations and social systems, and to present an integrative model for the POC.

The POC: A Social Construct

Companies are becoming more project-oriented. Projects and programs are applied in all industries and in the nonprofit sector. To perceive a company as a POC is a social construction. Any company (or parts of a company, such as a division or a profit center) that frequently applies projects and programs to perform relatively unique business processes of large scope can be perceived as being project-oriented. (To simplify the further reading, the term project will be used instead of "project and program." Many of the presented concepts apply to projects as well as programs.)

A POC can be defined as an organization that

- defines "management by projects" as an organizational strategy;
- applies temporary organizations for the performance of business processes of medium and large scope;
- manages a project portfolio of different project types;
- has specific permanent organization units, such as a PM office and a project portfolio group;
- applies a "new management paradigm"; and
- perceives itself as being project-oriented.

Observing the project orientation of a company requires that we put on a special pair of "project orientation" glasses to view the practices of project, program, and project portfolio management and to observe the organizational design and the personnel management practices to support these approaches. These observations are the basis for management interventions needed to optimize the maturity as a POC.

Organizational Fit of the Strategies, the Structures, and the Cultures of the Project-Oriented Company

According to the *organizational fit model*, a company can be described by its strategies, structures, and cultures. These have to fit in order to provide good-quality services and to be cost- and time-efficient.

The specific organizational strategy of the POC is one of "management by projects." Further, it is characterized by permanent and temporary organization structures, and by a culture based on a "new management paradigm." Projects as temporary structures can only be performed successfully if appropriate strategic and cultural provisions exist.

"Management by Projects": The Organizational Strategy of the Project-Oriented Company

Project-oriented companies consider projects not only as tools to perform business processes of medium and large scope but as a strategic option for the organizational design of the company. By applying a management-by-projects approach, the following organizational objectives are pursued:

- Organizational differentiation and decentralization of management responsibility
- Quality assurance by project teamwork and holistic project definitions
- Goal orientation by defining and controlling project objectives

- Personnel development in projects
- Organizational learning by projects

For the implementation of management-by-projects symbolic management measures, showing the importance of projects, are required. Such measures include the following:

- Showing in the organization chart of the company not only the permanent organization structures but also temporary organizations (see Figure 11.1).
- Including project-related functions in job descriptions of all managers and top managers
- Including a statement on the strategic importance of project management in the company mission statement
- Marketing and promoting project management.

Projects and Programs: Temporary Organizations of the POC

For business processes with different characteristics, different organizations are adequate. Line functions such as procurement or production will generally be responsible for routine

FIGURE 11.1. ORGANIZATION CHART OF A PROJECT-ORIENTED COMPANY.

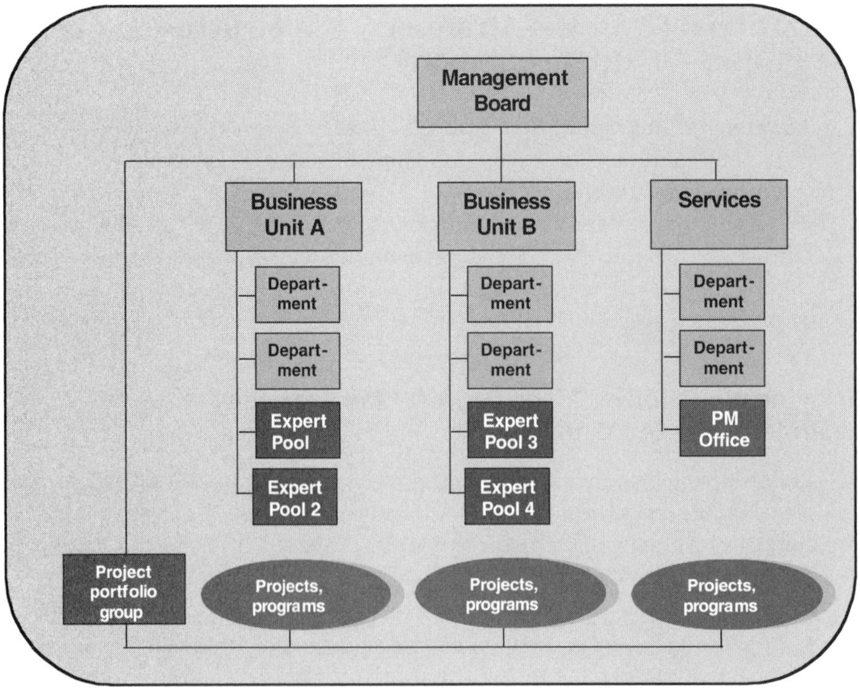

types of business processes. However, for relatively unique business processes of medium or large scope, and of short to medium duration, projects are the appropriate organizations. Projects can be defined for the performance of "contracts" for external clients, as well as for product developments, marketing campaigns, investments in the company infrastructure, or for reengineering activities for internal clients.

A program is a temporary organization for the performance of a business process of large scope. (See the chapters by Thiry, Jamieson and Morris, Archer and Ghamazadeh, Arttos and Deitrich, and Shenhar and Dvir, among others, for a further discussion of programs and program management.) A program consists of several closely coupled projects and activities. It has a time dimension and is medium or long term in duration. Typical programs are the development of a "product family" (and not of a single product), the implementation of a comprehensive IT solution for an international concern, the reorganization of a group of companies in a holding structure, and infrastructure investments considering several investment objects.

Projects and programs allow us to further differentiate companies. In addition to the permanent organizations of companies, such as divisions, profit centers, and departments, temporary organizations can also be added.

Clusters of Projects (and Programs)

For the integration of the different projects performed simultaneously in a project-oriented company, projects have to be clustered. Clustering according to the sequence in which projects are performed results in a "chain of projects." By relating projects to each other according to a defined criterion, such as the technology applied, a common client, or a geographic region, a "network of projects" results. By considering all projects performed by an organization, the "project portfolio" results. A *project portfolio* is defined as a set of all projects a POC holds at a given point in time and the relationships between these projects.

In a project portfolio, different project types, such as internal and external projects; unique and repetitive projects; marketing, contracting, organizational development projects; and infrastructure projects, might be included.

Supporting Projects by a "New Management Paradigm"

The project-oriented company is characterized by the existence of an explicit project management culture—that is, by a set of project management-related values and norms. For the project and the program management processes, specific procedures exist, creating a common understanding for the performance of these processes, the roles involved, and the management methods to be applied.

The application of this "new management paradigm" supports efficiency in the performance of projects. Traditional management approaches, based on a mechanistic management paradigm such as that of Taylorism, emphasize detailed planning methods, focus on the assignment of clearly defined work packages for individuals, rely on contractual agreements with clients and suppliers, and use the hierarchy as a central integration instrument. "New" management concepts, such as lean management, Total Quality Management, the learning organization, and business process reengineering, introduce new approaches. Among the common features of these "new" management approaches are the following:

- The use of the organization to create competitive advantage
- The empowerment of employees
- Process orientation and teamwork in flat organizations
- Continuous and discontinuous organizational change
- Customer orientation, and networking with clients and suppliers

Of course, projects can be performed within a traditional management culture. But this often results in costly, time-consuming, and for the project team members, frustrating experiences. The real benefits, the added values of project management, can only be achieved if some concepts of the new management paradigm are applied in the project-performing companies.

Management of Dynamics and Complexity in the Project-Oriented Company

POCs have dynamic boundaries and contexts. On the one hand, as the number and the sizes of the projects performed are constantly changing, permanent and temporary resources are employed and cooperations with clients, partners, and suppliers are organized in virtual teams. On the other hand, varying strategic alliances are established and relationships with different social environments of different projects have to be managed.

The greater diversity of projects that a company holds in its project portfolio, the more differentiated it becomes organizationally and the greater its management complexity will be. In order to support the successful performance of single projects as well as to ensure the compliance of the objectives of the different projects with the overall company strategies, specific integrative structures, such as a strategic center, expert pools, a PM office, and a project portfolio group are required.

Specific Business Processes of the Project-Oriented Company

The POC is characterized by specific business processes. The processes can be grouped by those that relate to single projects, those that relate to the project portfolio, and those that relate to the permanent organization. The processes of project management, program management, and assurance of the management quality of projects relate to projects. The assignment of projects, the project portfolio coordination, and the networking between projects can be defined as project portfolio management processes. The personnel management and the organizational design relate to the POC overall.

The specific business processes of the POC can be visualized in a spiderweb graph (see Figure 11.2), which can be used for the analysis of the maturity of a POC (see *Maturity of*

FIGURE 11.2. POC MATURITY MODEL BASED ON ROLAND GAREIS' MANAGEMENT OF THE PROJECT-ORIENTED COMPANY.®

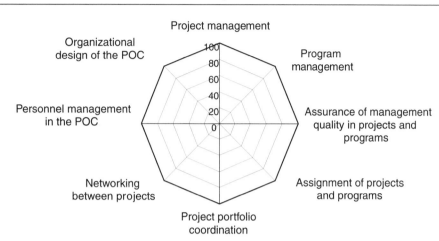

the Project-Oriented Company later in the chapter). For each of these processes, which will be described briefly in following paragraphs, the POC requires individual and organizational competences.

Project Management

The perception of projects influences the project management approach. The traditional perception of projects as complex, goal-determined, and risky tasks supports above all the use of project planning methods. Typical examples are the work breakdown structure, the CPM/network scheduling, CPM-supported resource and cost plans, and risk management.

When projects are defined as temporary organizations, the formal establishment of a project, its integration into the overall company organization, and the development of a project specific culture are emphasized. The perception of projects as social systems further promotes the emphasis on, and orientation toward, context in project management. The relationship of a project to company strategies to the other projects performed simultaneously, to the relevant social environments, and to the business case of the investment initialized by the project become a concern (see the chapters by Jamieson and Morris, and by Arttos and Deitrich in this regard). "Social" project controlling—that is, the controlling of the relationships to relevant project environments and the relationships in the project organization—is considered in addition to controlling the hard project facts (progress, schedule, costs).

ROLAND GAREIS Project and Programme Management® represents a such a systemic-constructionistic approach to project management. Projects are perceived as temporary organizations and social systems; the development of project plans is considered as a constructionistic process. The objects of consideration in the project management process are not only the scope of work, the project schedule, and the project costs, but also the project objectives, the project income, the project organization, the project culture, the project context dimensions—including relationships to the relevant environments, to other projects, and to the company strategies—and the business case.

Project management is defined as a business process of the POC, which includes the subprocesses: project start, continuous project coordination, project controlling, resolution of a project discontinuity, and project closedown. (A *project discontinuity* is a discontinuous development of a project. In the case of an existential threat to a project, we talk about a "project crisis." Another type of a project discontinuity, which also requires a change of the project identity—project objectives, strategies, organization, and culture—is the "project chance.") The project management process starts with the formal project assignment and ends with the approval of the project results by the project owner.

The project start is the most important project management subprocess, because in it the basis for the other project management subprocesses is established. The project plans, the project communication structures, the relationships to relevant environments, and so on are developed and defined in the project start process. For each project management sub-process, the objectives, functions, methods, responsibilities, and deliverables can be de-scribed; this enables the quality of the project management process to be measured.

The objectives of the project management process are to

- successfully perform the project according to the project objectives,
- contribute to the optimization of the business case of the investment, initialized by the project,
- manage the project complexity and project dynamics,
- continuously adjust the project boundaries, and
- manage the project-context relationships.

The project management objective, to contribute to the optimization of the business case of the investment, initialized by the project, is of great concern in product development and capital investment projects and of less concern in contracting projects. Decisions in projects can influence the business case of the investment—for instance, in a new office building—to a large extent. Therefore, the project manager and the project team have to take on the responsibility for the optimization of the business case of the investment too.

A project needs an appropriate degree of complexity to relate appropriately to its environment. It is a project management function to manage—to build up and to reduce—the project complexity. The differentiation of project roles, the creation of subteams, as well as the consideration of different functional disciplines and hierarchical levels in the project team, are organizational possibilities for building up complexity. The application of different project management methods (i.e., work breakdown structure, the schedule, the cost and resources plan, risk analysis, project environmental analysis, etc.) offers different perspectives of the project. This multimethod approach further contributes to the development of the project complexity.

A reduction of project complexity occurs by the application of project management standards and by agreements. The definition of project-specific rules and norms, the development of project plans, and agreements in project owner meetings and project team meetings provide an orientation for the project work.

The project boundaries define what belongs to the project and what does not. The project boundaries are determined by the project scope and by the project start event and end event. Defining the project start and end event allows the preproject phase and the post-project phase to become, or at least limit, the project context. The social context of a project is determined by its social environments. For a project, these are those environments that we can expect will influence the success of the project.

Projects evolve over time. In the course of this, we can differentiate between continuous and discontinuous developments. Discontinuous developments can take place when project crises or chances occur.

Program Management

A program, as we have seen, consists of several projects and activities that are closely coupled by common program objectives, strategies, and rules. Usually, some of the projects in a program are performed sequentially and some are performed in parallel.

Program management has to be performed in addition to the management of the single projects of a program. The program management process has the same structure as the project management process (though see the chapter by Thiry regarding different emphases in program management). This includes the subprocesses of starting, coordinating, controlling, and closing down a program, and possibly resolving a program discontinuity. Also, program management methods are similar to the project management methods—in other words, there is a program work breakdown structure, a program bar chart, a program environment analysis, and so forth.

To allow for autonomous projects on the one hand but to ensure the benefits of organizational learning, economies of scale, and networking synergies in a program on the other hand, a specific program organization has to be designed. Typical program roles are program owner team, program manager, program office, and program team (see Figure 11.3). The program owner assigns the program to the program manager, who is responsible for the program management. He or she is supported by the program team members and the program office. Typical program communication structures are program owner meetings and program team meetings. The function of the program organization is to integrate the different projects of a program, in order to fulfil overall program objectives and strategies.

The advantages of designing a program organization instead of defining a large "project" with several subprojects are as follows:

- A less hierarchical organization
- A clear terminology: a program manager and several project managers instead of one project manager and additional "project managers" of the subprojects

FIGURE 11.3. PROGRAM ORGANIZATION CHART.

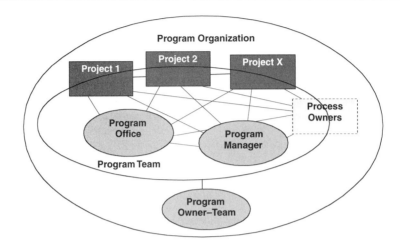

- Empowerment of the projects (of the program) by allowing for specific project cultures, specific relationships to social environments, and specific project organizations
- Differentiation between program ownership and ownerships for the different projects

Assurance of the Management Quality in Projects and Programs

Projects and programs are relatively autonomous organizations of the POC. The management of projects and programs is supposed to be performed according the general project and program management procedures of the POC. Management consulting and management auditing have to be performed in order to ensure the application of these general management procedures and to ensure the management quality in the projects and programs.

Management Consulting of Projects and Programs. The objective of management consulting of a project is to further develop the project management competence of this project. Not only are the competences of the project manager or the project team further developed (in which case this service would be defined as "coaching") but the temporary organization overall is consulted. The project as a temporary organization becomes the object of the management consulting service. This means that not only a permanent organization, such as a company or a profit center, but also a project can be the client in the consulting process.

By definition, projects are complex, risky, and dynamic. Therefore, they need high management attention and management quality. This can be ensured by involving project-external consultants. In a consulting assignment, any of the management subprocesses of a project (start, controlling, resolution of a discontinuity, closedown) or all of them can be considered. Specific management consulting services for programs might include the establishment of the program office, the development of a program marketing plan, or the definition and description of program- specific business processes.

The quality of the management consulting process can be measured. The key quality metric is the management competence of the project. It can be measured if there are improvements in the management competences of the members of the project organization, in the quality of the project meetings, in the project management documentation, in the project image, and in the relationships of the project to the relevant project environments.

The management consulting process starts with the initial contact between the potential consultant and the project. It includes analyzing the situation, moderating meetings and workshops, documenting these meetings and workshops, supporting the development and the updating of project management documentations, and so forth. Among the important intervention methods of consultants are interviews, documentation analyses, observations of meetings, and feedback.

The management consultants are assigned by the project owner. They are project-external people, but not necessarily company-external individuals. In many project-oriented companies, management consulting on projects and programs is considered an opportunity for job enlargement for senior (project) managers. Therefore, internal management consultants for projects and programs are developed in special training and coaching programs.

Management Auditing of Projects and Programs. Often a management audit on a project or a program is performed because of performance problems. Actually, management auditing is an instrument of quality management in projects and programs. It is an instrument of organizational learning of the POC.

The NEN-EN ISO 19011 (2002) defines auditing as "systematic, independent and documented process for obtaining audit evidence and evaluating it objectively, to determine the extent to which the audit criteria are fulfilled." Criteria for a management audit of projects are the project management approaches against which the projects are audited. These approaches are documented either in company-specific or in generic project management procedures.

While in a project audit the business processes for the contents as well as the project management process are considered, management auditing focuses on the project management competences only. The objects of consideration of a management audit are the project management subprocesses.

In the management audit of projects, we can differentiate the roles that the audit owner, auditor(s), and representatives of the audited project or program assume. The auditor(s) are project-external, but might be recruited from within the POC. The role of a management auditor of projects and programs is also sometimes seen to be an option for job enlargement for senior (project) managers of the POC.

Assignment of Projects and Programs

Companies invest in their infrastructure, in new products or services, in new markets, in the organization, or in their personnel. A project or a program might involve initializing such an investment. Therefore, an investment decision is often the basis for the decision to pursue a project or a program. For the investment decision, a business case analysis and initial project (or program) plans have to be developed. These documents are part of the investment proposal.

We must ensure the alignment of an investment with the company strategies. A formal instrument for this alignment is an investment scorecard. In such a scorecard, the investment decision criteria are documented (see Figure 11.4). Among the criteria to be considered for scorecards are financial data, customer relations, processes and resources, and innovations. (See also the chapter by Brandon on the balanced scorecard and project management.)

The investment decision is the basis for the decision about the appropriate organization for its implementation. A decision board has to decide if the investment is to be implemented by a project, by a program, or by organizational units of the permanent organization.

It is also necessary to analyze the fit of the project into the existing project portfolio. Each new project added to the portfolio means the mix has changed and a new portfolio is created. The relationships of the new project to the existing projects have to be considered and optimized. The decision about the project owner has to be made. The assignment of key personnel to the project, and the selection of partners and contractors for the project have to be made in consideration of the relationships to other projects in the project portfolio. (See the chapter by Archer and Ghasemzadeh.)

The authority to make the investment decision and the project assignment decision may either be divided between an investment decision board and a project portfolio group (PPG) or might be with the PPG only. The role of the PPG is described in *Organizational Design of the POC* later in the chapter.

FIGURE 11.4. INVESTMENT SCORECARD.

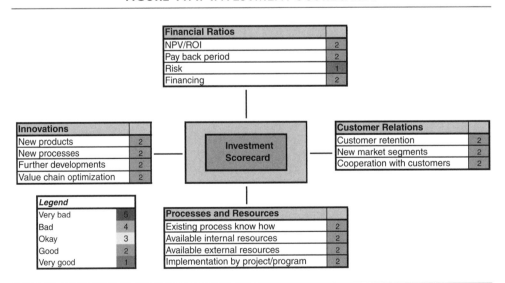

Project Portfolio Coordination

The objectives of the project portfolio coordination process are as follows:

- Optimization of the results of the project portfolio, and not of the single projects
- Definition of project priorities
- Coordination of internal and external resources
- Organization of learning of and between projects

The basis for the coordination of the project portfolio is a project portfolio database, which typically includes data about the project types, relations of projects to other projects, the project organizations, relevant project environments, and project ratios. The project portfolio database is not a project information system but contains aggregated project data only. It might be integrated in a project information system.

The project portfolio database allows the development of project portfolio reports. Typical project portfolio reports are the project portfolio bar chart, the project portfolio profit-

FIGURE 11.5. PROJECT PORTFOLIO SCORECARD.

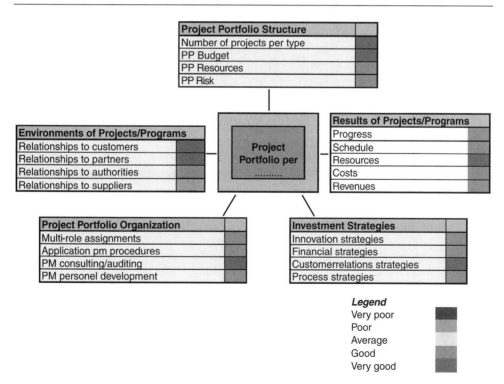

risk graph, and the project portfolio progress chart. An integrative project portfolio reporting tool is the project portfolio scorecard (see Figure 11.5). It shows how the actual project portfolio contributes to the implementation of the company strategies, reporting on the structure of the project portfolio and on the project portfolio status overall. Visualizing the project portfolio reports contributes to their acceptance as communication instruments for management and top management.

Networking between Projects

A set of closely coupled projects can be defined as a network of projects. Examples of the criteria which might relate projects in a network are a common technology applied, a common client, a common partner or supplier, or a common geographic region. The construction of a network of projects occurs at a point in time, in order to resolve a common problem or use a common opportunity. Therefore, a network of projects is not an organization with a common objective and a manager, such as a program, but it is an ad hoc communication structure.

The objective of constructing networks of projects is to identify synergies and potential conflicts between projects and to define strategies and measures to resolve the conflicts and to use the synergies. The networking between projects might result in a redefinition of the objectives of one or more projects of the network, in an assignment of common resources to two or more projects, or in renegotiations of contracts with the clients, partners, or suppliers.

The networking might be promoted by the PM office to generate the added value. The construction of the network of projects, the description of existing relationships between projects, and the establishment of new relationships is organized through ad hoc communications (meetings, workshops) of the project managers of the projects of the network. Also, project team members and representatives of relevant project environments, which might contribute relevant information, might participate. The communication between the projects requires a trust and openness between these partners.

An example of networking between projects is presented in Figure 11.6.

The demand for networking between projects of an Austrian consulting company, visualized in Figure 11.6, occurred because of a crisis in the project "eSupported PM-Seminars." In this project the cooperation with the IT supplier caused major problems. The project portfolio database showed that this IT supplier was also supplier or cooperation partner in other projects. Measures to resolve this problem could not be decided under consideration of the eSupported PM-Seminars project only, but the relationships to the other projects the IT supplier was involved in had to be considered too.

To analyze the consequences for all the projects the IT-supplier was involved in, a networking workshop with the project managers of the projects was organized, with the objective to analyze the relationships between the projects and to develop common strategies and measures to resolve the problem. The strategy agreed on was to further cooperate with the IT company but to reduce the dependency on it as partner and supplier. Measures agreed on were to resolve the crisis of the eSupported PM-Seminars project, to assign the

FIGURE 11.6. NETWORK OF PROJECTS PERFORMED WITH AN IT COMPANY.

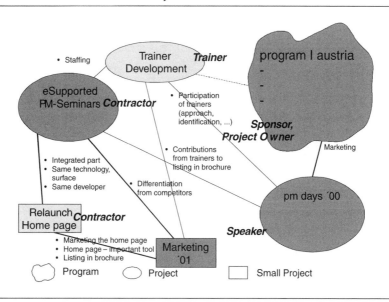

same developer to two projects, and to cancel the invitation of employees of the IT supplier to join the trainer team.

Personnel Management in the Project-Oriented Company

Personnel management processes in the POC include the recruiting, leading, developing, and releasing processes for project personnel.

In POCs a project management career path exists. This is based on definitions of competences for the different roles in the POC. *Competence* can be defined as knowledge, skills, behaviors, and experience required for the performance of a business process. The specific competences that are required in a POC relate to the performance of the specific business processes of the POC. The competences are required by individuals and by teams.

Competences of Project Management Personnel. The project management competences required differ according to the project roles to be fulfilled by individuals. The following project roles can be performed by individuals: project owner, project manager, project management assistant, project team member, and project contributor. The project contributor contributes to the performance of work packages but does not (compared with the project team member) participate in project team meetings. The project management functions to be performed by project personnel can be described in project role descriptions (see, for example, Figure 11.7).

FIGURE 11.7. PART OF THE DESCRIPTION OF THE ROLE "PROJECT MANAGER."

Role Description: Project Manager
Objectives
• Representation of the project interests
• Contribution to the realization of the project objectives and to the optimization of the business case
• Leading the project team and the project contributors
• Representation of the project towards relevant environments
Organizational Position
• Member of the project team
• Reports to the project owner team
Tasks
During the project assignment process
• Formulating the project assignment together with project owner team
• Nominating the project team members
During the project start process
• Know-how transfer from the pre-project phase into the project
• Development of adequate project plans
• Design of an adequate project organization
• Performance of risk management
• Design of project-context relations
• etc.
During the project controlling process
• Determination of the project status
• Redefinition of project objectives
• Development of project progress reports
• etc.

FIGURE 11.7. (*Continued*).

During the resolution of a project discontinuity
•   Analysis of the situation and definition of ad hoc measures
•   Development of project scenarios
•   Definition of strategies and further measures
•   Communication of the project discontinuity to relevant project environments
•   etc.
During the project closedown process
•   Coordination of the final contents work
•   Transfer of know-how into the base organization
•   Dissolution of project-environment relations
•   etc.

The project manager requires knowledge and experience not only to apply project management methods but also to creatively design the project management process. These design functions include the following:

- Selection of the project management methods appropriate for a given project
- Selection of the appropriate communication structures
- Facilitation of the different workshops and meetings
- Decision to involve a project management consultant
- Selection of the appropriate IT and telecom infrastructure
- Definition of the appropriate form for the project management documentations

The project management competence of a project manager is the capability to fulfill all functions specified in the role description. Besides the project management knowledge, skills, behaviors, and experience for a given project type, a project manager needs product, company, and industry knowledge. In international projects, cultural awareness and language knowledge are also prerequisites.

Competences of Project Teams. To perform a project successfully, a project team requires team competence. The competence of a project team can be defined as the competences of the project team members plus the social knowledge and experience of the team to create the "big project picture," to produce synergies, to solve conflicts, and to ensure learning in the team.

A project team cooperates in workshops and meetings. The application of project plans, such as a work breakdown structure, a schedule, a project environment analysis, and so on, are tools, to support the communication in the project team.

Organizational Design of the Project-Oriented Company

To integrate the different projects performed simultaneously, a POC requires specific permanent organization structures, such as a PM office, a project portfolio group, and expert pools.

The PM Office. To ensure that the different ongoing projects apply a common management approach, somebody in the POC has to take on the ownership for the project management process. The project management competences of the POC have to be institutionalized. The PM office is the organizational unit that can take on this responsibility. (See the chapter by Young and Powell for more on the PM office.)

The PM office is a permanent structure and is part of the base organization of the POC. It provides services for all projects of the POC and also for the project portfolio group. The PM office has to be differentiated from project offices and program offices, which are temporary, are part of the project or program organization, and provide services for one project or one program only.

The organization chart of a PM office (see Figure 11.8) includes the roles PM office manager and personnel for project, program, and project portfolio management services. The expert pools "PM-Personnel" and "PM-Trainer, PM-Consultants" might be coordinated by the PM office manager.

Services provided by the PM office might include the following:

- Development and maintenance procedures for project and program management, management auditing of projects and programs, and project portfolio management
- Development and maintenance of standard project plans (standard WBSs, work package specifications, milestone lists, etc.)
- Provision of project management support services
- Project and program marketing
- Organization of project management training, coaching, consulting, and auditing
- Promotion of project management as a profession by establishment of a project management career path, project management certification programs

FIGURE 11.8. ORGANIZATION CHART OF THE PM OFFICE.

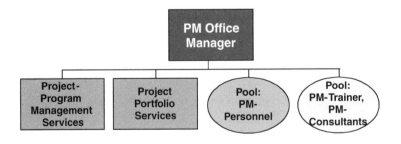

- Assurance of a project management infrastructure (meeting rooms, ICT tools, moderation tools)
- (Internal) benchmarking of the project and program management processes
- Maintenance of the project portfolio database
- Development of project portfolio reports
- Support of the meetings of the project portfolio group

The Project Portfolio Group. In a POC many projects are performed simultaneously. Synergies and possible conflicts between these projects have to be managed. The results of the project portfolio have to be optimized. Because of the high organizational differentiation of the POC, the management of the project portfolio might not be taken care of in the usual meeting structures, such as management board meetings, department head meetings, and so on. It might be preferable to delegate this responsibility to a specific communication structure, the project portfolio group.

The project portfolio group is a permanent organization structure of the POC. It could be considered as a staff or a line position reporting to the management board. Five to eight managers of profit centers and departments being strongly involved in projects and programs should be members of the project portfolio group. In a major Austrian telecommunication with some 1,500 employees and 70 to 80 projects at any given time in the project portfolio, the project portfolio group consists of the department heads for marketing, engineering, call center, IT, and PM office. The head of the group is the director of finance. The project portfolio group meets every week for two to three hours.

Depending on the duration and the dynamics of the projects in the portfolio, not more than 100 projects should be managed by one project portfolio group. In large POCs several project portfolio groups, differentiated by project types or by business units, might be required. The major services of the project portfolio group are the selection of projects, to be started and integrated into the project portfolio, the creation of synergies (common use of resources, economies of scale, organizational learning) and the resolution of conflicts within the project portfolio, and the quality assurance and the provision of early warnings in the project portfolio. Criteria for the assurance of the project portfolio quality are, for example:

- The number of projects for one project manager
- The number of projects for one project owner
- The number of projects per project type
- The project portfolio budget (per project type)

Expert Pools. Experts, to perform the work packages in projects, are required in POCs. Depending on the type of industry a POC is in and the type of projects it performs, different expert categories are required. Typical expert pools are pools of engineers (differentiated by mechanical, electrical, engineering, etc.), procurement experts, and marketing experts. These experts most of all need competence in the discipline they represent, but they also need project management competence in order to cooperate in teams. The roles in an expert pool are the expert pool manager, the experts, and exchange of experience groups and possibly supervision groups.

One important expert pool of a POC is that pertaining to "project management." The experts of this pool might be differentiated in relation to a project management career path into project management assistants, junior project managers, project managers, senior project managers, and program managers.

An expert pool manager has personnel management, knowledge management, and infrastructure management functions. He or she has to recruit, develop, and allocate personnel, has to further develop standards and provide ethics of work, and has to provide the infrastructure for the performance of the work packages. On the other hand, the quality control of the work packages, which are performed in the projects and programs, is not his or her responsibility.

Maturity of the Project-Oriented Company

Not only individuals but also organizations have the capability to acquire knowledge and experience and to store it in a "collective mind" (Senge, 1994; Weik and Roberts, 1993). Organizational principles, which might be stored in the collective mind of a POC, are project management procedures, project management templates, standard project plans, procedures for the management auditing of projects, as well as project portfolio management procedures, structures for a project portfolio database, and standard project portfolio reports.

Assessing and Benchmarking the Maturity of a POC

The organizational competences of the POC can be assessed with the *POC Maturity Model*. This model is based on a POC questionnaire and is visualized in a "POC spiderweb." The axes of the spiderweb represent the business processes of the POC. The maturity of a POC in the performance of each of these processes can be assessed by the application of a set of questions relating to the business process.

Questions relating to the project start process are grouped regarding the planning of project objectives, the project risk, the project context relationships, the project organization, and the project culture. They are not assessed for the application of a given project management method, but for the resulting project management documents.

The overall maturity of a POC can be represented by the area in the spiderweb (see Figure 11.9).

The maturities of POCs can be benchmarked and further developed. Instruments to develop the competences of individuals performing roles in the POC are (self-) assessments, trainings (classroom, on the job), and coachings. Instruments to develop the organizational competences of the POC are assessments, benchmarkings, and organizational development projects—for instance, to implement project portfolio management in order to establish a PM office.

Summary

"Management by projects" is the organizational strategy of the POC. The application of a "new management paradigm" supports the efficient application of projects in the POC.

FIGURE 11.9. MATURITY OF A POC.

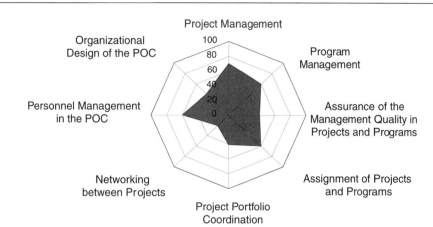

Specific organizational structures, such as expert pools, a project portfolio group, and a PM office, are required to perform integrative functions in the POC. For the performance of the specific business processes of the POC, specific competences are required by individuals, teams, and the POC overall.

It is important to differentiate between the project management process and other specific processes of the POC. Even though project management is the most important business process of the POC, competences for the performance of the other specific business processes of the POC have to be further developed too. The *POC Maturity Model* supports the assessment as well as the further development of these competences.

References

Ashby, W. R. 1970. *The process of model building in the behavioral sciences*. Columbus, Ohio: Ohio State University Press.

CCTA. 1996. *Managing successful projects with PRINCE2*. London: The Stationery Office.

Cleland, D., and R. Gareis. 1994. *Global project management handbook*. New York: McGraw-Hill.

Cleland, D., and L. Ireland. 2002. *Project management: Strategic design and implementation*. 4th ed. New York: McGraw-Hill.

Englund, R. L., R. J. Graham, and P. C. Dinsmore. 2003. *Creating the project office: A manager's guide to leading organizational change*. The Jossey-Bass Business & Management Series. New York: Wiley.

Gareis, R. (Hg.) 1994. *Erfolgsfaktor Krise: Konstruktionen, Methoden, Fallstudien zum Krisenmanagement*. Wien: Signum-Verlag.

———. 2000. Managing the project start. In *The Gower handbook of project management*, ed. J. R. Turner and S. J. Simister. Aldershot, UK: Gower.

Gareis, R., ed. 2002. *PM baseline knowledge elements for project and programme management and for the management of project-oriented organisations*, July, version 1, Prokekt Management Austria, Vienna, Austria.

Gareis, R., and M. Huemann. 2000. *PM-competences in the project-oriented organization*. In *The Gower handbook of project management, eds.* J. R. Turner and S. J. Simister. Aldershot, UK: Gower.

Gareis, R. and M. Huemann. 2003. *Project management competences in the project-oriented company*. In *People in Project Management*, ed. J. R. Turner, Aldershot, UK: Gower.

Hartmann, F. 2000. Don't park your brain outside. Newtown Square, PA: Project Management Institute.

Jones, M. O. 1996. *Studying organizational symbolism*. Thousand Oaks, CA: Sage Publications.

Kaplan, R. S. 1996. *The balanced scorecard: Translating strategy into action*. Boston: Harvard Business School Press.

Knutson, J. 2001. *Succeeding in project-driven organizations: People, processes, and politics*. New York: Wiley.

NEN-EN ISO 19011. 2002. *Guidelines for quality and/or environmental management systems auditing*. Brussels: European Committee for Standardization.

PMI. 2000. *A guide to the project management body of knowledge* (*PMBOK Guide*). Newtown Square, PA: Project Management Institute.

Schein, E. 1985. *Organizational culture and leadership*. San Francisco: Jossey-Bass.

Senge, P. 1994. *The fifth discipline fieldbook: Strategies and tools for building a learning organization*. New York: Doubleday.

Turner, J. R. 1999. *The handbook of project based management*. 2nd ed. London: McGraw-Hill.

Weik, A. and K. Roberts. 1993. Collective mind in organizations heedful interrelating on flight decks. *Administrative Quarterly* 38.

CHAPTER TWELVE

MANAGING PROJECT STAKEHOLDERS

Graham M. Winch

"Taurus meant an awful lot of different things to different people, it was the absolute lack of clarity as to its definition at the front that I think was its Achilles' heel."

PETER RAWLINS, FORMER CHIEF EXECUTIVE OF THE LONDON STOCK EXCHANGE, INTERVIEWED IN THE MARCH 7, 1995, EDITION OF THE *FINANCIAL TIMES*, ON THE DEMISE OF THE MASSIVELY OVERRUN AND UNSUCCESSFUL PROJECT TO COMPUTERIZE SHARE DEALING ON THE EXCHANGE.

The challenges for the project management team are growing more complex. This point is illustrated in many different ways throughout this book and is the fundamental insight of the "management of projects" perspective. The aim of this chapter is to address one of the more important elements in that complexity: the increasing diversity and power of *project stakeholders*. The chapter starts by briefly identifying some of the sources of this growing complexity, before formally defining the concept of project stakeholder. It then goes on to propose a framework for mapping the stakeholders on the project as a prerequisite for analyzing their ability to influence the definition of the project mission and to disrupt its execution. The framework presented here was used to analyze publicly sponsored construction projects in Winch and Bonke (2002). Here the analysis of power is developed further and the framework used to analyze the case of a private-sector-sponsored IT project for back-office settlement on the London Stock Exchange (LSE). It will be argued that one of the major differences between the failed project (TAURUS) and the successful project (CREST) was the effectiveness of stakeholder management. Some implications of the analysis are then drawn out for the effective management of project stakeholders by the project management team (PMT).

The Growing Complexity of Stakeholder Management

Projects have always had stakeholders, but they have usually been either the funders of the project as client or suppliers of the project as members of the project coalition. Inherently, these stakeholders have had an interest in the effective delivery of the project with the minimum capital investment for the functionality required by the business case, and the project management team could focus on this objective. However, long-run changes in the social, political, and economic environment of projects have meant that this is no longer necessarily the case, for a number of reasons:

- Since 1945, most projects were financed from the general revenue streams of the client organization—whether streams derived from profits on turnover or the raising of taxes. Increasingly, for both private and public sector clients, projects are financed by loans or equity raised by a special project vehicle (SPV) with the returns on that investment generated directly by the revenue stream from the asset created by the project (see the chapters by Ive and Turner). This immediately introduces financiers as a new class of project stakeholder, as well as creating a wholly new type of project actor in the SPV itself.
- Traditionally, the client—that is, the party with which contracts are made by the principal supply-side members of the project coalition—and the project sponsor were the same entity. This is no longer necessarily the case. In urban regeneration projects, for example, the sponsors may be local political elites, who then choose a public body to be the client, as in Manchester's sports-led schemes to host the Olympic Games and (successfully) the Commonwealth Games (Cochrane et al., 2002), or on the Central Artery/Tunnel (CA/T; Hughes, 1998) in Boston, Massachusetts.
- Regulators are growing ever more insistent on the project definition taking into account wider social objectives than the effective exploitation of the asset being created by the project. At least one authoritative study has concluded that the interventions of regulators are a principal source of budget overruns on projects (Merrow et al., 1981). The most obvious example here is environmental protection, institutionalized through environmental impact assessments, but operational safety, local purchasing and labor requirements, and land-use policy issues can also figure large in regulators' concerns.
- Direct action by environmentalists or local loser groups can also be highly disruptive for the project during the execution phase, and the only way to address this issue aside from confrontation is to address the concerns of such groups—to the extent that they can be considered as legitimate—during project definition.

These points indicate the diversity of parties that can be considered to be project stakeholders. Cleland (1998, p. 55) has defined "project stakeholder" thus:

> Stakeholders are people or groups that have, or believe they have, legitimate claims against the substantive aspects of the project. A stake is an interest or share or claim in a project; it can range from informal interest in the undertaking, at one extreme, to a legal claim of ownership at the other extreme.

There are two important aspects of this definition worthy of note. First, stakeholders only have to *believe* that they have a claim on the project to cause problems—that claim might be perceived as illegitimate by the client and PMT. As the sociologist William I. Thomas put it, "if men define situations as real, they are real in their consequences" (cited Coser,

1978, p. 315). Second, those claims are usually met by adjusting the project mission in some way, unless the claimant is too weak to press their claim. Even direct-action opponents whose claims are not considered legitimate by most parties can force changes in project definition by enforcing additional expenditure on site security and the like. As Fred Salvucci, a member of the Boston political elite sponsoring the CA/T, put it, these mitigations of stakeholder claims on the project are the modern equivalent of "delivering some chunk of mastodon meat back to the tribe" (Hughes, 1998, p. 221).

Project stakeholders are a diverse group—some are formally members of the project coalition, others not. The first group is usually defined as the primary or *internal* stakeholder group (Calvert, 1995; Cleland, 1998). They are defined by having a contractual relationship with the client or a subcontract from another internal stakeholder. They usually enter willingly into the project coalition, and are, by definition, positive about the project even if they negotiate toughly for their share of the value added by the project. Their claims are usually enforceable directly as breach of contract. The second group is usually defined as the secondary or *external* stakeholder group. They may have little choice about whether the project goes ahead and may be either positive or negative about the project. They rarely have a directly enforceable claim on the project and are therefore reliant upon regulators to act on their behalf, the mobilization of political influence either covertly or through public campaigns, or, occasionally, direct action.

Internal stakeholders can be broken down to those clustered around the client on the demand side and those on the supply side. External stakeholders can be broken down into private and public actors. This categorization, with some examples, is shown in Table 12.1.

On the demand side, a complex array of interests is indicated, and there can be no assumption that they are all aligned. One of the largest differences of interest within the demand side is often between the client and its employees. Particularly where the project is associated with reengineering business processes, employees may face significant changes in their work, or even lose their jobs, as a result of the project. Even where this is not the case, failure to adequately capture the needs of users as they perceive them can cause difficulties or even failure during the commissioning phase of the project. Similarly, the interests of financiers, clients, and sponsors may be divergent. Where project sponsors or

TABLE 12.1. SOME PROJECT STAKEHOLDERS.

Internal Stakeholders		External Stakeholders	
Demand Side	**Supply Side**	**Private**	**Public**
Client	Consulting engineers	Local residents	Regulatory agencies
Sponsor	Principal contractors	Local landowners	Local government
Financiers	Trade contractors	Environmentalists	National government
Client's employees	Materials suppliers	Conservationists	
Client's customers	Employees of the above	Archaeologists	
Client's tenants			
Client's suppliers			

Source: Adapted from Winch (2002) Figure 4.1. Used with permission of the Project Management Institute.

clients do not finance the project from their own resources, there is a great temptation to underestimate the costs and overestimate the benefits of the project (Flyvbjerg et al., 2003). For example, the finance of the Boston CA/T using the "10-cent dollar," where the U.S. federal government matches every ten cents put in by local taxpayers with 90 cents from federal taxpayers (Hughes, 1998), created a major misalignment of incentives. Much the same would appear to have happened on the West Coast Main Line project in the United Kingdom, which has suffered program and budget overruns similar to Boston CA/T. Where the project sponsors are on the supply side—as on the Channel Fixed Link project—this can cause great suspicion among financiers regarding the integrity of the decision making of the client (Winch, 1996).

On the supply side, a whole coalition of interests is arrayed. The supply side satisfies its claim on the project through the income stream generated by working on the project and the learning acquired through solving project problems (Winch, 2002). It is immediately clear that there is an inherent conflict of interest between the stakeholders on the demand side and those on the supply side as they compete to appropriate the income stream from the project—what Porter (1985) calls margin—in the project value system. Managing this conflict is the central task of project governance (Winch, 2001).

Among the external stakeholders, there is even more diversity. By and large, the internal stakeholders will be in support of the project, although there may be factions within the client that are backing alternative investments. External stakeholders may be in favor, against, or indifferent. In the private sector, those in favor may be local landowners who expect a rise in the value of their holding and local residents supporting a rise in the general level of amenity. Those against may also be local residents and landowners who fear a fall in amenity and hence the value of holdings. Such objectors are known as NIMBYs (not in my back yard) and can delay infrastructure projects such as airports for decades, if not stop them all together. Environmentalists and conservationists may take a more principled view than local losers, while archaeologists are concerned about the loss of important historical artifacts.

The public external stakeholders—in those situations where the public sector is not also the client—will tend to be indifferent. Their interest arises from the general level of economic activity, rather than from any particular project, and their claim is met through the taxes generated by this economic activity. Regulatory agencies that enforce regulatory arrangements such as those for urban planning, quality of specification, and heritage assets will tend to be indifferent to any particular project definition, so long as it complies with the codes. National and local government may, however, wish to encourage development, particularly in regeneration areas. At times, there may be conflicts of interest within the public sector between its role as a project sponsor and its responsibilities as a regulator, as happened with the Sheffield Arena project (Winch, 2002).

Managing Stakeholders

The successful management of stakeholders by the project management team requires the following processes (cf. Cleland, 1998):

- Identify those stakeholders with a claim on the project
- Specify the nature of each stakeholder's claim
- Assess each stakeholder's ability to press that claim
- Manage the response to that claim so that the overall impact on the definition and execution of the project are minimized.

Stakeholder mapping is proposed as a valuable aid to completing these management processes successfully. The first step in managing the stakeholders is to map their interest in the project (Winch and Bonke, 2002). This can be done using the framework illustrated in Figure 12.1. The focus of the approach is the project mission as represented by the asset to be created. Stakeholders can be considered as having a problem or issue with the project mission and as having a solution (tacit or explicit) that will resolve that problem. Where such solution proposals are inconsistent with the client's proposals, they can be defined as being opponents to the project. An important part of stakeholder management is to find ways of changing opponents to proponents by offering appropriate changes to the project mission and preventing possible proponents defecting to the opponent camp by offering to accommodate more explicitly their proposed problem solutions.

Once the stakeholder map has been drawn up, the power/interest matrix can be used to develop a strategy toward managing the different stakeholders (Johnson and Scholes, 2002). It consists of two dimensions: the power of the stakeholder to influence the definition of the project and the level of interest that the stakeholder has in that definition. The level

FIGURE 12.1. MAPPING STAKEHOLDERS.

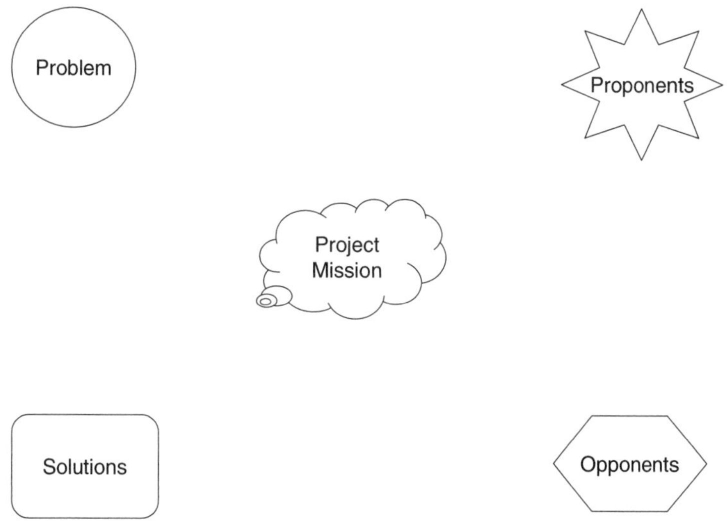

Problem

Proponents

Project Mission

Solutions

Opponents

Source: Bonke (1996).

of interest is conceptually simple—it is a function of the expected benefit or loss from the project. Power is a more slippery concept and is discussed more fully in the text that follows.

This matrix categorizes the stakeholders into one of four types, but the discussion here can only be indicative—where a particular stakeholder sits in relation to the project depends entirely on the specific context of that project. The first group is those who require *minimal effort*, such as the client's customers, or local and national government. A public relations approach to this group will often suffice, aimed at ensuring that those who might be opposed to the project stay in the low-interest category while those who are likely supporters are tempted to move to the high-interest category.

The second group are those who need to be *kept informed*. Groups who may be opposed to the project, such as local residents, conservationists, or environmentalists, need to be carefully managed. If such groups coalesce into well-organized movements and are able to mobilize the press behind them, they may well be able to move into the key player category, causing the project severe disruption, or even cancellation. To a certain extent, such groups can be bought off to prevent this happening, with inevitable consequences for the business plan. For instance, it is now standard practice for clients building in the City of London—which contains important Roman and mediaeval remains—to finance an archaeological dig prior to works commencing on site. Similarly, the concept of planning gain within the UK regulatory system is common, where a project promoter provides additional utility for the benefit of the local community to defuse potential opposition. Some groups—typically environmentalists—cannot, however, be bought off and can go on to disrupt physically the project during execution on-site. The impact of these groups is twofold:

- They upset the calculations underlying the business plan because of delays in the schedule and additional costs of security;
- They dissuade future clients from coming forward with similar projects.

Those who need to be *kept satisfied* usually fall into two main groups—regulatory bodies and the supply-side stakeholders—which require very different management approaches. Regulatory bodies are, in essence, the institutionalized interests of the low-power stakeholders. They provide forums in which local residents, landowners, and government can have their voice (planning enquiries), the safety of the client's or operator's employees, tenants, and customers is ensured (safety codes), and environmental and conservationist interests are heard (environmental impact assessments). The latter even allows the purported claims of stakeholders that do not yet exist—future generations—to be heard. The first task of the PMT is to ensure compliance with the regulatory requirements, supported by lobbying tactics where the requirements are open to interpretation. Supply-side stakeholders are placed in this category, rather than the key players category, for two reasons, First, most of them are mobilized after the project is defined; second, they will typically have a portfolio of projects at any one time—while their power to influence the outcome of any one project is very high, their interest in the definition of any one project is typically limited.

The final category is that of the *key player*. Here the client and sponsor are central; the analytic questions revolve around which of the other demand-side stakeholders are also in this category. Where finance is raised from the traditional sources—equity and debt secured through a floating charge for the private sector, and the taxpayers for the public sector—financiers are typically in the keep-satisfied category. However, where project finance techniques are used as the source of capital—as is increasingly common in both the public and private sectors—such financiers move into the key player category. In commercial property development where the asset is pre-let, the client's tenant can become a key player in definition, while in the provision of social housing this is rarely the case. The client's customers are also usually in the key player category, but through the proxy voice of the corporate marketing department; if they misunderstand the market for the facility, it is unlikely to be successful because those customers will simply use other competing facilities. Whether the client's employees are in the low- or high-power categories depends upon the internal organization of the client and its understanding of its business processes.

Understanding the Power of Project Stakeholders

Awareness of the importance of power relationships in project management has been growing (Pinto, 1996; and see the chapter by Magenau and Pinto later). A definitive analysis of power as a relationship is provided by Lukes (1974). He identified three facets of power, as illustrated in Figure 12.2.

- *Facet 1, overt power.* The ability of A to persuade B to choose the option A prefers
- *Facet 2, agenda-setting power.* The ability of A to set the agenda so that B's preferred option is "off the agenda"

FIGURE 12.2. THE THREE FACETS OF POWER.

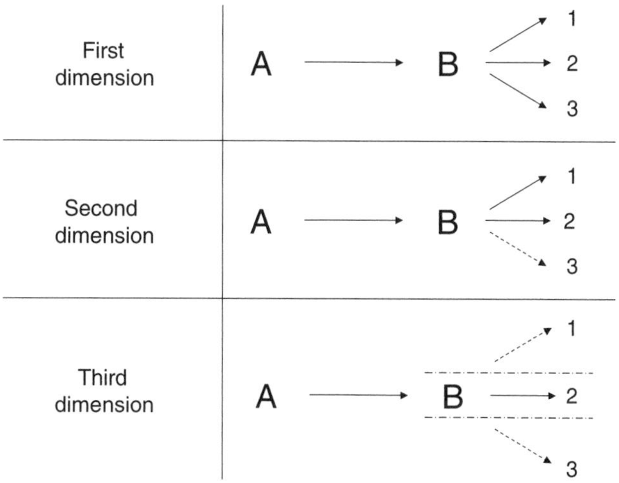

Source: Winch (2002). Used with permission of the Project Management Institute.

- *Facet 3, hegemonic power*. The ability of A to define the issues in such a way that B sees no alternative but to make choices favorable to A

There are five main sources of power in organizations (Handy, 1994), which can be illustrated with examples from the overt facet of power:

- *Physical* power. This is rarely used and of little relevance to project stakeholder management.
- *Positional power*. This is derived from the powerful actor's ability to deploy power derived from position in an organizational hierarchy or legal authority. For instance, the power of regulators usually derives from their backing in national legislation to protect the interests of relatively powerless external stakeholders. Similarly, the power of clients to have the last word on key decisions on the project is enshrined in the contracts used between the parties.
- *Resource power*. Those that are providing the resources for the project—especially capital—typically have considerable say in how those resources are used. Indeed, this can often amount to a veto on some options.
- *Expert power*. This is the power wielded by specialist advisors hired to provide expertise not otherwise available; their power is strongly linked to their reputation for competence.
- *Personal power*. This is the ability of charismatic individuals to sway opinions and win arguments simply by force of personality. This type of power should not be underestimated.

A telling example of the agenda-setting facet of power comes from the appraisal of privately financed public projects in the United Kingdom. The business case for all such projects must be subject to a public sector comparison to see whether private or public finance is the more cost-effective. However, should the comparison show that public finance is more cost-effective, it does not follow that public finance will be available for the project. Project sponsors are, therefore, tempted to manipulate the comparison to show that private finance is more cost-effective in order to obtain the support of financiers. An example of the hegemonic facet of power is the sponsorship of IT projects to tackle the so-called Y2K or "millennium bug" problem. This was the fear that many mission-critical IT systems would cease to function properly after 1999 because of shortcuts taken in the coding of many software programs. With the benefit of hindsight, the problem can now be seen to have been greatly exaggerated, and countries such as Italy that spent relatively little tackling the problem did not suffer significant adverse consequences (Finkelstein, 2000). The sponsors of Y2K projects successfully promoted them as TINA (there is no alternative) projects, and thereby distorted corporate capital budgeting.

The Case of Settlement on the London Stock Exchange

In March 1993, around £80m of investment in a computerized trading settlement system was canceled by the Board of the London Stock Exchange (LSE), followed by the resignation of the chief executive responsible. Many different analyses of this project failure are possible, for it is rich in lessons for project management. Here the focus is on the way in which poor stakeholder management led to project scope escalation to a level of unmanageable complexity, and hence project failure. The sources for this case are Drummond (1996) and Head (2001). Neither author bears any responsibility for the interpretation of their research presented here.

In October 1986, the so-called Big Bang had swept away the cozy, clubbable world of the LSE, constituted as a mutual organization for the benefit of its members. In essence, this was a government deregulation initiative designed to remove price fixing in trading commissions so as to ensure that the City of London retained its premier position in global trading markets. In preparation, the LSE had implemented nine separate IT projects for front-office trading, which were successfully launched on Big Bang day with a synchronized switch-on. IT was also seen as central both to minimizing back-office settlement costs and to speeding up the settlement process. The stock market crash of October 1987 had revealed the fragility of the existing paper-based system, and a systemic risk was identified of default during the lengthy settlement process. In November 1989, the Council of the LSE appointed a new chief executive, committed to reforming the LSE and ensuring that the "dinosaurs" retired to their clubs. However, like all stock exchanges, the LSE was simply a forum for its members:

- The "jobbers," who traded on behalf of brokers using an "open cry" trading floor and who were replaced by "market-makers" using computer screens
- The brokers, who dealt with investors and were rapidly being taken over by the major banks

The TAURUS (Transfer and Automated Registration of Uncertified Stock) system was aimed at computerizing the settlement system. This is the post-trade system by which stocks are exchanged for cash between seller and buyer. Settlement is far from the front-office buzz of trading, but an indispensable element in the process. Full computerization of settlement implies *dematerialization*, or the replacement of paper stock certificates, which can be placed in a bank vault or under the mattress, with asserted "rights" of ownership contained in an electronic register. This was a cultural change of immense significance, particularly to the private equity investors who accounted for around 75 percent of trades but only around 25 percent of value traded. In 1979, "Talisman" had been implemented and was making considerable progress in computerizing settlement. Instead of certificates being physically carried around The City from broker to broker, Talisman acted as an electronic clearing-house, and perhaps most importantly, centralized the clearing process through the LSE itself. It thereby came to provide 50 percent of LSE income. TAURUS represented the next stage of back-office computerization.

The agenda for TAURUS was set by the self-appointed Group of 30 (G30) senior City figures who had become increasingly concerned with back-office capabilities and were strong advocates of TAURUS. "The G30, with its exclusive membership of top-level policymakers and business leaders, is the most eminent of financial think-tanks" (January 24, 2003 edition of the *Financial Times*). The TAURUS project was launched in May 1986 with a schedule of 36 months and a budget of £6m (all budget figures are in then current prices). The crash of 1987 prompted the program to be speeded up, and a powerful committee—SISCOT (Securities Industry Steering Committee on TAURUS)—was established to oversee development. SISCOT decided to abandon TAURUS 1 in 1989. The main problem was the proposal to create a single central register of share ownership. This would have made the registrars, which had traditionally held such registers, redundant. While historically small firms, many of them had been recently been taken over by the major banks, which were exercising their muscle in the post-Big Bang environment and did not wish to see their investment wasted. The registrars thereby moved from a low-power to a high-power position.

However, the perceived urgency remained, and Coopers and Lybrand were appointed as project managers, led by a highly experienced and respected IT project manager who demanded complete autonomy for the PMT in August 1989. At the same time, major changes in the overall management of the LSE were taking place as it evolved from a mutual society to a corporate entity. The new, reforming chief executive took up his post in November 1989.

Although he tried to kill the project at this point because he could not see the rationale for it, he was advised that all was in order and that the newly appointed PMT was in full control. He therefore addressed himself to what he perceived as more fundamental problems with the LSE than back-office systems. In March 1990, the prospectus for TAURUS 8 was announced—a 19-month schedule costing around £47.5m. The decision to adapt an existing computer package, rather than develop a bespoke one, was then taken, and a contract was signed with Vista Corporation of the United States for their package, implemented on IBM mainframe hardware. Adaptations to the package were to be made on a cost-plus basis.

Learning from the experience of TAURUS 1 suggested sensitivity to stakeholder interests. The system was, therefore, designed from the outside in, so as to enable consultation with key stakeholders. This attempt to engage with stakeholders spawned over 30 committees around The City, and while "everybody was giving input to Taurus" and "everybody was being promised the earth" (all citations are from Peter Rawlins, LSE chief executive, 1989 to 93), it was not at all clear who was actually responsible for it on the client side. Formally it was overseen by a TAURUS board that met once a month for under two hours, while the external monitors that were intended to act on behalf of the LSE Council as on the Talisman project were sacked as part of a cost-cutting exercise. Building the system on the basis of the Vista package started in earnest in early 1990 with delivery due in October 1991. To move with the perceived urgency, information was sent to stakeholders so that they could start building their own systems, but as this information was received, stakeholders started to demand changes in the specification to suit their own interests, which then had knock-on implications for other stakeholders. SISCOT was the forum in which these debates were held and became known to those involved in the project as the "Mad Hatters' Tea Party."

As 1990 wore on, it became increasingly clear that dematerialization would require primary legislation. The regulatory body concerned—the UK Department of Trade and Industry (DTI)—produced a bill amounting to over 100 pages of legislation. It also required that the LSE indemnify investors for any errors in attribution of ownership caused by the system. It also refused to allow the LSE to make membership of the TAURUS system compulsory, which had been a major assumption of system design. The DTI, therefore, played the classic regulatory role of champion for low-power/high-interest stakeholders such as private investors.

The first public announcements of problems came in January 2001 as the delivery date was delayed, the blame for which was put on the problems with the regulator. However, the diversion of the project director, who now had to go round the country to sell participation in TAURUS now that it was voluntary, was causing problems of internal project management, despite the long hours and dedication of the project team. By the autumn of that year, budget sanctions had been increased and delivery was promised for April 1993. By mid-1992, 50 percent of the Vista package code had been rewritten, and the IBM servers for the state-of-the-art security interfaces had failed to work. The LSE refused to fund the secondment of more Coopers and Lybrand personnel to the PMT as the project slipped into crisis. Andersen Consulting had been commissioned in early 1992 to provide a general review of the LSE's trading and settlement systems. Formally, TAURUS was outside their brief, but in the October they were asked to include it in their scope. It was on the basis of their report that the chief executive took his decision to cancel the project in March 1993 and then resigned. Three-hundred and fifty other people also lost their jobs, and stakeholders lost an estimated £400m of associated investment.

The failure stimulated those stakeholders with a primary concern for the viability of The City to action. The Bank of England—the UK's central bank—set up a tight project team to develop what became known as CREST, to be run by a not-for-profit organization called CRESTCo. The central mission of the CREST project was the same as that of the TAURUS project—dematerialized back-office settlement. However, the mission was more restrictive in one crucial way: Membership of CREST was to be voluntary. It deliberately used outsiders to The City—career civil servants—to engineer the settlement processes that *had* to happen, rather than would be nice to have. When stakeholders lobbied against the proposed systems design, they were told to "take it or leave it." The team did not use a formal systems methodology but stuck rigorously to a design-then-build approach, supported by strong configuration management. Even here, not all stakeholders could be kept at bay. The UK Treasury had offered to give up £3bn of stamp duty on the purchase of stocks in order to encourage dematerialization with TAURUS, but reneged on this halfway through the CREST project. Nevertheless, the system was delivered on schedule and budget in July 1996. Two redesigns were required to meet system limitations identified during commissioning, thanks to higher-than-expected use of the system.

Apparently, the TAURUS PMT project did many things in accordance with project management best practice. Taking the ten critical success factors (Pinto and Slevin, 1998):

1. The original definition of the mission was clear—a dematerialized settlement system with a central register to enhance the international competitiveness of the LSE.
2. The project was publicly supported by everybody from the G30 down, and those who privately expressed reservations such as the incoming LSE chief executive were warned against interference.

3. Project scheduling and planning techniques were used.
4. Extensive user consultation was undertaken.
5. The LSE appointed a strong project manager and team with an exemplary track record in managing similar projects, and the team worked very hard to achieve project success.
6. There was no shortage of technical competence, and the PMT opted to use an existing package rather than designing a bespoke system from scratch.
7. Users were kept continually informed of progress through SISCOT.
8. Monitoring and feedback arrangements did exist, but the project manager had demanded complete autonomy so as to reduce the possibility of interference by the client.
9. Extensive communication with the user community was undertaken, and great efforts were made to accommodate their views.
10. Troubleshooting arrangements appeared satisfactory.

However, the implementation of these PM disciplines was vitiated by the inability of the client to manage the stakeholders in the project. Figure 12.3 shows the stakeholder map for the TAURUS and CREST projects, with the mission of achieving a dematerialized back-off settlement system at its heart. The essence of the problem was that the PMT was prepared to listen to too many stakeholders, each with its own agenda. This was in the context of TAURUS being defined as a TINA project by the G30, with vociferous backing from the financial press. Those who had doubts were reluctant to express them in this febrile context.

The TAURUS power/interest matrix shown in Figure 12.4 illustrates that the key players were a disparate group, and, as the case shows, had different agendas. In other words, the Council of the London Stock Exchange was not powerful enough in relation to its stakeholders (Drummond, 1996, p. 82).

The importance of this power relationship is shown by the success of the CREST project, as shown in Figure 12.5. Here the team was backed by a supremely powerful actor—the Bank of England—as opposed to a relatively powerless actor, the Council of the London Stock Exchange. The weakness of the Council was a larger problem for the LSE as it strove to meet the challenges of the post Big Bang era, and many of the broader organizational changes implemented by the CEO during the life of the project were intended to strengthen the executive against the members of the exchange. As a result, the strength of the LSE itself relative to its stakeholders had also grown because of reform efforts of the ousted chief executive, and the Council had been replaced by a Board by the time the CREST project was implemented.

Approaches to Managing Stakeholders

Drawing on the lessons of the preceding analysis, together with those cases presented in Winch and Bonke (2002), the following suggestions can be made for the improvement of stakeholder management.

The Alignment of Incentives. The most effective way of managing internal stakeholders is to make sure that their incentives are aligned—the problem of project governance (Winch

FIGURE 12.3. TAURUS/CREST STAKEHOLDER MAP.

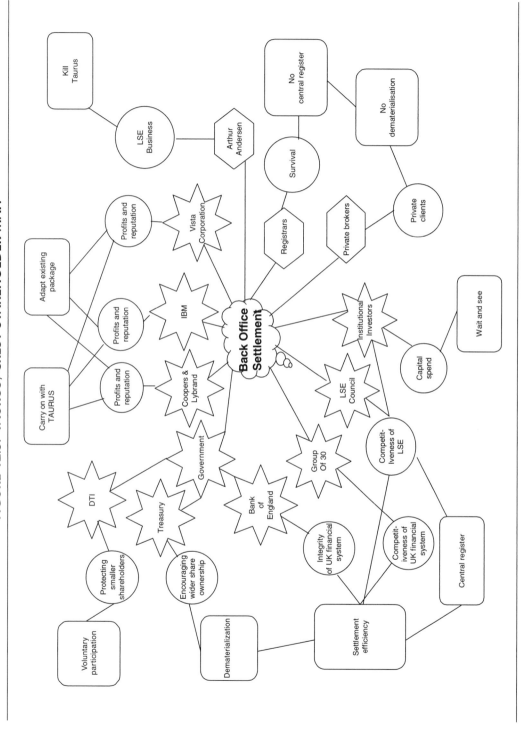

FIGURE 12.4. TAURUS POWER INTEREST MATRIX.

Level of Interest

	Low	High
Low	A: Minimal effort	B: Keep informed Supplier staff
Power to Influence		
High	C: Keep satisfied UK Government Group of 30 IBM Vista Corporation Private brokers Registrars	D: Key players LSE Members LSE Council Coopers & Lybrand Institutional Investors

FIGURE 12.5. CREST POWER INTEREST MATRIX.

Level of Interest

	Low	High
Low	A: Minimal effort supplier staff	B: Keep informed Registrars Private brokers Institutional investors
Power to Influence		
High	C: Keep satisfied Department of Trade & Industry HM Treasury Group of 30	D: Key players LSE Board Bank of England

2001; see also Miller and Lessard, 2000). On the demand side, this can be done by making sure that all key players have stakes in the project that move in step as project budget and schedule vary. A good model of this is where financiers, sponsors, clients, and facility operators all have an equity stake in an SPV—a good example of this approach is the successful Second Severn Crossing in the UK (Campagnac, 1996). Between the demand side and the supply side, this can be done through using incentive contracts as opposed to lump-sum or fee-based contracts..

Early Development of a Mitigation Strategy. Stakeholder mapping by the PMT should identify the claims that are likely to be made by external stakeholders and their power to press them. This should form the basis for a consistent strategy regarding which claims can be accepted without undermining the business plan and which are too costly. If likely claims by key players cannot be met without threatening the integrity of the business case, then not proceeding with the project is the wise option. The TAURUS project clearly lacked such a strategy and was overwhelmed by claims, while the CREST project adopted a take-it-or-leave-it strategy. The failure to identify the airlines as key players on the Denver International Airport project (Applegate, 2001; Montealegre et al., 1996) was, arguably, a major factor in the major schedule and budget overruns experienced.

Friends in the Right Places. This is often the responsibility of the project sponsors—for instance, the ability of members of the Boston political elites sponsoring the Boston CA/T to lobby effectively in Washington for funds was critical to its successful financing (Hughes, 1998). The project director of the Apollo mission was appointed for his contacts in Washington, not his technical expertise (Sayles and Chandler, 1993). Project management teams can be twin-headed, with one project director responsible for managing external stakeholders and another responsible for internal delivery, as on the Tate Modern project in London and Boston CA/T (Winch, 2002). One of the problems on TAURUS was that the project director was obliged to shift his attention to external stakeholder management, leaving a vacuum with internal delivery.

An Ethical Approach. Corporate social responsibility can be defined as the extent to which "an organization exceeds its minimum required obligations to stakeholders" (Johnson and Scholes, 2002, p. 216). One of the strongest arguments for an ethical approach to business is the damage to the brand that can occur when corporations fail to act responsibly, so much of the debate about business ethics has concerned the consumer goods sector. However, the issues are, arguably, equally relevant to the capital goods sector. In this spirit, perhaps *project social responsibility* can be defined as

> the extent to which the definition of the project mission exceeds the minima established in the NPV calculation and those required to obtain regulatory consents (Winch, 2002, p.78).

Explicit adoption of such an approach may well help to reduce the level of claims from external stakeholders who can see that their claims are being fairly and dispassionately considered. An ethical approach should avoid the bribe-and-ignore approach to consent management

Effective Consent Management. The management of consent (Stringer, 1995) within the regulatory framework is a strategic matter within project definition—for projects such as the fifth terminal at London's Heathrow airport, the management of consent amounts to a major project is its own right. Three basic approaches can be identified (Winch, 2002):

- Define-and-enquire
- Consult-and-refine
- Bribe-and-ignore

Where the regulations are unambiguous and prescriptive, then the define-and-enquire approach is appropriate—the codes are published and simply require to be interpreted. The consult-and-refine approach is more appropriate where codes are not prescriptive, or where there are significant uncertainties in their interpretation. The bribe-and-ignore strategy is, unfortunately, widespread—zoning codes are routinely ignored in many countries as recent tragedies where shantytowns have been engulfed by mudslides show. Recent earthquakes in a number of countries have indicated widespread ignorance of—or at least failure to implement—the codes relating to structural integrity. Its use is likely to undermine the other approaches to stakeholder management discussed here.

A Strong Client. A client that knows what it wants to achieve and how much it is prepared to invest to achieve it is more likely to be able to effectively manage stakeholders. A major difference between the TAURUS and the CREST projects was that the client was powerful in relationship to the project stakeholders on the latter and knew exactly what it was trying to achieve. A weakness of the Channel Fixed Link project was that it was "assembled round a hole like a Polo Mint . . . [there was] no client driving it forward with a vision of what the operator needed to have" (September 9, 1995 edition of the *Financial Times*). Where the client is selected by the sponsor on the grounds of political expediency rather than project management capability—as on Boston CA/T (Hughes, 1998)—then problems are likely to follow.

Getting the Concrete on the Table. The words of the project manager of the Storebælt Link in Denmark (cited Bonke, 1998, p. 10) are fundamentally wise: "we had to have the concrete on the table in a hurry." Stakeholders can change their minds about supporting the project—especially where political support is required—but the more capital that has been sunk in the project, the more likely it is to be pushed through to completion, even in the face of mounting doubts about the business case. This is the phenomenon of project escalation, and the tactic of making large spends early in the project life cycle on nonfungible assets can be used by project managers to keep their projects rolling. This was used on the Channel Fixed Link project to manage the risk of cancellation following a Labour win in the 1987 general election (Winch, 1996).

Public Relations. Keeping the external stakeholders informed using PR techniques can pay significant dividends, if only to avoid misleading rumors circulating. The press tend to focus on projects that go wrong, as the PMTs on the West Coast Main Line and Boston CA/T know well. Releasing good news in a systematic manner can help keep stakeholders on-side, both in press releases and on the project Web site. For example, the Web site for the C/AT (www.bigdig.com/) was exemplary, even if its stakeholder management overall was not.

Visualization. In earlier generations, the only way to visualize the asset being created was through artist's impressions or scale models. Digital techniques are making the visualization

TABLE 12.2. TAURUS AND CREST COMPARED.

Strategy	TAURUS	CREST
Client power	The LSE Council was weak in relation to the stakeholders.	The Bank of England was powerful in relation to the stakeholders. The LSE Council had also been replaced by the LSE Board.
Mitigation	A forum was set up—SISCOT—which encouraged claims to be made by stakeholders.	A take-it-or-leave-it approach made possible by voluntary membership of CREST.
Consent management	The project was hit by regulatory surprises, particularly from DTI, which undermined the business case.	The business case was not dependent on regulatory consent, and the need for such consent was reduced because of the voluntary nature of CREST membership.
Alignment of incentives	The principal contractors were awarded a cost-plus contract, and stakeholders were able to act as free riders.	Tightly managed process with strong in-house team. Free riding stakeholders (notably registrars and institutional investors) were moved to the low-power category.

task much easier. Regulators and other external stakeholders can see a digital image of the proposed facility in situ and can come to a more informed assessment of its real impact on their interests. User groups among the internal stakeholders can interact with virtual reality (VR) images of the proposed facility to provide their input to the design of a building, or rapid prototyping techniques can be used in the development of user interfaces for IT systems.

A number of strategies are, therefore, available to the PMT to help them in the management of project stakeholders. Table 12.2 summarizes the differences between the failed TAURUS project and successful CREST project in terms of their deployment of these suggested strategies.

Summary

"The only good stakeholder that I can recall is Professor Van Helsing in the Hammer horror films, who would drive a stake into the vampire's heart".

Alastair Ross-Goobey, an experienced fund manager and project financier cited in the *Financial Times* (February 21, 2003, edition), takes a rather jaundiced view of the claims of stakeholders compared to shareholders. However, at least so far the PMT is concerned, it is hoped that this chapter has convinced them that the Hammer Films approach to stake-holder management is less viable today than it ever was.

Projects are not very likely to keep all stakeholders happy. Projects are, fundamentally, vehicles of planned change in business and society (Winch, 2002), and changes nearly always benefit some stakeholders more than others. In many cases, this may simply means a number of employees losing their jobs; in a few, the perception may be that the irreparable destruc-tion of the natural environment or heritage assets is at stake. In all cases, project budgets have to be won and defended against the claims of sponsors of different projects competing for the same resources. Capital budgeting techniques are designed to allow these decisions to be taken rationally, but the inherent uncertainty associated with outturn capital costs and future returns from the assets created by the project leave plenty room for opinion and argument.

The growing complexity of project stakeholder management makes the balancing act required by the PMT between potential gainers and losers from the project ever more difficult. Mitigating losers too much may undermine the business case, while mitigating inadequately may lead to the mobilization of opposition through political pressure or direct action and oblige the cancellation of the project. The skills of project managers in internal delivery are well honed and can offer significant returns on investment for clients. However, projects still overrun program and budget, and fail to deliver a properly working asset. There is growing consensus that the problems lie in the setting of schedule and budget objectives and specifying the facility (Merrow, et al., 1981; Morris, 1994)—in other words, in the definition of the project mission. This chapter has argued that stakeholder management plays a vital part in defining the project mission, and has offered one way of aiding such management processes.

References

Applegate, L. M. 2001. *BAE automated systems (A) and (B)*. Harvard Business School Teaching Note 5-399-099.

Bonke, S. 1996. Technology management on large construction projects. London, Le Groupe Bagnolet Working Paper 4.

———. 1998. The storebælt fixed link: The fixing of multiplicity. London, Le Groupe Bagnolet Work-ing Paper 14.

Calvert, S. 1995. Managing stakeholders. In *The commercial project manager*, ed. J. R. Turner. London: McGraw-Hill.

Campagnac, E. 1996. La maîtrise du risque entre différences et coopération: Le cas du severn bridge. London, Le Groupe Bagnolet Working Paper 12.

Cleland, D. I. 1998. Stakeholder management. In *Project management handbook*, ed. J. Pinto. San Franciso: Jossey-Bass.

Cochrane, A., J. Peck, and A. Tickell. 2002. Olympic dreams: Visions of partnership. In *City of revolution: Restructuring Manchester*, ed. J. Peck and K. Ward. Manchester: Manchester University Press.

Coser, L. A. 1978. American trends. In: *A history of sociological analysis*, ed. T. Bottomore and R. Nisbet. London: Heinemann.

Drummond, H. 1996. *Escalation in decision-making: The tragedy of TAURUS.* Oxford: OUP.

Finkelstein, A. 2000. Y2K: A retrospective view. *Computing and Control Engineering Journal* 11:156–159.

Flyvbjerg, B., N. Bruzelius, and W. Rothengatter. 2003. *Megaprojects and risk: An anatomy of ambition.* Cambridge, UK: Cambridge University Press.

Handy, C. 1993. *Understanding organizations.* 4th ed. Harmondsworth, UK: Penguin.

Head, C. H. 2001. *TAURUS and CREST: Failure and success in technology project management.* Henley, UK: Henley Management College.

Hughes, T. P. 1998. *Rescuing Prometheus: Four monumental projects that changed the modern world.* New York: Vintage Books.

Johnson, G., and K. Scholes. 2002. *Exploring corporate strategy.* 6th ed. London: Prentice Hall.

Lukes, S. 1974. *Power: A radical view.* London: Macmillan.

Merrow, E. W., K. E. Phillips, and C. W. Myers. 1981. *Understanding cost growth and performance shortfalls in pioneer process plants.* Santa Monica, CA: RAND Corporation.

Miller, R., and D. R. Lessard. 2000. *The strategic management of large engineering projects: Shaping institutions, risks, and governance.* Cambridge, MA: MIT Press.

Montealegre, R., H. J. Nelson, C. I. Knoop, and L. M. Applegate. 1996. BAE automated systems (A): Denver international airport baggage-handling system. Harvard Business School Case 9-396-311.

Morris, P. W. G. 1994. *The management of projects.* London: Thomas Telford.

Pinto, J. 1996. *Power and politics in project management.* Newtown Square, PA: Project Management Institute.

Pinto, J., and D. P. Slevin. 1998. Critical success factors. In *Project management handbook,* ed. J. Pinto. San Franciso, Jossey-Bass.

Porter, M. E. 1985. *Competitive advantage.* New York: Free Press.

Sayles, L. R., and M. K. Chandler. 1993. *Managing large systems.* New Brunswick, NJ: Transaction Publishers.

Stringer, J. 1995. The planning enquiry process. In *The commercial project manager,* ed. J. R. Turner. London, McGraw-Hill.

Winch, G. M. 1996. The channel tunnel: Le projet du siècle. Le Groupe Bagnolet Working Paper 11.

———. 2001. Governing the project process: A conceptual framework. *Construction Management and Economics* 19:799–808.

———. 2002. *Managing construction projects: An information processing approach.* Oxford, UK: Blackwell Science.

Winch, G. M. and S. Bonke. 2002. Project stakeholder mapping: Analyzing the interests of project stakeholders. In *The frontiers of project management research,* ed. D. P. Slevin, D. I. Clelend, and J. K. Pinto. Newtown Square, PA: Project Management Institute.

CHAPTER THIRTEEN

THE FINANCING OF PROJECTS

Rodney Turner

This chapter describes the financing of projects. No project can take place without funding, and yet the raising of finance is something most project managers do not get involved in. However, the financing can have a huge influence on their project, and so project managers do need to contribute to the development of the financial strategy for their project, which in itself is a vital element of the overall project strategy.

The chapter is titled "The Financing of Projects," rather than "Project Finance." The vast majority of projects are paid for by the parent organization, out of revenue expenditure, as in the case of maintenance or research projects, or out of capital expenditure, as in the case of new investments. The capital is raised from the sources of finance described in the *Types and Sources of Finance* section, based on the reputation of the parent organization, and secured against its assets. The parent organization needs to repay the finance independent of the success or failure of the project. A few, usually large projects are financed as entities in their own right. This is called unsecured, nonrecourse, or off-balance-sheet financing. It is unsecured because it is only secured against the project's assets and its revenue stream. If the project fails, the lenders have no recourse against which to recover their money. It is off-balance-sheet, because if a parent organization invests in a project in this way, the capital invested does not appear in the company's balance sheet. Limited-recourse financing is a mixture of nonrecourse and recourse financing, where the parent organization invests some equity in the project, which will appear on its balance sheet and will be lost in the event of project failure, but the majority of finance will be loans secured only against the project's assets. Lenders prefer this because with some of their ownmoney invested in the project, the sponsor will have a greater interest in the success of the project. The term project finance is usually reserved for nonrecourse or limited-recourse financing. The other (usual) case is the financing of projects. This chapter considers both cases, and the sources of finance described in *Types and Sources of Finance* section can be used for financing the parent organization or an individual project. However, the chapter mainly focuses on the second case. In the majority of projects, the money will be made available by the parent organization. In this case, money may be borrowed to finance a specific project, but it will be guaranteed against other assets of the parent organization, and so will be available at lower interest rates and will appear on the company's balance sheet. The role of the project manager, champion, or sponsor is to justify the expenditure using investment appraisal techniques, (Aston and Turner, 1995; Lock, 2000; Akalu, 2001). It is only in the case of non- or limited-recourse financing that the project team will be involved in developing a financial package specifically for the project and approaching the financial markets to finance the project.

The financing of projects, including project finance, is a vast subject; whole books are devoted to the subject. Thus, in the limited space of a single chapter, we can only focus on a number of key issues. The next section begins by explaining characteristics of project finance. Then sources of finance are described. These sources are broadly the same for both the parent organization and individual projects. Types of finance are considered, and conventional and unconventional sources of those different types described. The next section describes project finance, the financing of individual projects on an unsecured, nonrecourse basis. Finally, the process of creating a financial package for a project to meet the investment needs while responding to the risk is considered.

Characteristics of the Financing of Projects

There are a number of features of the financing of projects that affect the design of a particular financial strategy. Some of these features relate to the financing of all projects; the latter ones refer more specifically to project finance. The features include the following:

- Finance is the largest single cost on a project.
- There is no project without finance.
- Financial planning begins at feasibility.
- The projects involved are often complex.
- Financial planning adds to project complexity.

Largest Single Cost

Finance is often the largest single cost on a project. On a large construction project, the material costs may be 30 percent of the total capital cost, construction costs 30 percent, and design, project management, commissioning, working capital, and contingency about 10 percent each. On the other hand, on a project that takes two years to build, on the day it is commissioned, the total cost of the financial package (interest paid to providers of debt and returns to equity holders) may amount to 20 percent, and 60 percent by the time the debt has been paid off. Finance is therefore twice as much as the next largest cost. On infrastructure costs taking longer to build, they will be even larger. Thus, the financial costs may be almost as great as the project costs, and so good financial management is at least as important as good project management.

This may not be obvious to many project managers, especially on projects financed by the parent organization, because the cost of project finance is not included in the estimate. It is allowed for in the investment appraisal process through the discount factor applied (see Aston and Turner, 1995; Lock, 2000; Akalu, 2001). Unfortunately, this may create distortions. Decisions will be made to minimize the capital cost, not minimize the financial cost or total cost. Usually, lower capital cost leads to lower financial cost. But sequencing the cash flow can make a difference. A lower capital cost solution with the expenditure front-loaded in the project may have higher finance charges than a higher capital cost solution with expenditure later in the project. Whole-life costing techniques are being developed to address this problem. For nonrecourse financing, the financiers and financial agents will be more concerned with optimizing the cash flows to minimize financial costs.

No Project without Finance

Consider three projects: the first one has the design package complete, but only 90 percent of the finance has been sourced; on the second both the design package and the financial package are 95 percent complete; and on the third, the design package is 90 percent finished, but the financial package is totally in place. Only the third project can receive the go-ahead to begin. No project can begin without the financial package in place. Yet many project managers will focus on design completion and almost ignore the need for the financial package. On projects financed by the parent organization, preparing the design package may be part of the investment appraisal process, and so there is no finance until the design package is complete. However, for nonrecourse financing, it may be necessary to obtain finance before design can start, and the financiers may want to influence the design solution adopted, preferring less risky solutions.

Financial Planning in the Feasibility Study

For this reason, financial planning for a project should begin at the same time as the technical solution or earlier. Regrettably, it is often begun at the last minute, when most other features are already determined. The most successful projects are ones in which the financial planning is a key part of the project strategy from the start and the aim of the project is to minimize the whole-life cost, including the financing charges and technical solutions adopted considering their impact on the financial solution. The lenders, whether the parent organization or external financiers, may have a view about which options should be selected. They may have a view about which options minimize the risk and therefore provide the safest haven for their money. They may not want the project's promoter to choose what appears to be the best value for the promoter or champion, but instead the one that provides the best value for them (the lenders), taking account of the risks. It is therefore useful to engage the financiers as early as possible, so options are not selected that have to be rejected.

Complexity

Problems with complexity will be particularly the case with projects using project finance, because they have several features that will make them especially complex:

- They will usually be very large.
- They may cross national boundaries.
- They often exceed the capacity of a single organization to plan, supply, and construct.
- They are technically complex, demanding skills not widely available.
- They are dedicated to a single purpose, a major project rather than a program.
- They are located at remote sites.
- They take place over long timescales, with the return on investment often taking decades.

Financial Planning Adds Complexity

Projects requiring project finance tend to be complex, but the finance planning adds to the complexity. Three issues that need to be emphasized, because they exert significant influence on the financial planning:

1. The sources of finance need to be identified before the technical specification of the equipment. The financial package often imposes a sense of compromise. But further, the lenders will impose constraint as to the level of risk they are willing to bear and may indeed look for higher rates of interest with higher-risk technical solutions.
2. The structuring and acceptability of the financial package is different from the perspective of lender and borrower. Each will take a different perspective of the risk, each giving a different emphasis to the risks and the compromises involved. The lenders may want higher returns for certain risks, and thus it is sensible to take account of their concerns in the project's feasibility and planning stages.
3. However, if you do involve the lenders early, be aware that the availability of finance and the terms on which it is available can be subject to significant and rapid change, for reasons beyond the control of the project sponsor or project manager. The impact of these changes needs to be tracked throughout the feasibility and planning of the project.

Types and Sources of Finance

This section considers types of finance and their sources. These types and sources may be used by the parent organization to finance itself, or they may be used in non- or limited-recourse financing to finance the individual project

Types of Finance

There are two main types of finance: equity and debt.

Equity. *Equity* is money subscribed by investors or shareholders. The shareholders get returns from dividends and capital growth in the value of their equity. With equity, there is no guarantee that a dividend will ever be paid, nor that the money itself will ever be repaid. A dividend can only be paid after the interest and scheduled repayments of loans have been paid. And the equity can only be drawn out of the company or venture after all the obligations to the providers of debt have been met. If the project performs badly, the providers of equity may receive nothing, but if it performs well, the returns may be huge. Because of the higher risks involved, the providers of equity expect higher returns on average than the providers of debt.

Mezzanine Debt. *Mezzanine*, or *subordinate*, *debt* is loans made by the holders of equity. It is sometimes treated as equity, especially in the calculation of debt/equity ratios and so is also sometimes called *quasi-equity*. It differs from equity in that it is repayable against a schedule of payments and the returns to investors are in the form of interest payments at a predetermined rate. However, those interest payments and scheduled repayments can only be made after all the obligations to the providers of senior debt have been met. Thus, mezzanine debt is higher risk that senior debt and so commands higher interest rates. Mezzanine debt can take the form of debentures, preferred stocks, and other instruments.

Senior Debt. *Senior debt* is money borrowed from a number of possible sources but particularly banks. It is repayable against agreed schedules of payment, including the predetermined interest payments. Senior debt must be repaid before all other forms of finance (hence, the name), and the providers of senior debt have the first claim on all assets of the venture if the borrower goes into liquidation. Debt can take the form of loans, bonds, and nonconvertible debentures. It can also take the form of equipment provided under supply contracts with repayment to be made over the life of the plant. There are two types of senior debt:

1. *Secured debt.* This is debt secured against assets or collateral easily convertible to cash. It must almost by definition be money lent to the parent organization, secured against its assets, and is repayable regardless of the performance of any projects for which it is intended. It commands a lower interest rate than unsecured debt—the interest rate only being dependent on the reputation of the parent organization and the security of its assets.
2. *Unsecured debt.* This is debt lent for a specific project, secured only against the assets of the project and its predicted revenue streams. It is unsecured, nonrecourse, off-balance-sheet project financing. Given that it is dependent on the project's success, the interest rate will be higher than for secured debt and will be linked to the riskiness of the project.

Eurotunnel, the construction of the channel tunnel between England and France, was an example of a limited-recourse financed project, involving a mixture of debt and equity. Equity was provided by the project promoters, mainly contractors involved in the construction of the tunnel, and by private investors. But the vast majority of the finance (in excess of 80 percent) was loans provided by banks. During construction of the tunnel, interest on the loans was added to the debt. But in the early stages, equity holders received some returns on their investment in that the share price steadily rose as the commissioning date, and hence the expected returns, became closer. In fact, the early revenue streams were not sufficient to cover the schedule of debt and interest payments to the banks. The shares now had no value, so the equity investors lost their investment, and the project effectively became nonrecourse finance. But there was no point the banks foreclosing on their loans, because the hole in the ground (the project's main asset) had no value if it was not being used. So the banks converted much of their loans to equity, and are now receiving their returns over the life of the project in the form of dividend payments.

Cost of Capital

Each form of finance has a cost associated with it: the cost of borrowing that type of capital. The *cost of capital* is the average cost of all forms of finance used by a project or company. The discount factor used when performing investment appraisal (Aston and Turner, 1995; Lock, 2000; Akalu, 2001) is the cost of capital inflated by any allowance for risk.

Cost of Equity. The *cost of equity* (returns gained by shareholders) is the dividends they receive plus capital growth of the equity. (During construction of the channel tunnel, the equity investors did not receive dividends, but there was growth in value of the equity, providing initial early returns.) The capital growth is not predetermined, and so in calculating the cost of equity, a guess has to be made as to its likely size. The most common model for calculating the cost of equity is now the Capital Asset Pricing Model (CAPM). According to this model, the cost of equity is

$$\text{Cost of equity} = \text{Risk free rate of return} + \text{Beta} \times \text{Equity risk premium}$$

where:

> *Risk free rate of return* is the most secure rate of return anyone could obtain from investing their money, for example, by investing in government bonds.
>
> *Equity risk premium* represents the excess return expected from investing in equity given that it is repayable after all other debt. It represents the average risk of the stock market. A value of 3 percent to 4 percent is often used, which in the past has underestimated returns.
>
> *Beta* represents the risk premium of a particular stock, measured (obviously) as the ratio of that stocks riskiness compared to the average of the stock market. For a particular company, it will represent the riskiness of the industry that company operates in and its own features that make it more or less risky than the average for the industry. For a project it will represent the riskiness of the project.

Cost of Debt. The cost of equity (dividends and capital growth) is payable out of untaxed income. The cost of debt (interest) is payable out of taxed income. Thus, the *cost of debt* is the interest rate minus an allowance for the tax not paid:

$$\text{Cost of debt} = \text{Interest} \times (1 - \text{Tax rate})$$

Cost of Capital. The total *cost of capital* is the weighted average of the different types of capital. (If there are several forms of debt, this sum obviously must be done over all the forms of debt.)

$$\text{Cost of capital} = \text{Ratio of equity} \times \text{cost of equity} + \text{Ratio of debt} \times \text{Cost of debt}$$

Conventional Sources of Finance

Shareholders. These are public or private investors, institutions, or individuals who provider the equity or quasi-equity in a company. Sources of equity include the following:

- Retained profit of a company
- Funds raised through the stock market
- Venture capital companies
- Joint venture partners
- International investment institutions such as the World Bank

Banks. Banks and other financial institutions are the main providers of debt. Commercial banks are the most readily available to most project investors. They split into retail banks, which provide finance in the local main-street and merchant banks. There is a large choice available for companies raising finance, and this has led to intense competition. In choosing a bank, the decision will not be so much based on the interest rate charged as on the following factors:

- The size of the bank
- The experience in financing that type of project
- Any support they may offer with the financial engineering

International Investment Institutions. Another form of bank is the international investment institutions, including the World Bank and other development banks (see Table 13.1). They will provide both debt and equity.

International Financial Markets. International financial markets offer an alternative to domestic markets, giving easy access to foreign sources of funds. There are many, but the two most important are the Eurocurrency and Eurobond Markets. The Eurocurrency markets are the most efficient in the world and provide for smooth movement of funds. They provide short-term finance at competitive rates but are primarily for large organizations. The Eu-

TABLE 13.1. INTERNATIONAL INVESTMENT INSTITUTIONS.

African Development Bank, AfDB
Asian Development Bank, ADB
Commonwealth Development Corporation, CDC
European Development Fund, EDF
European Investment Bank, EIB
European Bank for Reconstruction and Development, EBRD
Inter-American Development Bank, IDB
International Bank for Reconstruction and Development, IBRD, or World Bank
International Development Association, IDA
International Finance Corporation, IFC

robond markets provide promissory notes or bonds issued outside the United States. The bonds can be issued in small denominations, making them attractive to the small investor. They provide anonymity, so interest rates are usually lower than on domestic markets, making them the most competitive in the world.

Suppliers

Suppliers have been placed under conventional forms of finance because it is becoming increasingly common for suppliers to be paid out of the revenue stream of a large project rather than up front on supply of their piece of equipment. They may be paid under buyer or supplier credit, or by becoming part of the joint venture investing in a project.

Export Credit, Buyer and Seller credit

Most developed countries have saturated home markets and so encourage companies to export. On the other hand, developing countries need access to advanced technologies. Many OECD (Organisation for Economic Co-operation and Development) countries have established an export financial agency national interest lender to facilitate trade with developing countries. This agency often also provides insurance to cover export risks. The finance it provides may also be linked to the country's aid budget. Examples include the following:

- *Canada.* The Export Development Corporation (EDC)
- *United Kingdom.* The Export Credit Guarantee Department (ECGD)
- *United States.* The Export-Import Bank of the United States (EIBUS), the Private Export Funding Corporation (PEFCO), and the Overseas Private Investment Corporation (OPIC).

The United Kingdom's ECGD does not directly provide finance but provides a number of related services:

- Access to cheap finance, allowing companies to offer competitive terms and win contracts they might otherwise lose

- Insurance against political, legal, economic, and social risk
- Export credit insurance to insure a national supplier against nonpayment
- Support for a performance bond required by a national supplier to win a contract overseas

In providing or arranging finance for equipment for export, the export credit agency may lend money to the national supplier, known as *supplier credit* (see Figure 13.1), or to the overseas buyer, known as *buyer credit* (see Figure 13.2).

Unconventional Sources of Finance

There are several less conventional sources of finance.

Leasing. Rather than buying an asset, the project promoter leases it. The financier pays for the asset and receives their return in the form of rental for the asset. This makes the asset available to the project promoter through off-balance-sheet financing. Under the terms of the deregulation of the electricity industry in Ireland, this is now effectively the form of financing used for investment in the national grid. The national grid is operated by a privatized company, called EirGrid, which acts as promoter for new investment in the grid. However, investments are paid for and constructed by the generating company, the Electricity Supply Board, and EirGrid pays a lease for the assets.

Counter Trade. The seller accepts goods or services in lieu of cash. This can be expensive and cumbersome, especially as the seller has to sell the goods or services to receive their returns, and so is not well liked by project contractors and project financiers. However, it is quite common under buyer or seller arrangements (see preceding text) or project financing arrangement, as discussed in the next section, for all of the revenue from operation of the project's facility post commissioning to be paid into an escrow account under the control of the project's financiers. All debt and interest payments are the drawn from this account, before any surplus is made available to the project's promoter.

FIGURE 13.1. SUPPLIER CREDIT.

FIGURE 13.2. BUYER CREDIT.

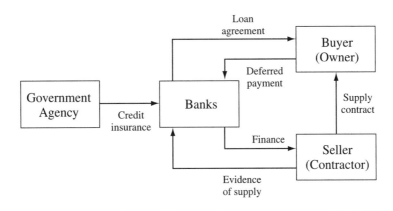

Forfaiting. Finance is made available through the sale of financial instruments due to mature at some time in the future. Finance is provided by trading in assets in the financial futures market. This can be a very expensive form of finance.

Switch Trade. Switch trade makes use of an uncleared credit surplus arising from bilateral trade arrangements. For example, if Country A has a credit surplus with Country B, exports from C to A can be financed with payments from B to C.

Debt/Equity Swapping. Debt/Equity swapping is designed to encourage investment in developing countries by owners of technology. A multinational company buys host country debt at a discount. This is redeemed in local currency at favorable rates of exchange and is used to set up local companies. These are used to transfer technology, generate foreign exchange, replace imports, and create employment.

Islamic Banks. These banks provide finance according to Sharia law, under which interest is banned. They provide finance and share in the profits of the investment. They appear to be like equity holders, except the payments are made against agreed-upon schedules and are repayable ahead of equity.

Project Finance

This section describes project finance—that is, unsecured, nonrecourse, off-balance-sheet financing of a project as a stand-alone entity. This has become popular recently under what is known as either the Private Finance Initiative (PFI) or Public-Private Partnership (PPP). Under this approach, a national government either works with the public sector or delegates responsibility to the private sector to design, construct, finance, operate, and maintain infrastructure projects that would normally be the responsibility of the national government working on its own (see Ive's chapter). The section begins by describing different forms of project undertaken in this way and then gives an overview of the structures commonly adopted, describing the roles of various parties and the contracts adopted.

Forms

Project finance is usually adopted for large infrastructure projects undertaken by the private sector on behalf of a national government, or working with the national government, under what is no known as the Private Finance Initiative (PFI) or Public Private Partnership (PPP). During the nineteenth century, it was common for infrastructure projects to be built by the private sector and paid for out of the revenues generated. In the early twentieth century, many national governments, under a sense of public spiritedness, took infrastructure development on themselves. However, during the 1960s and 1970s, companies in the mining and oil and gas industries began using limited-recourse financing for the following reasons:

- They had few assets other than the minerals in the ground with which to secure the loan.
- Off-balance-sheet financing offered tax advantages.

In the late 1970s and 1980s, many governments, especially developing economies, found they could not afford to undertake infrastructure development without involving the private sector. The Turkish government of the 1980s took a lead, but the decision to build the Channel Tunnel between England and France gave prominence to the approach. It became increasingly popular during the 1990s because governments

- saw it as a way of reducing public debt
- believed inherent efficiencies in the private sector would result in a cheaper product even though the private sector pays higher interest rates

PFI should be distinguished from privatization, which either

- transfers assets to the private sector that were previously owned by the public sector
- provides for services to be undertaken by a private company that were previously undertaken by the public sector

Project finance is used where a private sector company needs money for the construction of public infrastructure on the basis of a contract or license:

- An off-take contract where the public sector contracts to by the outputfrom the plant, as in the case of a power station
- A concession agreement in which the facility is constructed to provide a public service, as in the case of a hospital

- A concession agreement in which a facility is constructed to provide a service to the general public, as in the case of a toll road
- A license in which a facility is constructed to provide a new service to the public, such as a mobile phone network

Not all projects undertaken by the private sector on behalf of the government necessarily involve project finance. Some are paid for by the government. Others may be paid for out of the capital assets of the sponsoring company. These are likely to be smaller projects, which is why it was stated previously that it is larger projects for which project finance will be used.

Is there a difference between PPP and PFI? Some people treat them as synonymous. The generally agreed distinction is as follows (ECI, 2003):

- PFI is a subset of PPP and are projects financed by the private sector.
- PPP is all projects involving a collaboration between the public and private sector. The United Kingdom's treasury has issued a directive that all projects undertaken by government departments must be done either as PFI, prime contracting, or design and build, unless special dispensation is given. PFI and prime contracting are both PPP, but prime contracting is paid for by government funds. In the United Kingdom, design and build is treated as neither PFI nor PPP.

There are a number of different forms of such arrangements.

Build-Own-Operate-Transfer (BOOT). This method was the approach used on the Channel Tunnel. The private sector company constructs the project and owns and operates it for the concession period. It earns the revenues during that period, to repay the finance and make a profit, and at the end of the concession period, the company hands the asset over to the government.

Build-Own-Operate (BOO). Sometimes the asset is not handed over to the government at the end of the concession period. This situation may be in the case of a power station where at the end of the concession period, the plant has no residual value, or for a mobile phone network where the project company gets the benefit of the residual value.

The UK government prefers that all PFI projects use one of these two forms. For reasons of motivation and risk sharing, the promoter should own the asset through the concession period.

Build-Operate-Transfer (BOT), Design-Build-Finance-Operate (DBFO), and Build-Lease-Transfer (BLT). This form of financing is used when for some reason it is not appropriate for the private sector company to own the asset. The private sector company designs and builds the asset and obtains the finance to do so, and then operates it for the concession period, obtaining revenues to repay the finance and to cover its costs. Under BLT, the government raises the finance and then leases the asset to the project company. This circumstance was described previously under unconventional sources of finance and is used for construction of extensions to the national electricity grid in Ireland, as described previously.

Build- Transfer-Operate (BTO). This is identical to the previous form, except the project company owns the asset during construction. At commissioning, ownership transfers to the government, but the project company operates it.

Structures

Figure 13.3 illustrates a typical finance structure used for these types of project. At the heart is the project company, sometimes called the special purpose or project vehicle (SPV). One or more of the project stakeholders will be partners in this joint venture company. Almost certainly the operator will be a partner. All the others may or may not be partners.

Finance consists of a mixture of debt and equity. Project finance usually involves what would normally be considered high debt-to-equity ratios. A ratio of debt to equity of $4:1$ is common. For a company, the reverse, a debt equity ratio of $1:4$, is considered a large amount of debt. The higher the debt, the higher the return to the project sponsors. However, the banks insist that there should be some equity, because if the project fails, the equity

FIGURE 13.3. PROJECT FINANCE STRUCTURE.

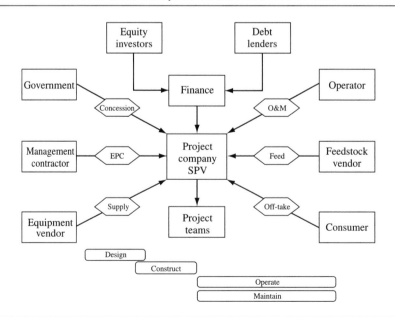

holders lose their money first, and so they will impose good management on the project company to maximize their own returns. The equity will primarily be provided by the partners to the SPV, but some equity may be raised from private investors.

Roles and Contracts

Figure 13.3 also illustrates some of the contracts involved. There are two main contracts that that enable the project to happen:

- The concession agreement with the government
- The off-take contract with the consumer of the project's outputs

In some cases the consumer will be the government, and so the off-take contract will be with the government. In other cases it will be with other bodies. In some cases the revenue from the off-take will be paid into an escrow account controlled by the lenders of debt. Interest and debt service in accordance with the agreed schedule of payments will be drawn from the escrow account before any surplus cash is paid to the SPV for payment of its costs and distribution to the equity partners. There are also a number of ancillary contracts:

- Design and build or EPC (engineering procurement and construction) contract with the managing contractor
- Equipment supply contracts with material suppliers
- Operation and maintenance agreement with the operator
- Fuel and other supply contracts with the feedstock suppliers

All of these people may be paid directly, or they may be given delayed payments, by either being given equity in the SPV or by having their payments converted into loans. If they are also equity holders, the loans will be subordinate debt to the bank loans.

There are also several support contracts:

- Permits and other rights
- Insurance
- A support agreement with the government

The support agreement with the government will impose a number of duties and obligations on both the government and concessionaire to facilitate the project completion.

The Process of Financial Management

You saw previously that the process of financial management begins at the feasibility study and continues right through the project to commissioning and operation. The key steps are as follows:

1. Conducting the feasibility study
2. Planning the project finance
3. Arranging the financial package
4. Controlling the financial package
5. Managing the risk

Sometimes the project champion, the project manager, and the organization they work for have little experience in the financing of projects. They will then be looking for considerable support from the lenders and may employ a financial consultant to support them.

Conducting the Feasibility Study

The financial assessment of the project begins at the feasibility study. The project will be assessed using financial appraisal techniques (Aston and Turner, 1995; Lock, 2000; Akalu, 2001). Options will be considered to maximize the value of the project. But the financial implications of those choices also need to be considered. Many decisions have financial implications, and those need to be taken account of from the start. During the feasibility, the financial objectives of the project will be set, and these will be different for the borrowers and the lenders, who will have different concerns. The sponsors concerns are as follows:

- Raising the necessary funds at times and in the currencies required by the project
- Minimizing costs and maximize revenues from the project
- Sharing risk between the project stakeholders, including the financiers
- Maintaining flexibility and control, including rescheduling repayments if necessary
- Being able to pay dividends to equity holders

On all projects, the lenders concerns are as follows:

- Repayments will be made in accordance with the schedule of interest and repayments.
- There is adequate security in the event of default by the borrower.
- Satisfactory dividends will be paid to equity holders.
- The risks are understood and will be managed.
- The basis of the cash flows, the nature of the business and market for products.
- There are adequate arrangements for insurance and maintenance to protect the value of the assets and maintain the cash flow.

In the case of project finance, the lenders will have a wider list of concerns:

- How robust are the project's cash flows?
- Do any other parties have claim on the cash flows?
- Are the project's assets dedicated only to the project?
- Can they be pledged as security?
- What is the market value of the assets in the case of project failure?
- Who are the project stakeholders?

Planning the Project Finance

Having undertaken the feasibility study, the project sponsor and their advisers will plan the financial package. This will involve the following issues:

1. From the feasibility study, the sponsor and their agents will have identified the total cost of the facility and the total amount of money that needs to be borrowed. Critically, this will include the scheduling of the cash flows, the time at which the money needs to be borrowed. It is also important to remember to include working capital and inflation.
2. Next, they will need to develop a financial strategy. They will need to plan the debt-to-equity ratio and develop a financial strategy involving a mixture of debt and equity to optimize profits while maintaining support of lenders.
3. They need to identify potential sources of debt and equity. They need to identify the sources of senior debt and determine the mixture of domestic and international finance. They also need to define the insurance required and the need for export credits. In the case of project finance, they need to consider who will be partners in the project company (SPV), how much equity will each provide, and will they provide any subordinate debt. They need to consider whether shares will be sold to private investors.
4. Having identified sources of finance, it can be worthwhile revisiting the cash flows to see if interest rates can be rescheduled to optimize interest rates.

Arranging the Financial Package

The financial package then needs to be assembled. The main supplier of senior debt will almost certainly be involved in this step as an adviser to the project promoter and will do most of the work. They will help the sponsor by:

- raising equity; a merchant bank will also be involved with this step.
- identifying additional sponsors and suppliers of subordinate debt.
- raising money through the Eurocurrency and Eurobond markets.
- liaising with government export agencies and arranging buyer or supplier credit.
- arranging insurance.
- arranging more sophisticated and less conventional financial packages.

Controlling the Financial Package

As the project progresses, the financial arrangements need to be managed, and again this will probably be done by the financial agent. This will involve monitoring progress against the plan and taking action to eliminate deviations from the plan. This may include the following:

1. *During project execution.* Monitoring expenditure to ensure it follows the predicted schedule and that debt and equity are drawn in accordance with the plan. It will be necessary to work with the cost controllers to forecast cost to complete, and make arrangements for additional finance if necessary.

2. *During commissioning.* Monitoring initial operating and maintenance costs and initial sales revenues to ensure they are as predicted and that the financial plan will be achieved. Variances need to be identified immediately and eliminated to ensure the project performs in accordance with its financial objectives.
3. *During operation.* Helping the sponsor with their management accounting to ensure the asset is operated at its peak profitability, and to provide reports to the lenders to avoid nasty surprises.

All of this is good project control, but it takes place at a level above the project. It draws on the project control and performance data, as discussed elsewhere in this book, to provide financial and business control for the project sponsor.

Controlling Risks

Risks on a project break into three broad categories:

1. Financial risks
2. Design and construction risks
3. Health and safety risks

The first of these is primarily the responsibility of the project sponsors and their agents. They will be manages using a risk management process, one of which will be described elsewhere in this book. However, a generic process follows six steps:

1. *Focus on risk management.* The planning of the financial packages should be done is such a way as to facilitate risk analysis and management. This has to begin during the project feasibility study.
2. *Identify risks.* Possible financial risks are listed in Table 13.2; most of these are self-explanatory.
3. *Qualitative assessment.* The risks will be assess in terms of likelihood and impact and prioritized for further assessment.
4. *Quantitative analysis.* The larger risks will be analyzed using techniques such as Monte Carlo analysis to determine the overall impact on the project and to help determine mitigation strategies.
5. *Develop a mitigation strategy.* A strategy for reducing the impact of the risks will be developed. For each significant risk, there is one of four possible actions:
   • Avoid the risk.
   • Reduce the likelihood or impact of the risk.
   • Pass the risk on to a contractor, insurance company, lender, or other stakeholder.
   • Develop a contingency plan.
6. *Monitor the risks.* The risks are monitored as the project progresses to ensure success of the mitigation strategy.

Finance is complex and the largest single cost on a project, so financial risk management is one of the most important success factors for a project.

TABLE 13.2. POSSIBLE FINANCIAL RISKS.

Type of Risk	Examples
Macroeconomic risks	Inflation
	Interest rates
	Currency and exchange
Political risks	Country risk
	Legislation and regulation
	Change of government
Commercial risk	Viability and feasibility
	Cost and schedule completion
	Performance and operation
	Revenue
	Availability, reliability, and maintainability
Contractual risks	Management
	Equipment supply
	Feedstock supply
	Concession and licenses
	Sales agreements

Summary

The cost of providing finance on a project is one of the largest single costs, being equal to or even greater than other major costs such as materials, design, or implementation over the life of the project. Further, as the arrangement of the financial package can have a direct impact on many of the design and technological choices on the project, it is necessary for the financial strategy to be considered as an integral part of the overall project strategy, beginning at the feasibility study. In this chapter, we considered the financing of projects, looking at the financing of projects (recourse finance) and project finance (non- or limited-recourse finance). We considered the features of project finance, and the sources of finance to companies, for both recourse and nonrecourse financing of projects. Both conventional and unconventional sources were described. The specific case of project finance was then described—that is nonrecourse, off-balance-sheet finance for projects, particularly as it applies to public-private partnership, with the private sector undertaking infrastructure development on behalf of governments. Finally, we considered the process of financial management on projects and how the financial strategy can be made an integral part of the project strategy, to minimize the cost of finance on the project.

References and Further Reading

Akalu, M. M. 2001. Re-examining project appraisal and control: developing a focus on wealth creation. *International Journal of Project Management* 19(7):375–384.

Aston, J., and J. R. Turner. 1995. Investment appraisal. In *The commercial project manager, ed.* J. R. Turner. London: McGraw-Hill.

Dingle, J., and A. Jashapara. 1995. Raising project finance. In *The Commercial Project Manager*, ed. J. R. Turner. London: McGraw-Hill.

ECI. 2003. *Public private partnerships: A review of the key issues.* Loughborough, UK: European Construction Institute.

Lock, D. 2000. Project apppriasal. In *The Gower handbook of project management*, ed. J. R. Turner and S. J. Simister. Aldershot, UK: Gower.

Merna, A. 2000. Managing finance. In *The Gower handbook of project management*, ed. J. R. Turner and S. J. Simister. Aldershot, UK: Gower.

Nevitt, P., and F. J. Fabozzi. 2000. *Project financing.* London: Euromoney Institutional Investor.

Turner, J. R., ed. 1995. *The commercial project manager.* London: McGraw-Hill.

Walker, C., and A. Smith. 1995. *Privatised infrastructure: The BOT approach.* London: Thomas Telford.

Yescombe, E. R. *Principles of project finance.* San Diego: Academic Press.

CHAPTER FOURTEEN

PRIVATE FINANCE INITIATIVE AND THE MANAGEMENT OF PROJECTS

Graham Ive

The first section of the chapter defines what is meant by the Private Finance Initiative (PFI). We then proceed to explore the special or key challenges that PFI as a structure of organization of the project process poses for the management of projects. This is done using the framework of generic project processes proposed by Winch (2000; 2002), namely: (1) defining the project mission, (2) mobilizing the resource base, (3) riding the project life cycle and leading the project coalition, and (4) maintaining the resource base.

In PFI there is a project management challenge both for the public sector client and its advisors and for the private sector provider. Accordingly, this chapter looks at the issues from both perspectives. However, where shortage of space makes it impossible to be comprehensive, priority has been given to the client's problems.

Throughout, issues will be illustrated by reference to a "stylized" and simplified account of an example of a PFI project—a 40-year contract to design, build, finance, and operate a major new teaching hospital, called the Gower Street Redevelopment Project, for University College London Hospitals (UCLH) National Health Service (NHS) Trust. The outline business case for this project was developed in 1994 and involved consolidating four hospitals (UCH, Middlesex, Tropical Diseases, EGA for Women), occupying many buildings, into a single new complex.

While the chapter tries where possible to give normative advice, its main aim is to describe the tensions and trade-offs that PFI project owners and managers must resolve, given their own priorities and their project's attributes.

What Do We Mean by PFI?

The name PFI (Private Finance Initiative) is the term used in the United Kingdom, but also elsewhere, for what, for project management purposes, would more clearly be labelled as DBFO (Design-Build-Finance-Operate) procurement and contracting.

DBFO is potentially applicable to any kind of project for delivery of a service under-pinned by provision of a specific capital asset or set of assets (facility or facilities). In the United Kingdom, for instance, it has been used for public sector projects requiring new or refurbished buildings or infrastructure assets, military vehicles and equipment, and IT sys-tems. It can even be used for projects with private sector clients, though the contract period will then tend to be much shorter than the 25-to-40-year operating periods found in pubic sector DBFOs.

PFI and Concession Contracts Compared

PFI is a form of a broader phenomenon, BOOT (build-own-operate-transfer) contracting. The acronyms DBFO and BOOT are sometimes used as interchangeable synonyms. I pro-pose that it is more useful to restrict the use of DBFO (and thus PFI) to mean the following version or variant of BOOT and to distinguish it clearly from concession contracts.

Concessions. In non-DBFO BOOT contracts, the essence of the matter is that the role of the public sector becomes restricted to that of granting and regulation of a *concession* (see Campagnac and Winch, 1997), which gives the concession company monopoly rights to use and appropriate income from a specified asset or set of assets for a defined period. There may in addition be either a public subsidy paid to the concession company or the reverse, participation by a public body in the distribution of profits arising, but neither of these is essential. What is essential is that the concession company obtains most of its revenues by charges to users of the services it provides. What distinguishes this from the ordinary com-mercial production and sale of private goods and services is that in this case the producer requires either the use of public property or the exercise on its behalf of rights and powers belonging solely to the state; and that, in return, it accepts more or less close regulation by an agency of that state, the form of which regulation is articulated in the concession agree-ment between the public and private sector parties.

DBFO. In PFI or DBFO, on the other hand, the essence of the matter is that the public sector (and not the users) purchases the services and pays the PFI company (the private provider) for providing to a set of users (which can be the general public or a defined subset thereof); they remain therefore, in a political-economic sense *public services*. The financing of the assets required in order to provide those services becomes private (hence, private *finance* initiative). But the *funding* of the service remains public—proximately, government pays the PFI company and it is taxpayers, not users, who ultimately pay for provision (HM Treasury Task Force, 1997).

In comparison with conventional public services, two things have changed: (1) instead of both purchasing and providing those services to the public, the public sector now only purchases some or all of them, while the PFI company does some or all of the providing; and (2) the public sector no longer directly purchases (and therefore owns) the assets required to provide those services—instead, ownership rights in those assets are split between the PFI company (which can, as a slight simplification, be thought of as owning those assets for the length of its PFI service-provision contract with the public sector) and the public sector body (to which ownership of the assets will normally pass, without further payment, at the end of the PFI contract).

In the great majority of UK PFI projects, service provision is split between public and private sector bodies. A public sector body continues to provide *final* services to users (for example, education services, clinical services) and therefore to employ the producers of such services and control the main interface with users, while the PFI company provides *intermediate* or *support* services (some or all of the range of what can be described as facility management or FM services). The exceptions to date are highway and custodial projects. In these cases the PFI company provides the final service and there is no public sector involvement in provision.

UK-type DBFO projects are found around the world, but most especially to date in countries with "Anglo-Saxon" political, legal, and business cultures, such as Ireland, the United States, Canada, Australia, New Zealand, and South Africa, but also in such diverse countries as Chile and the Netherlands. Many international banks, service providers, and management, legal, and technical advisers with experience gathered mostly to date in the United Kingdom are actively selling the "UK model" around the world, and the UK government and its agents are actively advising many other governments on its use. Thus, a study such as contained in this chapter, based on UK experience and practice, is of far more than parochial relevance.

In the United Kingdom and elsewhere, PFI has been used by government where it does not wish to transform public services and public assets into private services and private assets, as would be achieved by a full marketization of services and privatization of the service provider, but where it does nonetheless wish to introduce an element of private sector provision of services (on efficiency or comparative value-for-money grounds) and private sector finance (on macroeconomic, reduction of public sector borrowing requirement grounds).

Thus, PFI has been applied mostly in the field of core public services such as defence, custodial, justice, police, education, and health services, rather than in the standard BOOT fields of utilities.

If we stick to this clear and useful distinction, then we must say that UK projects such as the Severn, Skye and Thames Crossings (toll bridges), and city mass transit systems with passenger fares (Docklands Light Railway, various tram and light-rail systems), though undoubtedly UK examples of BOOT or concession, are not instances of what is new and unique about PFI or DBFO. They therefore lie outside the main focus of this chapter, though many of its arguments apply to them (and all concession projects) as well as to PFI projects.

There is a third type of arrangement also sometimes classed together with PFI. In this case, there is a public sector provider of marketed goods or services (a "nationalized industry") that acts as client by placing a BOOT contract. This would include, for example,

BOOT projects placed by London Underground Ltd. or by Scottish (publicly owned) utility providers. In project management terms, this has much in common with "narrow" PFI (payment by a public client for asset-based services delivered by a private provider under an output specification). Consequently, almost all of the following analysis also applies to such projects.

For project management purposes, the difference is clear between a private concession company as the client for the project in one case (with government and regulator as non-client stakeholders) and a public sector body as the client for the project in the other, with this client defining and specifying the services to be provided and for which it will pay.

Aims and Objectives of Parties to PFI Contracts

The basic idea of PFI, as defined by the National Audit Office (National Audit Office, 2002; see also HM Treasury Task Force, 1997) is that, while the public sector still pays for the project, they do so

- by entering into a long-term contract to purchase services, not facilities
- by leaving to the private sector the procurement, ownership, and operation of the underlying facilities required to deliver the preceding service
- by specifying the services to be purchased in terms of outputs, not inputs
- by linking service payment to delivery
- by defining delivery in terms of performance measures set out as part of the output specification (performance specification).

UCLH

In the case of UCLH, for example, the NHS Trust and the Department of Health used PFI because they hoped thereby

- to get expenditure on a capital project approved by the Treasury (ministry of finance), in a context of tight limits on approvals for capital spending requiring public borrowing
- to get better value-for-money-spent (VFM) than they would from conventional procurement by getting a price for provision of services that reflected lower private than public sector costs of provision
- to get greater certainty at the precontract stage about construction time and price
- to get greater certainty of long-run service delivery

Meanwhile, the successful PFI bidder, Health Management Group (HMG), at bid stage a joint venture between AMEC plc (a major construction and engineering services group) and Building and Property Group Ltd (a major UK facility management service provider), hoped, as putative owners of the PFI project company, the special purpose vehicle (SPV), would get

- a high return on equity capital invested in the project
- a long-run stable flow of profit income
- an increased presence in and experience of a fast-growing market

As construction and FM companies, respectively, they hoped to get

- large and profitable contracts from the SPV
- experience in a fast-growing market
- opportunities to find innovative, lower-cost ways of providing construction and FM solutions that would raise their profit margins

Consequences of Using Project Finance

PFI almost invariably involves *project* rather than corporate finance. That is, a special purpose vehicle or company (SPV) is set up whose only asset and only revenue is the PFI contract with the public sector body and the revenues due under that contract. These revenues are "ring fenced" and kept strictly separate from the other revenues and expenditures of the companies owning the equity of the SPV. Thus, they cannot be used to cover losses or liabilities elsewhere in the parent firms. Private finance is provided through this SPV, in the form of equity stakes (normally held by subsidiaries of the same parent companies that own the firms that will take the principal construction and operating contracts from the SPV) and debt (either bonds issued by the SPV or loans to the SPV by banks). Project finance debt is in principle distinguished from corporate finance debt by its "nonrecourse" character. That is, the lender has no recourse to the other corporate assets or revenues of the equity investors in the SPV in the event of a financial default by the SPV. However, in practice this distinction is blurred by guarantees given by those parent companies (these become "contingent liabilities" for the companies giving the guarantees; unlike simple liabilities to service debt, these need not appear on the balance sheets of the companies in question). The SPV is usually a "non-stick" entity, retaining neither risks nor profits. All profits are passed to the companies owning its equity, and all risks transferred to it from the public client by the PFI contract are promptly passed on to the companies with which it enters into supply contracts, namely, the principal constructor and principal operator, or to other third parties. One exception is that sometimes the SPV retains responsibility for replacement of components needed during the life of the contract, though sometimes this is passed on to the operator.

One further defining characteristic of UK-style PFI is the debt-to-equity ratios of the project finance structures. It is unusual for equity to amount to as much as 15 percent of the capital structure. Moreover, the main equity-providers normally hedge their exposure to financial risk arising from fluctuations in interest rates. It is widely asserted by practitioners that these high proportions of debt and hedging (a) have the effect of reducing the weighted average cost of capital (in part, but not only, because interest payments attract corporation tax relief) and thus the amount of gross profit that must be covered by PFI contract prices and (b) reflect the unwillingness of PFI company sponsors to risk more equity capital, or to expose that capital to greater financing risk.

However, such high levels of financial gearing leave lenders exposed if there are substantial risks affecting projected revenue and cost streams (small proportional reductions in revenues or increases in costs could wipe out the equity, put the SPV in financial default, and leave lenders to bear any further losses). Thus, the lenders seek to assure themselves that the probabilities of such risks are very low. Increasingly, therefore, PFI has come to be seen as suitable for low-risk projects and unsuitable for projects with a high intrinsic total risk, unless the majority of that risk is borne by either the public sector client or the parent company owners behind the SPV, by means of additional guarantees to the lenders. This somewhat militates against one of the perceived rationales for the public sector adopting PFI, which is the transfer to the private sector of "significant" project risks.

PFI and Project Risks

PFI project failures, for all stakeholders, to date have been greatest in projects with high intrinsic risk:

- High "demand risk" ('if we build it, will they come?'); for example, the Royal Armouries Museum
- High "technical risk" (a technical specification at frontier of technical capabilities of the provider); for example, the National Physical Laboratory

PFI has been more successfully applied to projects with only "normal" levels of construction risk and no demand-risk transfer to the private sector. The PFI company, under direction from its lender, manages these normal risks by aiming for certainty of construction price and time. It achieves this certainty by

- specifying "tried and tested" solutions
- exerting strong "change control"
- transferring "all construction risks" in its contract to the design-and-build (D&B) supplier

"Time" risks are then controlled by direct monitoring of the contractor's performance and by incentives (penalties for late completion), and "price" risks are controlled by contract and by guarantees to lenders from parent of the D&B company.

UCLH

The lender, Abbey National Treasury Services (ANTS), and the SPV, Health Management Group (HMG) expressed concern that the innovative heating and cooling system proposed by HMG's designers had not been used previously in a UK hospital and perceived that unproven technology could increase the likelihood of availability deductions. To overcome this concern, HV plant for the entire building were "pooled," so that all plants in effect served all areas of the hospital. This meant that

> if one plant failed, it would merely reduce overall capacity of the system but not cause unavailability in any one part of the hospital. Once convinced that several plants would have to fail simultaneously for there to be non-availability, ANTS and HMG approved the solution. This raised costs but increased revenue certainty.

During project development, the SPV is carefully monitored by advisers appointed by, and reporting to, the principal lender. The project monitors' task is to help identify key risks in advance and then monitor the PFI company's performance in managing each risk, advise the PFI company on what it has to do to keep the bank happy, and report (or not) to the bank that performance is good enough for the banks to release their next tranche of finance. By controlling the flow of project finance in this way, the banks seek to manage design and construction risks. The banks, and their monitors, are not particularly concerned about risks of cost overrun, because the SPV will have passed these risks to their D&B contractor, whose balance sheet should be strong enough to bear them. Rather, their main concerns are with anything that could delay the start or halt the flow of service payments by the public sector client. This could be either a construction delay or a risk that the project would fail to pass quality (safety; fit-for-purpose) tests at commissioning stage, or that it subsequently becomes "unavailable" to any significant extent.

It is characteristic of PFI (and concessions) that the SPV's cash flow position is at its worst during construction. Once service delivery payments (or, in concessions, user payments) begin, monthly operating cash flows are strongly positive, and sufficient to cover negative financing cash flows (interest payments). But the SPV initially has only just enough finance (and then only if the bank releases tranches of finance as scheduled) to cover construction expenditure (principally, payments under contract to its D&B contractor). Interest nominally due on the construction loan normally has to be "capitalized" (i.e., added to the amount of loan principal that has eventually to be repaid, and on which interest is charged). Thus, a delay in the start of service delivery payments (because the project is late reaching operational stage) is likely to be fatal to the SPV's ability to service its debt and force it to seek a rescheduling or debt-to-equity conversion from the bank. Later, the SPV may be able to sue the D&B contractor (probably its principal initial shareholder) for substantial damages, but by that time the ownership of the SPV will probably have passed to the bank, and the other equity owners (besides the constructor that failed to perform) will also have lost their investment.

BOOT thus creates one of the strongest incentives to the PFI or concession contract holder to deliver on time of any procurement mechanism yet devised. The SPV and the bank will not simply rely on the contract passing responsibility for risk of delay to the D&B contractor but will actively manage this risk.

Note the asymmetry in the treatment of construction budget and schedule risks. If construction costs overrun, the contractor may in theory attempt to sue to reclaim these later from the SPV (in effect, from the other shareholders in the SPV), on the grounds that these overruns were caused by changes in the SPV's requirements. However, in the meantime the completed facility is in the hands of the SPV, and it is collecting operating revenue. Thus, the SPV will probably survive.

Construction-based PFI projects have achieved an excellent track record of becoming operational at or before the contracted date (National Audit Office, 2003). We do not have public information on cost-control performance. There will have been a number of cases where construction costs have exceeded contract price agreed between an SPV and its D&B contractor. In such cases, the normal consequence will be for the parent company of the D&B contractor to absorb the loss.

The PFI contract transfers project delivery risks to the PFI company. However, it has proved better (for value-for-money) not to transfer to a risk-averse private firm risks that the latter is in no position to manage or control (HM Treasury Task Force, 1997). Thus, for example, general inflation risk is retained by the public sector (payments due are indexed to changes in the Retail Price Index or some similar inflation index). Also, the public sector bears the risk of the impact of decisions under its control, such as new "specific" (though not general) legislation or regulations affecting PFI, or the price of changes it requests in the specification of the service after the contract is signed. However, other construction risks, and many operating risks, are transferred under the terms of the PFI contract, the exact allocation of risks varying from the PFI sector norm in each project and being negotiated normally after appointment of the preferred bidder.

PFI usually involves transfer of ownership of the asset, at a nominal price, at expiry of the contract, though it needs to be noted that much earlier official guidance suggested it might not be the case (Private Finance Panel, 1996D and 1997). More exactly, the public sector normally has an option to take over the assets at the end of the contract, which it need not exercise if it judges the facilities to have become a net liability (to have negative net future value). Since the assets are usually highly specialized, the client's *ex ante* estimation of their residual value-in-use at contract expiry normally exceeds the bidder's *ex ante* assessment of possible residual value to them, and it is thus optimal to contract for ownership to revert.

In what follows I shall use the specialized terminology of the PFI procurement process. For fuller definitions see Construction Industry Council (1998) and HM Treasury Task Force (1999a).

Defining the Project Mission

Here the key and distinctive challenges of PFI are all consequences of the use of output and not input specification (Private Finance Panel, 1996c; Public Private Partnerships Programme, undated; Department of Health, 2002).

Assessing Value-for-Money

In UK PFI it is an HM Treasury policy requirement that the final proposal of the winning bidder for each PFI project be appraised for value-for-money against a public sector comparator (PSC). Only if it is deemed to offer the public sector better VFM than the PSC (after adjusting for the value of risks transferred under the PFI but retained by the public sector under the PSC) should the proposal be accepted and a PFI contract signed.

The PSC is an attempt to estimate the price of a "conventionally procured" solution to meet the same business case need and output specification. It therefore should comprise both a construction price estimate for the design and construction solutions that would be produced by following current public sector procurement best practice and a public sector operating cost estimate for the same period as the PFI contract, converted to present value terms.

The PSC should ideally be based on a realistic "reference project", and this should be a "real" alternative option that could be developed, publicly financed, and chosen instead of the PFI bid. However, as the National Audit Office reports into PFI have pointed out, often one or both of these conditions are not met.

On the other hand, it is equally important that the PSC should be for a truly comparable level and quality-of-service delivery over the project life. Often, the costing of the "operating" element of the PSC is subject to query on these grounds.

In the absence of a price for a PSC that is a real alternative, *ex post* assessment of VFM (for example, by the National Audit Office) becomes a matter of assessing the effectiveness of competition achieved in the PFI project procurement process and of identifying where (and how securely) the projected benefits or DBO cost savings are coming from, in a particular project case, to offset the extra financing costs and procurement costs of PFI (CIC, 2000).

In practice, to date few PFI proposals have been rejected at final business case stage on grounds of lack of VFM compared to the PSC, after putting a value on risk transfer. Rather, more projects have fallen on "affordability" grounds, or from inability to agree an allocation of risks acceptable to all parties.

UCLH

The outline business case was based on the financial (net present value of savings in operating expenditure and capital receipts from land, less capital costs) and nonfinancial (improved healthcare, education, and research provision) benefits. Eight alternative project solutions were appraised. Highest ratio of benefit-to-cost was found to be for centralization of all accommodation in the locality of the existing UCH site. This version of the project, known as the 4box solution, from the four new buildings envisaged, was put to NHS Executive for project approval.

The Trust then initiated a PFI procurement, using estimated costs and benefits of its designers' proposals for the 4box as the public sector comparator—the basis for judging whether the best PFI bid was better value-for-money than conventional procurement—and including it in bid documentation as a point of reference for bidders, who were allowed, however, to submit variant proposed solutions.

Writing the Output Specification

The first challenge for the public sector's project managers is to turn the project mission statement and outline business case into an output specification. The difference between output and input specification is part of the basic idea of PFI (see *Aims and Objectives of Parties to PFI Contracts*). This goes beyond "performance" or "functional" specifications, as those might be found in the brief for a turnkey design-and-build project, because they are not primarily specifications of performance or functionality to be achieved at the point of hand-over or commissioning of the project, but standards to be achieved throughout the n-year life of the PFI contract. Thus, in addition to an output specification of the facility required, there are specifications of operating services to be provided. Moreover, in the fields in which UK PFI has been chiefly applied (for example, schools, hospitals, prisons), there is no recent UK history of the type of "turnkey" or "full" design-and-build procurement that is based on performance specification of facilities. Recently, in UK public sector building, if any form of integrated procurement has been used, it has been "develop and construct" (where the client's consultant prepares concept drawings, site layout, building dispositions, and plan forms) rather than full design-and-build.

Capturing users' requirements in a brief is always one of the hardest parts of a project to accomplish. In conventional building procurement there is at least a feedback process when architects' interpretation of and proposed solution to that brief can be reviewed by users. It is often pointed out that users may find it difficult to "read" and fully understand the implications of architectural drawings. Might a written/numeric specification be easier for lay persons to understand? On the other hand, what if the problem lies not in the medium of communication (drawings or written specifications of outputs) but in the articulation of implicit knowledge or in inappropriately "rational" models of thinking that actually proceeds by use of image? We all know how we like our coffee, but many of us would find it hard to describe what we are looking for in terms of attributes (units of bitterness, and the like) and would find it easier to say "like a real Italian espresso" and rely on an "architect" sharing our understanding of what we mean by that. Likewise, army generals know what they mean by "consistent with the military ethos" but find it hard to put it into words in an output specification. Elephants are easy to recognize but hard to define.

Note also that in the competitive stages of PFI procurement there are only very limited opportunities for bidders' designers to consult with users. Partly this arises from constraints on the time available to develop designs at tender stage, known as invitation to negotiate (ITN) stage, and partly from the procedural requirement for even-handed treatment of all bidders.

UCLH

> UCLH proposed a radical shift from grouping in-patients by clinical speciality or department to grouping provision according to level of clinical need and dependency. In its output specification, it provided a draft set of operational policies for healthcare provision. It also included reference to a set of design guidance documents and principles. It indicated the number of acute (537) and low-dependency (60) beds that it thought it required bidders to provide. However, the final project company's proposals, accepted by the Trust, were for 669 acute and no low-dependency beds.

Assessing Affordability

The next challenge is to produce accurate budget estimates and thus assess affordability of the project at an early enough stage to avoid either: (a) late aborting of projects, with consequent write-offs of costs incurred to that point; or (b) late-stage downward rescoping of projects. Here the problem to be avoided is as follows. If, when bids are received, the lowest bid is higher than the public sector client can afford, either the project must be aborted or a preferred bidder appointed with whom a price for a reduced-scope specification will then be negotiated. The latter involves the client, in effect, in negotiating a service-content and a price with an incumbent monopolist, not under competitive pressure from rival bidders. Official advice is that the client should only proceed to negotiate scope and price with a preferred bidder if the bid price is "within negotiating distance" of affordability. The preferred situation for the public sector is that only relatively small details of risk allocation should be negotiated after receipt of competitive bids.

It is open to the client to publicize, at ITN stage, the amount of annual service payment they can afford and thus discourage non-affordable bids. However, if, in the view of bidders, the client has "specified a Rolls-Royce" but "provided a budget for a Ford," mere announcement of the budget limit will hardly solve the problem. In this situation, the choices open to a bidder are these: withdraw from bidding so as to avoid further unrecoverable bidding costs on a project that may be aborted; take the output specification at face value and ignore the budget limit; or take the budget limit seriously and "reinterpret" the output specification.

UCLH

> The Trust made explicit to bidders both the design solution of its public sector comparator and its annual affordability constraint.

Whereas project managers can draw on a formidable body of information and experience in order to estimate market prices for an input specification for a building (with all its corresponding design drawings and technical descriptions), the price estimation, by clients' technical advisers, of output specifications for construction-based PFI services has involved a higher degree of uncertainty and error and yielded ranges rather than single-value estimates.

Ideally, projects should be checked for affordability before going out to the market, that is, at outline business case (OBC) stage. The OBC should involve "a clear definition in terms of service delivery of what is sought: the output specification" and "should incorporate a Reference Project (RP), i.e. a particular possible solution to the output requirement. . .which is worked-up in sufficient detail to provide full and adequate costing. . .including quantification of key risks. . .and hence is *prima facie* affordable" (HM Treasury Task Force, 1999). However, at this stage the specifications are often in practice extremely sketchy and therefore impossible to price accurately. It seems that a common practice is instead to estimate the price of a reference project without a detailed specification, simply using (updated by indexation) historic averages for simple descriptors (such as "construction cost per bed place" plus annual facility management cost per place in existing facilities). Cost estimates derived in this way are, of course, subject to a high degree of error. (As yet unpublished research I've directed has found evidence of wide discrepancies, for UK schools' PFI projects, between technical advisers' price estimates for the DBO costs of the output specification and the actual prices contained in lowest bidders' financial models.) Thus by deferring design until after calling for bids, PFI projects lose the advantage of relatively accurate prebid estimates (permitting early checks for affordability) offered by the traditional construction process of design-estimate-then-tender.

The process of formation of an output specification for a PFI project is too often in practice one in which initially, as more users are more fully consulted, more and more desiderata are added in to the specification, only, once bids have been received, for more and more desiderata to be cut out as unaffordable.

Securing Stakeholder Consent

The final distinctive challenge posed by PFI to the process of defining the project mission concerns the management of stakeholder consent. PFI is, of course, politically controversial. In locally devolved projects, stakeholders will include councillors in local authorities and members of boards of trustees or governors who may be either opposed to the whole idea of PFI (as "creeping privatization") or at least suspicious of it. Government has sought to overcome opposition of this kind by offering local authorities Hobson's choice—a PFI project or nothing. There have been a few cases (the pathfinder Pimlico School project is perhaps the best known) where local representatives have chosen "nothing." This way of forcing PFI on reluctant stakeholders has not been good for its reputation.

> "He who's convinced against his will
> is of the same opinion still"

A relatively recent development in the PFI market has been the "bundling" of smaller projects, especially refurbishments and partial rebuilds of groups of schools. This increases the number of stakeholders in the project enormously, and there are several important instances where the project client coalition fell apart (for example, the Brent secondary schools project) before a contract could be agreed.

Perhaps the strongest opposition to PFI has come from the trade unions representing public sector employees. Opposed to PFI on principle, these unions are most particularly concerned when existing union members will be transferred to become employees of the private sector. From this perspective, PFI projects fall into two kinds: where the business case is essentially to expand total capacity; and where the case is essentially to replace existing capacity. In the former, the PFI companies essentially recruit new employees. In the latter, they take over the existing public sector employees. UK legislation says that any transferred employees must work on wage levels, terms, and conditions that are "no worse" than they previously enjoyed. For private sector project managers, this often creates a problem of managing a two-tier workforce, comprising a tier of transferees and a tier of new recruits, potentially doing similar jobs on different pay and conditions, and may reduce the firms' ability to introduce operating regimes based around "flexible working."

Although client requirements are meant to be captured by the output specification, both the client and other public regulatory stakeholders will have technical requirements embodied in technical guidance notes and codes, to which they must be convinced the project conforms before they will give consent. Often in the United Kingdom these standards, requirements, and codes are prescriptive in form and therefore simply require to be interpreted, rather than be consulted or negotiated upon. However, their prescriptiveness may greatly reduce the opportunity for the PFI consortium to develop unusual or innovative solutions.

Finally, in the United Kingdom, PFI projects, like any building project, require first outline and then detailed planning (zoning) consent. The normal practice is for the client to obtain outline consent on the basis of a maximal "footprint," mass and floor area, before proceeding to ITN. This is done to reduce the risk to bidders that consent may not be forthcoming. After appointment of the preferred bidder (PB), it becomes their responsibility to obtain detailed consent for their proposed solution. Nevertheless, it is not unknown for PFI bidders to propose solutions on sites other than those owned and proposed by the client.

There are sometimes, therefore, effectively two interdependent projects for the PFI company to manage: the PFI project itself, as defined by the client, and a property development or dealing project, involving some acquisition and some disposal of sites. If proceeds from sales of sites are to be used to fund part of the PFI project, this may mean that separate "bridging finance" has to be arranged to cover the period until sites are vacated and can be sold. Otherwise, in the simple case where the boundary of the site required is identical to the boundary of the site owned and proposed for use by the public sector, PFI requires that the site be leased by the PFI company.

UCLH

> The Trust went out to ITN with a four-site planning consent (the so-called 4box proposal, which had also been used for the OBC), only for the winning bidder to propose a single-site two-towers-and-podium solution, not requiring some of these sites but requiring some adjacent land not included in the 4box, on which the bidder had obtained an option to buy from its private owner. In such cases the bidder shares the risk that planning consent will not be forthcoming, but, of course, reduces competitors' ability to match their proposal.

Mobilizing the Resource Base

Under this heading we will consider the forming and the motivating of the project coalition.

The client's procurement problem is synonymous with the principal/agent problem identified in economic theory, and it is that theory that I will use here. The principal/agent problem arises when an agent (in this case, a PFI company) knows more about its real competence or about the effort it will expend than does the principal (in this case, the public sector client). The theory of principal and agent represents the main attempt by economists to generate advice to clients about how to design the *reward structure* of their contractors, such as a PFI company, so as to obtain optimal outcome. The rest of the problem can be decomposed into the *adverse selection* problem and the *moral hazard* problem—the former to do with selecting and the latter to do with opportunistic behavior by a supplier (Milgrom and Roberts, 1992; Douma and Schreuder, 1998).

Devising the PFI Co.'s reward structure

One of the most important tasks, therefore, for the client and their project manager, is to think carefully about just how to link payment to the PFI company to project outcomes, using the output specification to define those outcomes.

In PFI, payment to the agent is a mixture of a fixed element (agreed in advance and unlikely to vary according to outcomes or outputs achieved) and a variable element that is explicitly tied, via specific performance measures, to the quantity and quality of service-delivery outputs achieved.

PFI payment mechanisms divide the client's unified payment for services into two, or sometimes three, parts:

- A part for availability of the facility or of major parts thereof, and therefore with deductions for nonavailability for use
- A part for level of performance of the facility management (FM) or other service functions, with deductions therefore for performance below a set standard
- Sometimes, a part linked to a variable volume of use (as with early "shadow toll" highways, but also, say, laundry services payments made pro rata to volume of bed use in hospitals).

The availability payment invariably dominates the mix, not only in PFIs for "accommodation service" (where the costs of constructing and then maintaining the facility are the majority of the PFI company's costs, so that the structure of payments is broadly aligned to the structure of costs), but even in PFIs for "final service" (like prisons) where the cost structure is very different. Until commissioning is achieved, the timing and amount of the expected availability payment must be regarded as subject to risk. Once commissioning is complete, however, the client has agreed that the project is, at that point, meeting all its availability requirements. Thereafter, this part of the payment can only vary (downwards) if all or part of the facility subsequently seriously deteriorates in condition. Thus, the risk, to the PFI company, becomes a matter of preventing complete and minimizing partial nonavailability through deterioration leading to failure to remain available for use. The initial questions facing the client are as follows:

- By how much, and for how long, should a measured condition have to deteriorate in order to constitute "nonavailability" in terms of the payment mechanism?

- Shall payment be linked mainly to "aggregate" availability or separately to the availability of each part of the facility?
- How big or small a part of the availability payment will be deducted in such a case?

The PFI Co and its lenders want to be able to regard the availability element of the payment as "virtually fixed", once commissioning has been achieved. Then, on the basis of the security offered by this "fixed" stream of revenue, low-risk/low-interest debt finance can be arranged. Given that the client, too, should expect to share in the savings from lower interest costs, they are likely to answer the questions posed above in ways that allow this treatment of availability payments as being "mostly" fixed.

The economics of optimizing incentives via payment mechanisms is, in principle, straightforward (Private Finance Panel with HM Treasury, 1997; p. 8), but in practice fiendishly difficult. Penalties for nonperformance or nonavailability should be set so that the monetary value of the penalty is just less than the monetary imputed difference of value of the benefits to users between the two states (of availability and nonavailability, or of performance to standard X and performance to level Y). If this penalty is greater than the cost to the PFI company of remedying the poor performance or nonavailability, then the company will be induced to undertake the remedy; and if not, then nonremedy is preferable to remedy, because marginal cost of remedy exceeds marginal value of the resulting benefit (I have assumed, for simplicity, that the cost of remedy is the same as the cost of prevention; if not, the prescription becomes somewhat more complex).

This requires, first and foremost, that the client knows the value of each marginal benefit. Lack of such knowledge is one of the biggest impediments to effective design of incentive-payment systems. If this problem is overcome, there remains the problem for the PFI company of obtaining knowledge of the whole-life cost and revenue implications of each of its design and operating choices, without which it will be unable to respond rationally to any structure of incentives embodied in a payment mechanism.

However, the client may feel that it is simply unacceptable for some events of nonavailability to occur at all (such as failure of power supplies to operating theaters) and thus penalize such events so heavily that the contractor has very strong incentives to avoid it happening (by paying for the cost of backup systems, for example). Whereas, for other nonavailability events, the incentives will be set weaker (failure of power supply to office areas, for example) and therefore lower relative to the costs of remedy. Such "strong" incentives need to be used sparingly, partly because they will significantly add to overall cost and thus price (and by adding to the SPV's risk may make the project nonbankable, i.e., unattractive to lenders) and partly because they run the risk of distorting the PFI company's allocation of resource and managerial effort.

How many and what kind of separate outputs should be measured and linked to the payment?
The "equal compensation principle" of agency theory requires that

if the agent is required to allocate time between two or more activities, then [for an incentive contract to work] the marginal return to the agent from time spent on each activity must be equal, or else the activity with the lower marginal return will receive no time (Millgrom and Roberts, 1992; p 228).

Simplifying more than a little, the moral of the equal-compensation principle story is that the client should ensure that each and every desired output or outcome is given its own place in the payment mechanism. However, as you shall see, there are countervailing arguments in favor of simpler (or different) payment mechanisms.

The first of these arguments derives from the "informativeness principle." This states that the client should exclude from the PFI company's compensation formula any performance measures that partially reflect factors outside the agent's control and include in the formula any measure that increases the likelihood that the size of payment will relate to the actual effort of the PFI company.

One corollary might be that it would be desirable to benchmark each PFI company's performance against other comparable PFI projects and link payment mainly to benchmark (i.e., relative) scores. This would be so *if* "factors outside the PFI Co.'s control" that impact project outcomes (such as the common procedures and practices of the client) are likely to impact also on all those other PFIs, while differences in performance between SPVs are more likely to relate to differences in quality and quantity of their effort. This has hardly as yet been attempted in PFI.

How large should the deductions for nondelivery of each output be?

To answer this, project managers and clients need to think about the costs of monitoring. The issue can be thought of in terms of the "monitoring-intensity principle." This states that the level of monitoring of performance or output must be proportionate to the intensity of the incentive. If contracts set up large deductions in relation to an output, then the client must be prepared to spend more in total on the costs of monitoring of that output. Other things being equal, the more *costly* it is to monitor accurately a particular output, the weaker the strength of the incentive that should be created by linking payment to that output. Ideally, the amount of measurement and the intensity of incentives should be chosen together, as a package. Where outputs are heterogeneous and intrinsically hard to measure (because they are services, not commodities), as in PFI projects, and *a fortiori* if they are intermediate rather than final services, again as in most PFIs, so that accurate measurement is very costly, it may therefore be best to set up only relatively weak incentives to produce those outputs.

In practice, most PFI clients use "weak" incentives for service quality and rely more on there being a shared understanding of what constitutes "good provision" with, and an organizational culture of quality management in, their providers. It is normal for the providers themselves to be asked to do most of the monitoring of quality of service, because it is much less costly for them to undertake.

There are some further concerns for clients. What if methods of delivering outcomes are to a large extent technologically fixed, so that the same methods will be used despite creation of incentives? What if the things the client can measure are only more or less crude approximations of or proxies for the things they really wish to be provided? What if the PFI company (or, in a PFI context, their provider of capital, i.e., the bank) is *highly* risk-averse? Or, finally, what if the PFI company, or one of its principal suppliers, is, in terms of organizational behavior, conservative and habitual rather than progressive and innovative, again despite creation of incentives?

All the preceding issues are key matters a PFI client has to judge before deciding on an incentive structure. To guide such judgments, we need research to clarify what payment mechanisms work best in practice in what PFI project contexts. A highly provisional summing-up might be this: Beware overly strong incentives unless outcomes are few and simple to measure.

In practice, it may be better in PFI projects to decouple the issues of incentive and replacement. Incentives alone (because "weak" and incomplete) may be inadequate to prevent serious nonperformance. The real remedy, in extreme cases, will lie in replacement. It should not be the case that performance measures' importance is judged solely by the size of impact on payment. Any performance measures, however, need to be objective and accurate enough to be used to begin moves to replace the agent, if necessary. Replacement of FM providers is often possible, within a PFI, because the latter may have only relatively short-term contracts from the SPV. However, it can be very difficult indeed to make sure that the performance measurement of the FM provider only measures their own effort and not that of the D&B provider.

Adverse Selection

In a construction context, adverse selection means that suppliers who intend to offer inferior (but hidden) quality or to behave opportunistically with respect to claims and variations have the greatest incentive to charge the lowest prices. Thus, simple reliance on price competition does not work and, if insisted upon by the client, poor-quality and opportunistic suppliers will drive out good quality and trustworthy ones. The classic solution in construction was to divide procurement of the design from that of construction, on the basis that, while quality in design could not be assured in advance and defects could be hidden, and variations in client requirements during design development could not be prevented, the same would be less true of construction, if the construction contract consisted of an agreement to execute, under inspection, a completely predeveloped design. Designers therefore had to be selected in ways that assured the client that they could be trusted, whereas contractors did not and could be selected by lowest-price tender competition, within only weak parameters of pre-qualification to eliminate the egregiously insolvent or incompetent. Designers might well work in an in-house department of the client or, if not, would be appointed on the basis of reputation, track record, and an ongoing relationship with the client over past and projected future projects (Winch, 2002).

Now, PFI defies the logic of this approach, by making the client choose the designer, constructor, and operator all together, by the same method, and as a combined (take-it-or-leave-it) set. At the same time, it rules out use of the in-house provision solution for the client. This means, in effect, that the client needs to have the same level of confidence in their PFI consortium (confidence that they will not try to deliver and conceal inferior quality, and confidence that they will not exploit opportunities to extract monopoly rents) that they classically had to have in their designer. Alternatively, if they do not have such confidence but are still determined to use PFI, they must develop much more effective *contractual* means of inspecting and enforcing quality, and of ensuring that they do not require variations or changes not explicitly provided for in the contract.

However, as will be shown, contracts alone are unlikely to be efficacious in securing client objectives. Trust, and also a shared conception of project mission and a collaborative rather than confrontational climate between client and PFI company, is the *sine qua non* for effective PFI projects.

PFI uses a form of selective competitive tendering, followed by negotiation with a preferred bidder. Normally three (never more than four) consortia or firms are short-listed to bid. These are the firms that score highest in prequalification evaluations of firms that responded to the invitation to express interest (EOI). Track record in PFI projects is one key criterion for short-listing. Ideally, but in practice by no means always, clients should only short-list firms that are not and will not become too busy with other PFI projects, to avoid uncompetitive bids or bidders withdrawing. A review of the literature of PFI project case studies reveals at once that in a large proportion of cases, firms that were part of a short-list of two or three bidders subsequently withdrew, or submitted uncompetitive, nonconforming bids. Thus, it has been common for the client to have only two, or even only one, bid(s) to consider (see NAO PFI project case study reports, various dates).

UCLH

> The Trust chose a short-list of three consortia. One then dropped out because it became overcommitted on other PFI projects. Another collapsed after its FM partner withdrew following a conflict with its contractor. Thus, the Trust received just one bid.
>
> The unitary payment proposed in this bid exceeded the Trust's announced affordability constraint. However, the variant nature of its proposed solution was judged to yield further savings in operating costs to the Trust of healthcare provision, thus increasing the amount the Trust would have available for the unitary payment. Thus, the Trust judged the bid to be both affordable and better VFM than the PSC, because it provided more benefits in terms of healthcare provision, though at higher annual cost.

Bid evaluation should either proceed in two stages (with stage 1 being a technical appraisal of bids for ability to meet requirements, with only bids passing the standards required proceeding to stage 2, selection of lowest-price bid, where lowest-price is judged across all scenarios, allowing for differences in risks accepted or contract period proposed) or comprise a "weighted" evaluation in which price, risk acceptance, robustness, and scores on other quality criteria are all combined into a weighted composite score (HM Treasury Task Force, 1999b). Where the weighting method is used, invariably in practice by far the greatest weight is given to price (at most, to this author's knowledge, bids with a high score for "quality" but a higher price have been selected over bids with a merely satisfactory score for quality if the price difference is of the order of 5 percent or less).

Moral Hazard: "Holdup"

In principle, negotiations with the preferred bidder (PB), appointed following clarification of initial bids and evaluation of variant bids departing from proposed "standard" commercial terms and risk allocations, should be

> limited to fixing the final detail of the documentation and satisfying the reasonable requirements of the supplier's financiers . . . by insisting the PB's financiers have indicated their comfort with the risk allocation embodied in their bid at a stage where there is still the lever of competition (HM Treasury Task Force, 1999a).

In practice, however, it is common for one or other party to seek materially to amend the terms of the proposed contract. On the client's side, this is most likely to occur because of problems of nonaffordability (see preceding text). On the PB's side, it may reflect either changes proposed by their financiers, once the latter come to examine risks in detail, or an attempt to take advantage of the easing of competitive pressure. Public sector guidance recommends that the client should, where possible, keep a "second choice" bidder ready in reserve, as a counter to such opportunism, and reserve the right to require the PB to conduct a funding competition at this stage (Office of Government Commerce with Partnerships UK, 2002) if the source of the problem lies with their proposed financier.

The costs of preparing contracts designed to guard against later holdup of either party by the other in PFI have been much higher overall than in any other mode of construction procurement (Construction Industry Council, 1998 and 2000). The publication recently of standard forms of contract for use in some of the main PFI market sectors, as well as standard specifications and payment mechanisms, may help to reduce these transaction costs (Gruneberg and Ive, 2000, Chapter 5). However, this will depend on the extent to which clients and bidders feel obliged to vary from these standard terms.

It has proved very expensive, both in terms of advisers' fees and of the impact on contract price, for PFI clients to attempt to foresee as a contingency every relevant "state-of-the-world" and every possible change in their requirements over a period of several decades ahead. Thus, in practice, PFI clients are likely to find it better to commit to a loss of flexibility in their PFI-contracted requirements, leaving their non-PFI projects to buffer and absorb any such changes.

A large part of the fee costs—fees paid by the client to technical, legal, and financial advisers—arise from the perceived need to prepare complete and unambiguous tender documents. If the client is trying to achieve a complete contract and depend upon their ability to enforce this contract in the event of disputes, they must incorporate ways of measuring performance and specify methods of resolving disagreements over actual levels of performance.

In some cases, clients appear to have lost sight of the big picture in a mass of detail, paying for the development of very elaborate performance measures and payment mechanisms that actually are unlikely to achieve any key client goals, because of the small value (and perhaps also contestability) of the penalties resulting, relative to the unified service charge or to the likely costs of providing remedies.

In the case of one hospital PFI (not UCLH), a very complex payment mechanism was developed. It comprised 14 components in the payment, with some components based on actual or "notional" bed occupation units (BOU) and non-bed attendances (NBA) for each area of the hospital, including marginal rates of £3.67p per NBA, along with a complex system of deductions, retentions, and deficiency points with seven performance bands. All of the bands led to a potential difference between most optimistic and most pessimistic feasible scenarios, of an amount equal to 1.44 percent of the total payment stream to the SPV (Parker, 2000).

Riding the Project Life Cycle and Leading the Project Coalition

The PFI company is in a position potentially to take a holistic view of project costs and revenues across, if not the whole life of the facility, at least the life of the PFI contract. However, various barriers stand in the way and need to be overcome by its management of the project before the opportunities offered can be seized.

First, the project must be run by an integrated coalition and led by a project director keen to optimize the project's returns over its contract life cycle, and not prepared to accept subgoal optimizing behavior by members of that coalition (where the contractor tries to minimize construction cost regardless of operating costs or impact on revenues, for example). Each organization involved in the coalition needs to be responsive and adaptable to the fact that it has become a shareholder in its own client (the SPV) and thus, in addition to its "usual" interest (in maximizing profits from a construction or FM contract) has another interest, in maximizing profits for the SPV. This means, first, contributing to improving the chances of its consortium of winning the bid competition and, second, contributing to the *whole-life* PV (discounted present value) of the net revenues (profits) expected from that bid if successful. This latter is, of course, the PV of the difference between service payments receivable from the client and the whole-life costs of providing that service.

Devising solutions so as to maximize profit over the whole life of the contract, however, requires that the bidder's project management team have good information about the WLC (whole-life cost) and revenue implications of their choices of design solution and operating regime. If the managers lack confidence in the quality of that information, they may regard options that will certainly increase initial construction cost but that are "expected" to reduce subsequent operating costs as, in effect, "riskier" than the alternative, and thus reject them.

By improving the feedback flow of information about costs-in-use, actual replacement lives of components, and the like, well-managed project organizations can in principle reduce the uncertainty with which WLC forecasts are regarded. Stable consortia, operating in the same market and comprising the same partners working together over many successive projects, ought to begin to receive detailed feedback from their operator partners or departments on actual operating and replacement costs, in forms sufficiently robust to convince their bankers. However, this data will only provide a clearer picture of the "path most traveled." The other paths will continue to disappear into dark forests of ignorance, unless there are some pioneer PFI projects that are allowed to explore those paths.

Maintaining the Resource Base

It is under this heading that you can best consider the intersection of issues regarding the management of a single project with issues of strategic resource planning. In the context of the management of *programs* of projects, especially, resource-maintenance activities, including maintaining a sufficient number of resource-holding firms, become a key part of the successful management of projects.

Maintaining the Number of Firms

All government departments funding PFI programs have had to concern themselves with first the creation and then the maintenance of a population of provider firms that own or employ the relevant specialized resources and possess the relevant organizational experience and knowledge. All PFI markets in the United Kingdom have to date shown strong tendencies toward supply-side concentration. Economies of experience appear to be high. Barriers to "late" entry appear to be substantial. In a market comprising only three or four firms (consortia), exit by even one firm would pose major problems for client procurement strategy, which is to rely upon competition at tender stage to obtain prices that give value-for-money. A similar, though temporary, problem arises if one or more of such a small number of suppliers are too loaded with other projects to be able to bid keenly for a particular project.

The program client must deliberately develop the market in ways that minimize the risk of a firm exiting. This may require it to disadvantage temporarily any firm that is threatening to achieve market dominance.

Since there will be PFI markets in which competitive pressure on bidders is sometimes weak, clients cannot afford to rely solely on competition, but have to be able to negotiate effectively, from a basis of knowledge of the costs of an efficient provider and from the bargaining power of a real alternative option. For this, public clients need to continue to put some projects of each type out to "conventional" procurement.

Encouraging Investment in the Resource Base

Opportunistic behavior all round is understood to be kept in check mainly by the adverse impact such behavior would have on a player's profits in future plays of the game as other players retaliated (dangers of exclusion of a consortium from future short-lists, dangers of exclusion of a partner from future consortia). The greater the threat or possibility of opportunistic behavior, undoubtedly the more the damage that will be done to the efficient maintenance of the resource base, as all become reluctant to make contract-specific investments in resource development. It is therefore important from this perspective that PFI is perceived to comprise a program that will continue indefinitely.

Acknowledgments

I am grateful to Kai Rintala for permission to draw on his case study doctoral research into University College London Hospital PFI project. This research will be published in 2004 by VTT (Valtion teknillinen tutkimuskeskus) of Finland as "The Economic Efficiency of Accommodation Service PFI Projects." I am also grateful for comments by the editors, and by Chris Field, Kai Rintala, and Hedley Smyth.

References

Audit Commission. 2003. *PFI in schools: The quality and cost of buildings and services provided by early PFI schemes.* London: Audit Commission.

Campagnac, E., and G. M. Winch. 1997. The social regulation of technical expertise: the corps and the profession in France and Great Britain. In *Governance and work: The social regulation of economic relations in Europe*, by R. Whitley and P. H. Kristensen, Oxford, OUP.

Construction Industry Council. 2000. *Role of cost saving and innovation in PFI projects.* London: Thomas Telford.

Construction Industry Council. 1998. *Constructors' key guide to PFI.* London: Thomas Telford.

Department of Health. 2002. *Standard output specification.* London: Department of Health.

Douma, S., and H. Schreuder. 1998. *Economic approaches to organisations.* London: Prentice Hall.

Gruneberg, S., and G. Ive. 2000. *Economics of the modern construction firm.* Basingstoke, UK: Macmillan.

HM Treasury Task Force. 2000A. *Technical note no. 7: How to achieve design quality in PFI projects.* London: HM Treasury.

HM Treasury Task Force. 2000B. *Technical note no. 6: How to manage the delivery of long term PFI contracts.* London: H M Treasury.

———. 2000c. *Technical note no. 5: How to construct a public sector comparator.* Series 3—technical notes. London: HM Treasury.

———. 1999a. *A step-by-step guide to the PFI procurement process.* Revised version. Series 1: Generic guidance. London: HM Treasury.

———. 1999b. *Technical note no. 4: How to appoint and work with a preferred bidder.* Series 3—technical notes. London: HM Treasury.

———. 1999c. *Technical note no. 3: How to appoint and manage advisers to PFI projects.* London: HM Treasury.

———. 1999d. *Technical note no. 1: How to account for PFI transactions.* Series 3—technical notes. London: HM Treasury.

———. 1997. *Partnerships for prosperity: The private finance initiative.* Series 1—generic guidance. London: HM Treasury.

Milgrom, P., and J. Roberts. 1992. *Economics, organization and management.* Englewood Cliffs, NJ: Prentice Hall.

Mumford, M. 1998. *Public projects, private finance.* Welwyn GC, UK: Griffin.

National Audit Office. 2003. *PFI: Construction performance.* Report by Comptroller and Auditor General, HC 371, Parliamentary Session 2002-3.

National Audit Office. 2002. *PFI and value for money.* Conference presentation by Mr Jeremy Colman, Assistant Auditor General.

Office of Government Commerce and Partnerships UK. 2002.

OGC guidance on certain financing issues in PFI contracts. London: Private Finance Unit, OGC.

Parker, M. 2000. *The importance of the payment mechanism in the allocation of risk in PFI projects.* MBA diss., University of Warwick.

Private Finance Panel with Highways Agency. Undated. *DBFO: Value in roads.* London: Highways Agency.

Private Finance Panel with HM Prison Service. 1996a. *Report on the procurement of custodial services for the DCMF prisons at Bridgend and Fazakerley.* London: HM Prison Service.

Private Finance Panel with HM Treasury. 1997. *Further contractual issues.* London: Private Finance Unit, HM Treasury.

———. 1996b. *PFI in government accommodation.* London: Private Finance Unit, HM Treasury.

———. 1996c. *Writing an output specification.* London: Private Finance Unit, HM Treasury.

————. 1996d. *Risk and reward in PFI contracts.* London: Private Finance Panel.

Private Finance Panel with HM Treasury. 1995. *Private opportunity, public benefit.* London: Private Finance Unit, HM Treasury.

Public Private Partnerships Programme (4Ps). Undated. *Output specification for PFI projects: A 4Ps guide for schools.* London: 4Ps. www.4ps.co.uk/publications.

Winch, G. M. 2002. *Managing construction projects: An information processing approach.* Oxford, UK: Blackwell Science.

————. 2000. The management of projects as a generic business process. In *Projects as business constituents and guiding motives,* ed. R. A. Lundin and F. Hartman. Dordrecht, Netherlands: Kluwer.

INDEX